Contributions to Management

More information about this series at http://www.springer.com/series/1505

António Carrizo Moreira •
Luís Miguel D. F. Ferreira •
Ricardo A. Zimmermann
Editors

Innovation and Supply Chain Management

Relationship, Collaboration and Strategies

Editors
António Carrizo Moreira
Department of Economics, Management,
Industrial Engineering and Tourism
University of Aveiro
Aveiro, Portugal

Luís Miguel D. F. Ferreira
Department of Mechanical Engineering
University of Coimbra
Coimbra, Portugal

Ricardo A. Zimmermann
Department of Economics, Management,
Industrial Engineering and Tourism
University of Aveiro
Aveiro, Portugal

ISSN 1431-1941 ISSN 2197-716X (electronic)
Contributions to Management Science
ISBN 978-3-030-08960-3 ISBN 978-3-319-74304-2 (eBook)
https://doi.org/10.1007/978-3-319-74304-2

Printed on acid-free paper

This Springer imprint is published by the registered company Springer International Publishing AG part of Springer Nature.
The registered company address is: Gewerbestrasse 11, 6330 Cham, Switzerland

Preface

Innovation plays an important role in a firm's competitiveness. Its role has been extensively studied at product, process, organizational, and marketing level. It has also been analyzed using a multifaceted perspective, with implications for a firm's performance. However, although innovation processes are important from the firm's perspective, the role of interorganizational networks must not be overlooked. How a firm shares innovation processes throughout the supply chain (SC) underpins a firm's competitiveness as much as the innovation processes inside the firm.

Systematic and discontinuous innovation has a pervasive role in spreading change. When firms embrace innovation in their interorganizational processes, in developing new products jointly with their partners, they involve multiple innovation processes in upstream and downstream activities in the supply chain. In the context of new technologies that threaten to alter the configuration of supply chains, this book discusses the key issues, challenges, opportunities, and trends in the relationship between innovation and supply chain management (SCM).

In our recent experience as professors, researchers, and consultants, we have witnessed the challenges that both innovation and supply chains have faced and the opportunities they have offered separately and altogether. The idea of this book has evolved from these perceptions. We endorse and complement Zinn and Goldsby's statement in a recent editorial from the *Journal of Business Logistics*: What a great time in history to be contributors to the fields of innovation management and supply chain management!

This work builds upon the conclusions of Zimmermann, Ferreira, and Moreira in a recent article published in the *Supply Chain Management: An International Journal* as it seeks to identify and explore the intellectual structure of the intersection of innovation and supply chains (especially in Part I) and explore the different ways that the topic is addressed in the literature. Many of the authors who contribute to this book are engaged in this intellectual pursuit and have devoted their wisdom to this area for several years.

The book is composed of state-of-the-art contributions from innovation and supply chain management scholars from all over the world (UK, Portugal, Brazil,

Italy, the Netherlands, USA, Denmark, Sweden, Canada, France, and China), whose contributions are of added value to academic researchers and practitioners, providing some of the most advanced research, concepts, experiences, and case studies in order to improve firms' competitiveness.

The book also presents some of the most recent developments and best practices in the fields of innovation and supply chain management. This book is intended and designed for a broad audience that includes practitioners and managers, as well as academics and postgraduate students who seek readers regarding relationships, collaboration, and technology involving innovation throughout the supply chain. In this respect, this book is unique as it encompasses applied research, concepts, and practical experience organized in 15 chapters that have been grouped into four different parts.

We hope you enjoy reading the book as much us we enjoyed being the editors and working with our colleagues!

Aveiro, Portugal — António Carrizo Moreira
Coimbra, Portugal — Luís Miguel D. F. Ferreira
Aveiro, Portugal — Ricardo A. Zimmermann

Introduction

The book encompasses applied research, concepts, and practical experience organized in 15 chapters that have been grouped into four different parts. Part I describes the intellectual structure of the relationship between innovation and supply chain management. Part II deals with strategies and implications for innovation in the supply chain when involving suppliers. This part covers the importance of coordination, cooperation, and collaboration in new product development (NPD) throughout the supply chain, as well as how small and medium-sized firms (SMEs) differ from large firms. This part also contributes to the debate about supplier-enabled innovation in complex projects and how the Product Innovation Charter (PIC) needs to be addressed by introducing suppliers to the PIC. Finally, this part also addresses the intricacies and practicalities of supplier involvement in NPD.

Part III, titled “Strategies and Implications for Innovation,” embraces and explores different topics, such as purchasing involvement in discontinuous innovation, the importance of culture in information sharing among manufacturing firms, risk allocation and supplier development, and the importance of supply chain innovation. This contributes to our understanding of the importance of the purchasing department in stimulating innovation in the supply chain. The importance of culture in information sharing among industrial firms features strongly in this part, although it has previously been under-researched. While Part II focuses on qualitative studies, as well as personal points of view, Part III offers several quantitative studies, which is a clear indication of the diversity of approaches used.

Finally, Part IV addresses some very exciting topics for the future innovative outlook of supply chains: new technologies and their importance for firms’ future competitiveness. Among the most important topics we can refer Industry 4.0, technological innovation, advanced supply chains, and the role of big data and predictive analytics. They are certainly game changers for most firms.

Part I: The Intellectual Structure

Part I is composed of a single chapter that describes the intellectual structure of the relationship between innovation and supply chain management.

In chapter "*The Intellectual Structure of the Relationship Between Innovation and Supply Chain Management*", Zimmermann, Ferreira, and Moreira analyze the intellectual structure of the relationship between innovation and supply chain management. Starting from the importance and complexity of this relationship, and using the principles of systematic literature review to identify the papers to be analyzed, the authors develop a bibliometric analysis of the topic. The results show the relevance, the topicality, and the all-embracing character of the theme. Citation analysis was used to identify the most influential studies in the area, and co-citation analysis made it possible to identify the knowledge base of the topic and its intellectual structure. The 35 articles identified as the intellectual base of the topic are divided into four clusters: papers that focus on the structural characteristics of the supply chain network, papers that are predominately characterized by the study of supply chain trust and collaborative advantage, papers that highlight the importance of the long-term integration of suppliers and customers, and papers that deal with some miscellaneous trends in the topic. The chapter contributes to theory by identifying the different approaches that address the relationship between innovation and supply chains in the literature.

Part II: Strategies and Implications for Innovation

Part II is composed of seven chapters dealing with supplier–client relationships, new product development, complex projects, early supplier involvement, and the product innovation charter. Hilletofth, Reitsma, and Erikson authored chapter "*Coordination of New Product Development and Supply Chain Management*", which deals with a specific but important topic: the coordination of supply chain management and new product development. In a case study, they analyze why and how NPD and SCM should be coordinated. To that end, they explore the critical success factors (CSFs) for NPD involving the market, product, strategic, and product characteristics. Hilletofth, Reitsma, and Erikson conclude that a strong focus on the demand side to develop premium products will necessarily drive high demands on the supply side of the company (SCM), which leads to the development outcome of coordinating NPD and SCM. In this way, the company can ensure that it is able to develop new products and that its supply chain can deliver innovative solutions. Time to market is guaranteed, not only during the NPD phase, but also the product is moved rapidly to the market. Consumer preferences are respected in relation to new products, as well as in lead times, service levels, and supply chain solutions. NPD processes need to identify costumer-oriented solutions, well beyond mere technological improvements, which requires that the company understands its consumers, as well as having supply chain solutions that provide proper consumer services. Holistic solutions in the NPD processes need to involve marketing, product development, R&D, and manufacturing representatives as traditionally occurs, but also involve representatives from sourcing and distribution, in order to coordinate NPD and SCM from the very beginning.

Chapter "*An Investigation of Contextual Influences on Innovation in Complex Projects*" deals with supplier-enabled innovation in the context of complex products, where Kavin and Narasimhan propose a framework for the analysis of innovation-fostering practices to address innovation performance. Based on the unique characteristics of complex products, they argue that an open approach to innovation with external partners needs to be undertaken in order to internalize external knowledge. Flexible management practices that employ network-based solutions are necessary. Risk-taking behavior needs to be ingrained in the organizational practices together with a well-developed absorptive capacity, so that innovation thrives in the firm. For innovation-fostering practices to succeed, Kavin and Narasimhan state that organizational incentives and infrastructural governance practices are needed that foster collaboration and knowledge sharing practices among internal and external stakeholders. This implies that communication must be based on trust and commitment in order to build confidentiality among external partners. They conclude that if firms are to succeed in complex product contexts, they need to follow a relational, network-based approach in which transparency and effective communication encourage commitment and knowledge sharing.

Chapter "*Necessary Governing Practices for Success and Failure of Client-supplier Innovation Cooperation*", by Servajean-Hilst, is about governance practices for the success of supplier–client innovation cooperation. Based on the necessary condition analysis (NCA) of 160 supplier–client relationships, he concludes that, to succeed, firms need to manage their supplier–client relationships and portfolios strategically. The involvement of the supplier's top management is a necessary condition for success, and lack of involvement of the client's top management is a necessary condition for failure. Moreover, involving purchasing and R&D functions is essential for the supplier–client relationship to work positively. Encouraging attitudes and the absence of threats are necessary conditions for success. Defining governance roles and responsibilities are also crucial if relationships are meant to last. The chapter ends with a list of critical government practices that are necessary, and those that should be avoided, to promote the flourishing of suppliers and clients in innovation cooperation.

Chapter "*Collaborative New Product Development in SMEs and Large Industrial Firms. Relationships Upstream and Downstream in the Supply Chain*", by Silva and Moreira, addresses collaborative new product development (CNPD) involving upstream and downstream relationships with suppliers and clients, taking into account both SMEs and large firms. Based on a set of eight case studies—where they analyze the type of collaboration, CNPD focus, CNPD objectives, and types of suppliers and clients—they seek to answer two research questions: How does CNPD differ in upstream and downstream relationships? How do firms intervene in CNPD according to their size and the innovation created?

This chapter concludes that CNPD is asymmetric, more often actively engaging suppliers than clients, because interaction involving industrial suppliers is more frequent and intense than interaction with industrial clients. Moreover, firms generally involve their suppliers to diversify their product portfolio. Silva and Moreira also conclude that CNPD is not restricted to large firms but also occurs when SMEs involve large firms as suppliers. Silva and Moreira demonstrate that CNPD is influenced by the technological intensity of the industry in which firms operate. In general, collaboration between firms operating in the same industry results in product differentiation, whereas CNPD carried out between firms operating in different industries creates diversified products or promotes increased efficiency in the firms' activities.

Silva and Moreira conclude that although CNPD is normally carried out between large firms operating in high-tech industries with large-scale production, SMEs operating in high-tech industries involve large firms in CNPD. Moreover, upstream and downstream CNPD is influenced by the technological intensity of firms' operating industries, and firm size affects their intervention in CNPD only when a high scale of production is required.

Product Innovation Charters are the focus of chapter "*It's Time to Include Suppliers in the Product Innovation Charter (PIC)*", where Roy explores the importance of the mission statement of innovation to managers in influencing how and when to involve suppliers. Roy explains the importance of the Product Innovation Charter and argues that suppliers need to be explicitly included in the charter. For that, firms need to be aware of the roles and capabilities, not only of existing suppliers but also of new potential suppliers, as they can be a new source of ideas and technology. In order to balance the innovative potential, whether incremental or radical, the Product Innovation Charter needs to include the management/incorporation of new technologies as well as intellectual property concerns throughout the whole product development process in order to avoid intellectual property leaks and to encourage active/participative supplier involvement. At the end of the chapter, Roy sets out a set of guidelines for framing supplier relationships in a Product Innovation Charter.

In chapter "*Mission Impossible: How to Make Early Supplier Involvement Work in New Product Development?*", Van Weele reports insights from his personal reflection on PhD research projects he has supervised, dealing with the obstacles and difficulties with early supplier involvement in new product development. This chapter builds on the premise that problems of effective supplier involvement are related to the manufacturer organization, the supplier organization, and the supplier–manufacturer relationship. Van Weele supports the idea of using timely supplier involvement rather than early supplier involvement, where timeliness is matched to key processes—prioritizing, mobilizing, coordinating, timing, and informing—

when dealing with supplier interface management involving development management activities, project management activities, and product management activities with suppliers. Effective supplier collaboration must involve strategic, operational, and collaborative management processes, which need the exchange of information between all parties involved and human capital to generate time-tuned group dynamics among firms. Van Weele concludes that, although early supplier involvement may result in disappointments, in order to embark on joint collaborative product development activities, it is important to address the human perspective with sufficient resources and adequate governance rules. Joint project teams need to be aware of the project mission, project objectives, and the project work plan and be aware how investments will be recorded, how both parties deal with intellectual property, and how progress is assessed regularly so that teams are committed to the relationship.

Part III: Strategies and Implications for Innovation

Part III is composed of four chapters that deal with four different topics: purchasing involvement in discontinuous innovation, the importance of culture in information sharing among manufacturing firms, risk allocation and supplier development, and the importance of supply chain innovation.

Calvi, Johnsen, and Picaud address purchasing involvement in discontinuous innovation in chapter "*Purchasing Involvement in Discontinuous Innovation: An Emerging Research Agenda*". This is an important and under-researched topic dealing with the role of the purchasing department in the organizational structure and its influence on product innovation involving discontinuous change. After a systematic literature review of an initial sample of 287 papers that resulted in the analysis of 22 articles, they conclude that a common theme across the research is that radical/discontinuous/breakthrough innovation leads to the need to change supplier relationships, which is at odds with the typical steady-state behavior of most purchasing departments that seek stable relationships that maintain cost and integrate responsibility of their suppliers over the entire product life cycle. Calvi, Johnsen, and Picaud put forward three propositions to help purchasing departments deal with discontinuous innovation. First, the purchasing function needs to go well beyond the existing supply chain, so that it can complement the delicate balance of long-term collaborative relationships with the flexibility of being constantly aware of breakthrough technology from potential new suppliers. Second, the purchasing function needs to handle the challenge of discontinuous innovation through an ambidextrous approach; it is important to organize the purchasing department with skills and competences to develop strategic sourcing activities and at the same time to explore the possibility of embedding new technologies from new sources. Third, the purchasing department, as an innovation-oriented organization, needs to develop absorptive capabilities in order to acquire, assimilate, transform, and exploit external knowledge from the supply chain to help the firm to develop and produce brand new products and generate discontinuous innovation. As recommended in chapter "*It's Time to Include Suppliers in the Product Innovation Charter (PIC)*" by Van Weele, the purchasing department needs to play an important facilitator role with R&D and engineering departments in technology scouting if the firm is to generate new discontinuous products.

Golini, Mazzoleni, and Kalchschmidt study the national culture as an antecedent for information sharing in supply chains in chapter "*National Culture as an Antecedent for Information Sharing in Supply Chains: A Study of Manufacturing Companies in OECD Countries*". Using an interesting approach, Golini et al. investigate the relationship between national culture and the willingness of a company to invest in information sharing with both their suppliers and customers. The main focus of the chapter is the specific role of the country's cultural peculiarities of power distance and individualism-collectivism in influencing the extent of external supply chain information sharing. The authors used data from the fifth edition of the International Manufacturing Strategy Survey, a project that studies manufacturing and supply chain strategies through a detailed questionnaire administered simultaneously in different countries. A set of 392 companies from 16 countries belonging to OECD were used in the analysis. For Golini et al., the results indicate a significant and complex relationship between individualism–collectivism and power distance and the amount of investment that a focal company is willing to make in information sharing with its supply chain partners. The chapter provides an interesting contribution both to theory—extending the debate on supply chain integration at global level—and practice—helping managers to recognize the cultural implications of cross-cultural collaboration.

In chapter "*Risk Allocation and Supplier Development in Automotive Supply Chains: A Study of Nissan Europe*", Camuffo investigates the case of Nissan Europe—in the context of the merger with Renault—to discuss risk allocation in Original Equipment Manufacturer (OEM)–supplier relationship, a topic that is critical to ensure innovation and competitiveness. Camuffo analyzes vertical interfirm relationships and explores the extent to which Nissan shares risk with its suppliers and how the level of risk sharing relates to suppliers' financial, structural, location, and technological characteristics. The data used were collected from a variety of sources, including data provided by Nissan and structured interviews, as well as information on the supplier relations of the Nissan Europe Barcelona plant with 113 companies. These companies supply about 80% of the total purchasing volume for the car models produced at the plant, and for these suppliers, Nissan represents a significant share of their business, up to 60% of their revenues. The results show that the OEM absorbs more risk (a) the greater the supplier's environmental uncertainty, (b) the more risk averse the supplier, and (c) the less severe the supplier's moral hazard. The analysis also shows that Nissan absorbs risk from their suppliers to a non-negligible degree, but that global pressure to reduce costs, technological changes, and organizational changes related to the alliance with Renault moved the company toward a more competitive configuration.

Chapter "*Does Supply Chain Innovation Pay Off?*" investigates the relationship between supply chain innovation and firm performance among 187 Danish manufacturers. Stentoft and Rajkumar separate supply chain innovation into three main components—business process, network structure, and technology. Firm performance is measured in terms of market performance and operational performance. They conclude that, when analyzed as a single construct, supply chain management exerts more influence on operational performance than on market performance. This indicates that although firms understand the development of a market-oriented supply chain, firms must be forced to manage their supply chain if they are to improve their competitive behavior. When the supply chain management construct is decomposed into its three main components, the results are somewhat different. Business process, network structure, and technology all influence operational performance, but only network structure influences market performance. The results clearly indicate that firms associate supply chain management more with operations than with market issues. However, firms need to be aware that if they want to be competitive and succeed in the market place, they must intertwine business processes, network structures, and technology at both market and operational levels.

Part IV: Information and Technology

Part II is composed of four chapters that address technology-based issues that are important for the future competitiveness of firms: Industry 4.0, technological innovation, advanced supply chains, and the role of big data and predictive analytics.

In chapter "*Technological Innovations in Supply Chains*", Druehl, Carrillo, and Hsuan offer an overview of a set of emerging technologiesl—3D printing, virtual reality, autonomous vehicles, drones, and the Internet of Things (IoT)—that can be applied in many stages of the supply chain (SC) and that offer tremendous potential to improve SC transparency, reduce costs, and increase convenience for consumers. Druehl et al. focus on those five technologies to achieve a more profound view of each, their impact on the SC, and interesting future research questions. They highlight not only the stages in which these technologies are but also their potential impact. They discuss each technology, identifying where and how each can be used in the SC. They identify managerial, IS/IT, and policy implications including benefits, risks, existing research, and potential future research areas. Druehl et al. argue that there is still a great deal of uncertainty about these technologies as they are still under development, the regulatory landscape is evolving, and dominant designs and platforms are not yet established. All the technologies will require changes to public or corporate infrastructure such as factories, SC networks, highways, and communication networks. Most require integration with existing SC information systems, as well as with suppliers' and customers' systems, to gain the full benefits. Moreover, the technologies and their uses raise some fundamental questions about data safety and privacy. As standards and dominant designs emerge, there will probably be a period of consolidation in each industry and its supporting industries. More interesting is the question of how these technologies will combine. Combinations seem likely at some future date as these technologies address different needs in the SC, and where they can potentially interact, they seem to reinforce one another, each making the other more useful.

In chapter "*The Role of Informational and Human Resource Capabilities for Enabling Diffusion of Big Data and Predictive Analytics and Ensuing Performance*", Mishra, Luo, and Hazen discuss the role of information and human resource capabilities for enabling diffusion of big data and predictive analytics (BDPA) and ensuing performance. Mishra et al. claim that meaningful information cannot be extracted by just applying analytical tools to data. It requires intense collaboration between analysts and managers exploiting data and analytic tools to discover new knowledge. Innovations like BDPA have the potential to improve customer response times, lower inventories, shorten time to market for new products, improve decision-making processes, and enable a supply chain visibility. However, to realize these benefits of innovation, it is not sufficient to simply adopt the innovation. Instead, it must be accepted, routinized, and assimilated to some extent within the organization. The authors argue that the current knowledge on BDPA regarding how it might link organizational capabilities and organizational performance remains unclear, and knowledge of the support human resources (HR) might give this linkage is even more limited. Drawing from the resource-based view, Mishra et al. propose a model to examine how information technology deployment (strategic information technology flexibility, business–BDPA partnership, and business–BDPA alignment) and HR capabilities affect organizational performance through BDPA. A survey mainly targeted at Indian firms was conducted and 159 usable responses were obtained. Mishra et al. conclude that strategic information technology flexibility, business–BDPA partnership, business–BDPA alignment, and HR capabilities have a direct impact on BDPA diffusion, whereas these constructs have an indirect impact on organizational performance. Those findings provide guidance and assurance that BDPA usage can benefit organizations.

In chapter "*Adoption of Industry 4.0 Technologies in Supply Chains*", Dalmarco and Barros discuss how supply chains may benefit from the adoption of I4.0 technologies by their partners and highlight some of its implementation challenges. I4.0 is a concept used to characterize the new strategic positioning of German industry, based on a flexible Internet-based production system that uses communication improvements that allow a more decentralized production process, integrating sensors and actuators through Internet connection. Dalmarco and Barros analyze eight technologies that cover most I4.0 applications and claim that, at an individual level, technologies such as additive manufacturing, collaborative robots, visual computing, and cyber-physical systems establish the connectivity of a certain company. However, the integration of the whole supply chain, based on the principles of I4.0, demands that information provided by each company (big data) is shared through a collaborative system based on cloud computing and IoT technologies. To share useful information safely, cyber security techniques must be implemented in individual systems and cloud solutions. Summing up, even though the adoption of I4.0 demands an individual initiative, it will only raise the supply chain's competitive advantage if all companies adapt their manufacturing and supply chain processes. The main advantage foreseen here is based on an improved communication system for the whole supply chain, bringing consumers closer to the production process. To assist companies and researchers interested in I4.0 for supply chains, this chapter summarizes the main technologies applied to I4.0 and examples of their adoption by different industries.

Dalmarco and Barros argue that, besides improving the productivity of the supply chain, the adoption of I4.0 technologies adds the possibility of new business models. The integration and expansion of the supply chain and the combination of products and services available to other companies and to the final customer are some of the possibilities available. The development of innovative projects among supply chain companies is also easier when partners are already digitally integrated. In the end, the use of the Internet to share and absorb data is the new trend of the Internet-based society, and the adoption of technologies related to I4.0 is the first step supply chains should take to stay competitive.

To conclude, in chapter "*Advanced Supply Chains: Visibility, Blockchain and Human Behaviour*", Kharlamov and Parry discuss one of the most recent and potentially most significant technologies: Blockchain technology. Blockchain technology is secure by design and can enable decentralization and visibility, with applications in cryptocurrency transactions, historical records, identity management, traceability, authentication, and many other areas. Blockchain technology is a great invention of the digital age with a multitude of possible applications in supply chains. However, successful adoption of such technology requires that the people, process, and technology are ready. Kharlamov and Parry propose a conceptual framework where the concept and technology can balance between positive and negative manifestations depending on human behavior, therefore determining the success of Blockchain technology application in supply chains. Kharlamov and Parry claim that, while both the concept and technology are relatively ready, human behavior is a challenge, as it is known that people suffer from habits and perform poorly when exposed to large volumes of data. The list of biases is extensive with the respective debiasing methods that can potentially help to correct for error. Therefore, any implementation of the Blockchain technology in the future should consider the behavioral aspect in order to ease its implantation, acceptance, and use. The authors claim that much of the possible future of supply chains depends on the readiness of human psychology to accept automated and decentralized systems.

Acknowledgments

First, a word of gratitude to all coauthors who embarked with us in this exploratory journey of intertwining not only topics of innovation with supply chain but also their knowledge and experience creating an insightful book that otherwise would not be possible.

Last but not the least, to all our beloved ones.

Thanks to Alice for "steering the boat" in my absence. Without her patience and support, it would have been harder. To my son and daughter, Luís and Isabel, a word of gratitude. They just don't know that I miss them twice as much as they miss me.

Thanks to Isabel, companion of a life, thank you for your immense patience, support, and love. For my dear children, Joao and Joana, who often wonder why I spend so much time in the office, I dedicate this book to them too.

Thanks to Patrícia for being the "coauthor" of my life now and ever. Thanks to Miguel for being a great son and a little big friend. And thanks to my daughter Lara, who was born during this journey and has been a new reason to carry on and go further.

António Carrizo Moreira
Luís Miguel D. F. Ferreira
Ricardo A. Zimmermann

Contents

About the Editors

António Carrizo Moreira obtained a bachelor's degree in electrical engineering and a master's degree in management, both from the University of Porto, Portugal. He received his PhD in management from UMIST—University of Manchester Institute of Science and Technology, England. He has a solid international background in industry leveraged by working for a multinational company in Germany as well as in Portugal. He has also been involved in consultancy projects and in research activities. He is assistant professor at the Department of Economics, Management, Industrial Engineering, and Tourism, University of Aveiro, Portugal, where he headed the bachelor and master degrees in management for 5 years. He is member of GOVCOPP research unit.

Luís Miguel D. F. Ferreira is currently Assistant Professor of Logistics and Supply Chain Management at the Department of Mechanical Engineering, University of Coimbra, Portugal. He obtained a bachelor's degree in mechanical engineering from the University of Coimbra. He received his master's degree and PhD from Instituto Superior Técnico—University of Lisbon, Portugal. His research interests include topics related to supply chain management, supply chain risk management, sustainable supply chain management, and international purchasing. His research has been published in *Supply Chain Management: An International Journal*, *Production Planning and Control*, among others. He has also been deeply involved in consultancy projects with public institutions and private companies. He is member of CEMMPRE research unit.

Ricardo A. Zimmermann is a research fellow at the Department of Economics, Management, Industrial Engineering, and Tourism in the University of Aveiro, Portugal, and is a member of the Research Unit on Governance, Competitiveness and Public Policies (GOVCOPP). His research interests have been focused on supply chain management, innovation management, and strategic management.

Ricardo has work experience in companies in Brazil and Portugal in areas such as strategic management, quality management, risk management, project management, budget and costs management, and corporate governance. He has also experience as a consultant and as an assessor in quality awards.

Part I
Innovation and Supply Chain Management

The Intellectual Structure of the Relationship Between Innovation and Supply Chain Management

Ricardo A. Zimmermann, Luís Miguel D. F. Ferreira, and António Carrizo Moreira

Abstract Innovation is recognised as an important source of competitive advantage by both academics and managers. Nowadays, supply chain partners play a crucial part in driving many aspects of innovation, from the definition of the product concept to the launch to the market. This chapter analyzes how the relationship between supply chain management and the innovation process is addressed in the literature and discuss ways to improve the performance by means of this relationship. A bibliometric analysis—including citation and co-citation analysis—is carried out to study the intellectual structure of the topic. In the end, four literature clusters were identified, and their characteristics are discussed.

1 Introduction

Innovation is a complex process that is becoming more and more important for businesses as markets are becoming more competitive than ever (Jean et al. 2012). Addressing changes in customer needs, new technologies and trends and performing proactively are all crucial. Supply chain partners play a crucial role in driving innovation forward, both downstream and upstream, from the outset of the product concept phase to the launch of the product to the market. A number of studies refer the importance of supply chains and their actors in the innovation process (Roy and Sivakumar 2010; Golgeci and Ponomarov 2013; Narasimhan and Narayanan 2013; Arlbjorn and Paulraj 2013; Zimmermann et al. 2016).

Innovation enables the development of unique products and services leveraging firms in their quest for competitive advantage (Hilletofth and Eriksson 2011; Blome et al. 2013; Bellamy et al. 2014). As firms' ability to innovate is the result of internal

R. A. Zimmermann (✉) · A. C. Moreira
University of Aveiro, Aveiro, Portugal
e-mail: ricardoaz@ua.pt; amoreira@ua.pt

L. M. D. F. Ferreira
University of Coimbra, Pólo II da Universidade de Coimbra, Coimbra, Portugal
e-mail: luis.ferreira@dem.uc.pt

A. C. Moreira et al. (eds.), *Innovation and Supply Chain Management*, Contributions to Management Science, https://doi.org/10.1007/978-3-319-74304-2_1

and external factors (Roy et al. 2004; Berghman et al. 2012; Fawcett et al. 2012), great innovators depend on external actors to secure most of their advantage when it comes to innovation (Fawcett et al. 2012). Many companies rely on their supply chain partners for innovative input (Koufteros et al. 2007; Zimmermann et al. 2016) and "the development of supply chain management capabilities focusing on innovation is seen as a key competitive weapon" (Blome et al. 2013, p. 60). However, integrating suppliers in product and process development involves significant risk, time, and financial resources from both parties (Koufteros et al. 2007; Silva and Moreira 2017).

A growing body of literature suggests that, to improve their performance, including innovation performance, firms need to deepen the extent of their supply chain integration, cooperation and collaboration, which involves multiple business processes upstream and downstream involving their suppliers, customers and their internal functional units (Petersen et al. 2005; Fawcett et al. 2012; Blome et al. 2013).

Taking these facts into account, this chapter analyzes how the relationship between supply chain management and the innovation process is addressed in the literature. In other words, the study has the objective of analyzing the intellectual structure of the topic by means of a bibliometric analysis. The following research questions are addressed:

- When and where were studies about the relationship between innovation and supply chain published?
- What is the intellectual structure of the literature?
- How has the diffusion of the topic through research literature taken place?
- What are the main themes addressed in the literature on the topic? Is it possible to identify different clusters? What differentiates the clusters?

2 Methodology

A bibliometric analysis was performed as a way of mapping and profiling the literature on the relationship between supply chain management and innovation. The papers were identified using the principles of the systematic literature review method, as presented by Denyer and Tranfield (2009), and were analyzed with the intention of providing useful results for researchers and practitioners. The combination of the two methods is called Systematic Literature Network Analysis (Strozzi et al. 2017). In the first phase the papers are selected and evaluated, and the output of this phase is a set of selected papers. In the second phase the articles are analyzed to answer the research questions.

The ISI Web of Science database was chosen as the source of research. This strategy is used in other reviews of literature in the area (Strozzi et al. 2017). To search for studies to be analyzed, three categories of keywords were defined: (1) Words related to innovation: innovation, innovate, innovativeness. We decided to use the term innovat* to cover all possibilities; (2) Words related to supply chain:

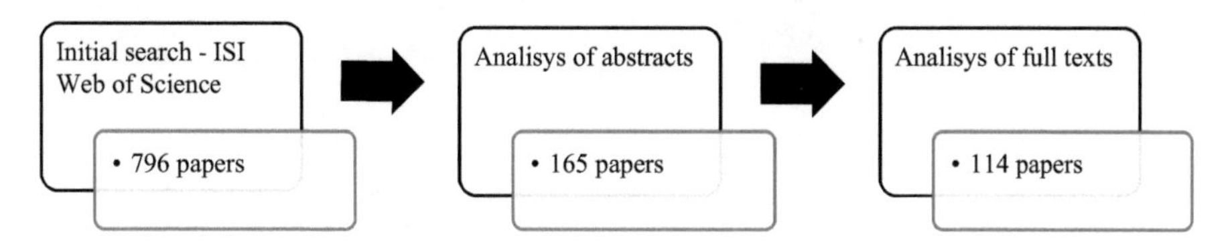

Fig. 1 Location and selection of the articles

supply chain, SCM; (3) Words related to alignment/relationship/partnership: we decided again to use the asterisk in the following terms: align*, partner*, coordinat*, collaborat*, relation*.

The search was based on all possible combinations of the three groups of keywords, using the "Topic" field to search. Only journals (articles and reviews) were searched, limited to the areas of "Business Economics", "Engineering" and "Operations Research Management Science". There was no restriction on the date of publication.

The abstracts and keywords of the articles were read to identify the focus on the relationship between the supply chain and the innovation process of organizations. Finally, the articles were fully read and, using the same criterion, 114 articles were selected (Appendix). The search was conducted in March 2017 (Fig. 1).

Following the suggestion of other studies, and as a way to increase the reliability of the selection, the articles were evaluated simultaneously by the three researchers and doubts and disagreements were discussed until consensus was reached. The articles were only included if all reviewers agreed.

3 Bibliometric Analysis

Gerdsri et al. (2013, p. 404) define bibliometric analysis as "a method that uses statistical and mathematical methods to analyze the literature of a target discipline by investigating the pattern in its bibliographies". In this chapter, the main idea is to get a broad and thorough view of the global context on the topic.

Bibliometrics comprises various methods, usually grouped as citation or co-citation analysis (Charvet et al. 2008). Citation analysis is based on the direct counts of references made to, or received from other documents. Co-citation analysis exploits paired citations as a measure of association between documents, or sets of documents. According to Charvet et al. (2008, p. 48), "one of its major applications is the discovery of intellectual linkages amongst (scholarly) communications and the creation of science maps". Co-citation analysis has been widely used across disciplines, including marketing, operations management, and strategic management.

The program BibExcel was used to conduct the bibliometric and statistical analyses from the 114 articles identified. BibExcel is the software most commonly used for performing bibliometric analysis in management and organizations (Charvet et al. 2008). The data source file used as the input to BibExcel was in a

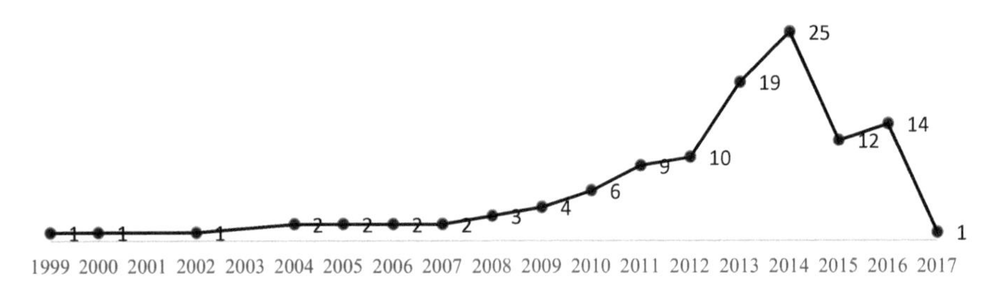

Fig. 2 Number of articles per year

plain text format and contained bibliographic information on the articles. The analysis focused on authors, titles, journals, years of publication, keywords, affiliations and references.

The open source software package Gephi was used to carry out the network analysis and graphical investigation. It uses a 3D render engine to develop illustrations of large networks in real-time and assist in speeding up the exploration process. In the graphs generated, the published papers are shown as nodes and citations are represented by the arcs and between the nodes (Fahimnia et al. 2015).

3.1 *When and Where?*

Initially, the data from the articles were used to help answering the first research question, which is "When and where were the studies about the relationship between innovation and supply chain published?" The answer to this question should clarify the breadth of interest and the potential for emerging, alternative perspectives on the topic. The aspects observed were year of publication, publication source and location of authors.

Figure 2 shows the evolution of the topic in the literature since 1999, when the first article was published. About 70% of the articles were published in the last 5 years (since 2012), which shows that the theme is relatively new in the literature.

When it comes to the journals where the papers were published, there is a clear indication of the relevance and the all-embracing character of the theme, as the articles have been published in 40 different Journals. However, it is clear that the journals in the field of operations management have paid more attention to the topic than the journals in the areas of management, innovation and strategic management. Accordingly, the journals with the largest number of articles are the International Journal of Production Economics, followed by the Journal of Supply Chain Management, and Supply Chain Management: An International Journal. Table 1 presents the main publishing journals.

Table 1 Main sources of publication

Journal	1999	2000	2002	2004	2005	2006	2007	2008	2009	2010	2011	2012	2013	2014	2015	2016	2017	Total
International Journal of Production Economics									1	2	1	2	2	3		2	1	14
Journal of Supply Chain Management										1		1	5	2				9
Supply Chain Management: An International Journal					1			1		1			2	2		1	1	9
Journal of Operations Management					1	1	1		1					2				6
Research Policy			1					1					2			1		5
Production Planning & Control													1	1	2	1		5
International Journal of Production Research													1	3				4
Industrial Management & Data Systems							1				3							4

(continued)

Table 1 (continued)

Journal	1999	2000	2002	2004	2005	2006	2007	2008	2009	2010	2011	2012	2013	2014	2015	2016	2017	Total
Production and Operations Management												1	1	1	1			4
Journal of Purchasing and Supply Management													1	1	2			4
Journal of Product Innovation Management								1		1				1	1			4

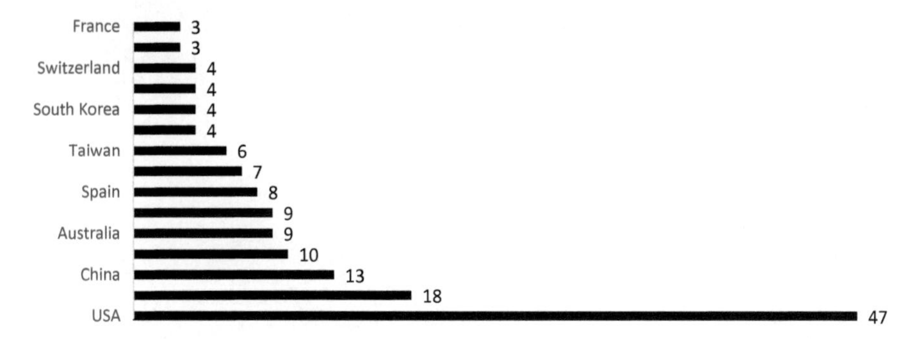

Fig. 3 Countries with the largest number of publications

Finally, the articles are also widely dispersed geographically (authors from 32 countries were identified), demonstrating that the subject is of global interest, as Fig. 3 shows.

This first analysis of the literature shows that the topic has aroused the interest of researchers from different parts of the world in recent years and that the theme has potential for continuous growth.

3.2 Keyword Statistics

Using the data extracted from the papers, an analysis was conducted to identify the most frequently used words and terms in article titles and keywords, respectively. The most frequently used words in paper titles were "supply", "innovation" and "chain". On the other hand, the most popular keywords are "innovation", "supply chain management" and "supply chain". Considering the search terms used to find the articles, there was no surprise in the main words used in titles and keywords.

However, it is important to highlight the use of the word "performance" among the most used words in titles. The high number of papers that uses this word in the title reveals the contribution of the topic to the improvement of firms' performance. Concerning the keywords, it is important to highlight the word "integration", which was used together with the terms "supplier" and "supply chain", and "trust" (Table 2).

3.3 Citation Analysis

To evaluate the relevance of each publication, a citation analysis was conducted, which counts the number of times a paper is cited in other publications. Citation analysis is frequently used to evaluate or compare articles, journals, academic programs and institutions (Charvet et al. 2008). In this case, we use citation analysis to compare the papers and to identify the most influential studies in the area.

Table 2 The most frequently used words in paper titles and keywords

Word in titles	Frequency	Keyword	Frequency
Supply	62	Innovation	33
Innovation	56	Supply chain management	25
Chain	53	Supply chain	13
Product	34	New product development	8
Performance	25	Product development	8
Development	21	Supplier integration	6
Supplier	20	Innovativeness	5
New	20	Supply chain integration	5
Integration	17	China	5
Relationships	12	Product innovation	5
Role	12	Trust	5
Knowledge	11	Game theory	3
Management	11	Smes	3
Effects	10	Open innovation	3
Firm	9	Absorptive capacity	3
Empirical	8	Supply chain performance	3
Collaborative	8	Performance	3
Innovativeness	7	Collaboration	3
Industry	7	Dynamic capabilities	3
Chains	7	Structural equation modeling	3

The BibExcel citation analysis results shows that the 114 articles in the sample cited each other 134 times. The most cited papers in the core sample are shown by number of local citations in Table 3.

3.4 Co-citation Analysis

A co-citation analysis was developed to identify the intellectual structure of the theme. Co-citation analysis is used in the majority of bibliometric studies in management and organizations and citation practices to connect documents, authors, or journals (Zupic and Cater 2015). When co-citation is applied to the cited articles, it is able to identify the knowledge base of a topic and its intellectual structure. The knowledge base of a field is the set of articles most cited by the current research. These publications are the foundations on which current research is being carried out and contain fundamental theories, breakthrough early works, and the methodological canons of the field (Zupic and Cater 2015).

Based on the co-citation analysis, 39 articles emerge as the core sample, as they are the studies which have been cited by the others. However, four articles were removed as they appeared as remote nodes (Fig. 4).

Table 3 Articles from core sample with the highest number of local citations (only those articles with 3 or more)

Article	Local citations
Petersen, K., 2005, V23, P371, J OPER MANAG	21
Roy, S., 2004, V32, P61, J ACAD MARKET SCI	13
Koufteros, X., 2007, V25, P847, J OPER MANAG	10
Soosay, C., 2008, V13, P160, SUPPLY CHAIN MANAG	8
Bhaskaran, S., 2009, V55, P1152, MANAGE SCI	7
Craighead, C., 2009, V27, P405, J OPER MANAG	7
Choi, T., 2006, V24, P637, J OPER MANAG	6
Ettlie, J., 2006, V37, P117, DECISION SCI	4
Jean, R., 2012, V43, P1003, DECISION SCI	3
Kim, B., 2000, V123, P568, EUR J OPER RES	3
Chong, A., 2011, V111, P410, IND MANAGE DATA SYST	3
Narasimhan, R., 2013, V49, P27, J SUPPLY CHAIN MANAG	3
Panayides, P., 2009, V122, P35, INT J PROD ECON	3
Salvador, F., 2013, V49, P87, J SUPPLY CHAIN MANAG	3
Wynstra, F., 2010, V27, P625, J PROD INNOVAT MANAG	3

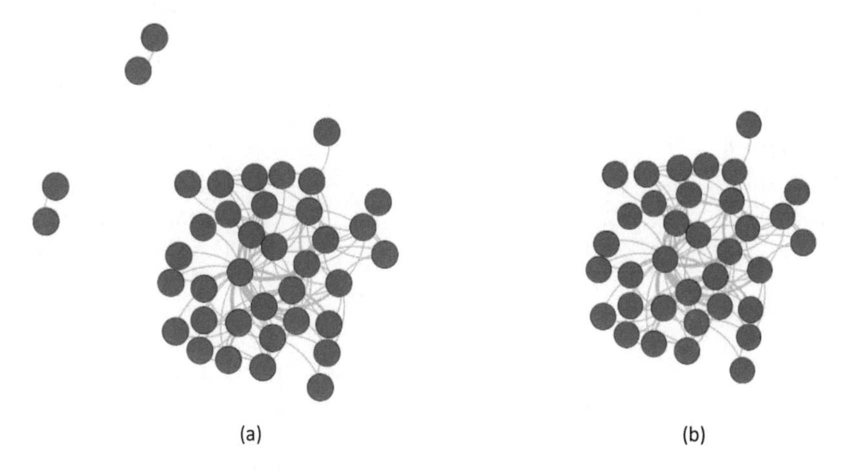

Fig. 4 Co-citation network with and without remote nodes removed. (**a**) The initial 39-node co-citation network. (**b**) The 35-node co-citation network after removing the with remote nodes remote nodes

The 35 papers remaining articles can be understood to be intellectual base of the topic (Table 4).

Table 4 Intellectual base of the topic based on the co-citation analysis

Author	Year	Vol	Journal
Koufteros XA	2007	V25	J OPER MANAG
Petersen KJ	2005	V23	J OPER MANAG
Choi TY	2006	V24	J OPER MANAG
Bhaskaran SR	2009	V55	MANAGE SCI
Ettlie JE	2006	V37	DECISION SCI
Lau AKW	2007	V107	IND MANAGE DATA SYST
McIvor R	2004	V32	OMEGA-INT J MANAGE S
Wagner SM	2014	V32	J OPER MANAG
Jayaram J	2013	V51	INT J PROD RES
Billington C	2013	V22	PROD OPER MANAG
Bellamy MA	2014	V32	J OPER MANAG
Roy S	2004	V32	J ACAD MARKET SCI
Soosay CA	2008	V13	SUPPLY CHAIN MANAG
Roy S	2010	V63	J BUS RES
Jean RJ	2012	V43	DECISION SCI
Seo Y-J	2014	V19	
Wang LW	2011	V134	INT J PROD ECON
Panayides PM	2009	V122	INT J PROD ECON
Pero M	2010	V15	SUPPLY CHAIN MANAG
Blome C	2013	V49	J SUPPLY CHAIN MANAG
Cao M	2010	V128	INT J PROD ECON
Fawcett SE	2012	V55	BUS HORIZONS
Chong AYL	2011	V111	IND MANAGE DATA SYST
Hilletofth P	2011	V111	IND MANAGE DATA SYST
Modi SB	2010	V46	J SUPPLY CHAIN MANAG
Wynstra F	2010	V27	J PROD INNOVAT MANAG
Koufteros X	2012	V48	J SUPPLY CHAIN MANAG
Caridi M	2012	V136	INT J PROD ECON
Craighead CW	2009	V27	J OPER MANAG
Narasimhan R	2013	V49	J SUPPLY CHAIN MANAG
Salvador F	2013	V49	J SUPPLY CHAIN MANAG
Oke A	2013	V49	J SUPPLY CHAIN MANAG
Kim B	2000	V123	EUR J OPER RES
Wong CWY	2013	V146	INT J PROD ECON
He YQ	2014	V147	INT J PROD ECON

3.5 *Data Clustering*

Finally, in order to understand how the literature deals with the different themes that are part of the main topic "supply chain management and innovation", a data clustering analysis was conducted. Cluster analysis is a frequently used technique for finding subgroups inside a topic (Zupic and Cater 2015). The nodes of a network

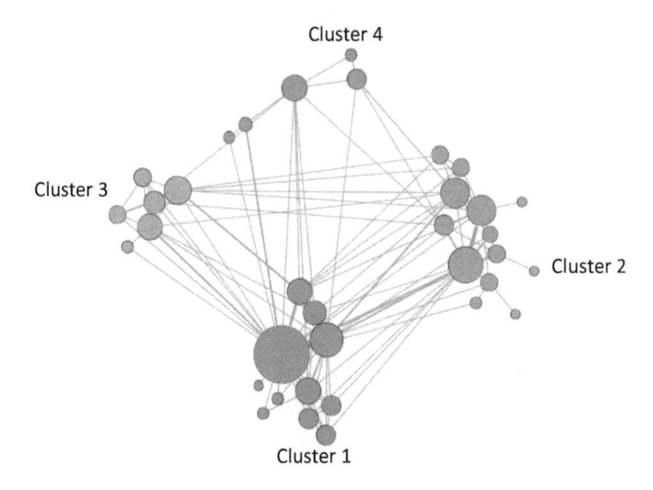

Fig. 5 The position of the four literature clusters

can be divided into clusters where the density of edges is greater between the nodes of the same cluster than those of the others (Fahimnia et al. 2015). A cluster can be seen as a group of well-connected articles in a research area with limited connection to papers in another cluster or research area.

From the intellectual base of the topic, the literature mapping and network analysis identified four clusters. The papers that are part of Cluster 1 focus on the structural characteristics of the supply chain network, with a special focus on the supply base. Cluster 2 is predominately characterized by the study of supply chain trust and collaborative advantage. Authors in Cluster 3 highlight the importance of supplier and customer long term integration. Cluster 4, which was the last cluster to emerge, is composed of a set of papers which approach some trends in the topic, mainly related to strategy. Figure 5 shows the position of the four clusters.

Figure 6 shows the evolution of the clusters over time. It stands out that Cluster 1, 2 and 3 have emerged since the beginning while Cluster 4 emerged later, in 2009. Although Cluster 3 has the first article published on the theme (in 2000), the other papers were published from 2013 onwards, providing evidence of the recent interest in its approach.

Table 5 shows the number of articles published each year in each cluster and Table 6 shows the articles that belong to each cluster.

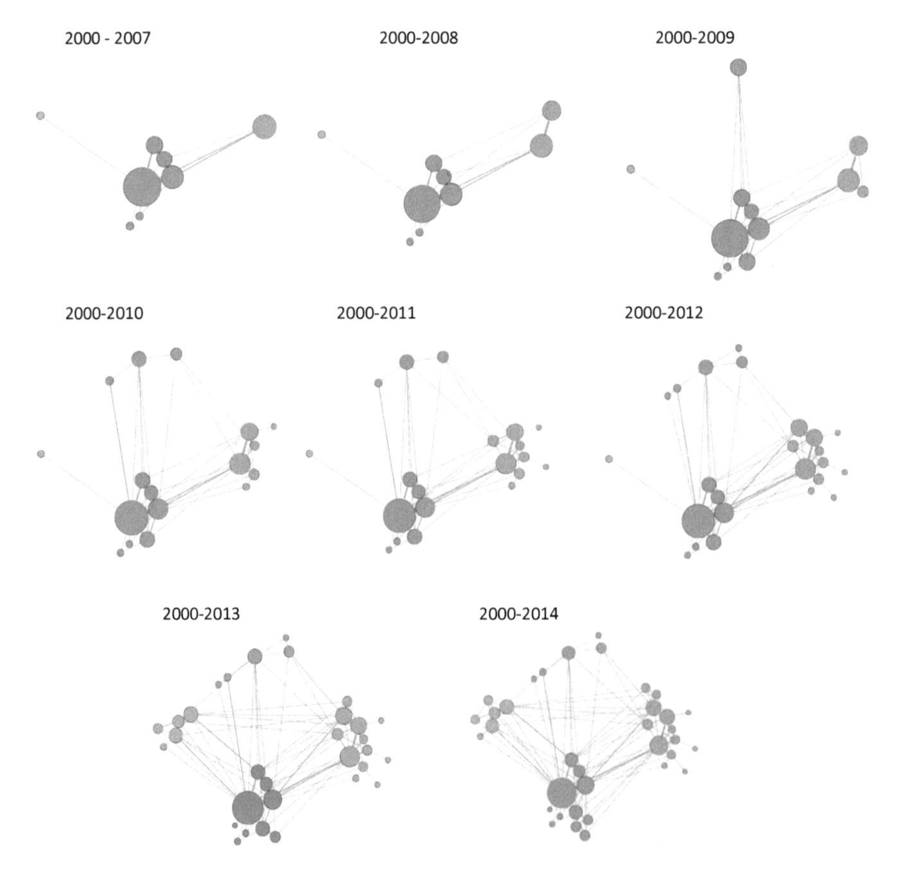

Fig. 6 Evolution of the research areas/clusters over time

Table 5 Number of published papers per cluster

Year	Cluster 1	Cluster 2	Cluster 3	Cluster 4	Total
2000			1		1
2004	1	1			2
2005	1				1
2006	2				2
2007	2				2
2008		1			1
2009	1	1		1	3
2010	1	3		1	5
2011		3			3
2012	1	2		1	4
2013	2	1	4		7
2014	2	1	1		4
Total	13	13	6	3	35

Table 6 Papers belonging to each cluster: co-citation PageRank measure

Cluster 1	Cluster 2	Cluster 3	Cluster 4
Petersen K, 2005, V23, P371, J OPER MANAG	Roy S, 2004, V32, P61, J ACAD MARKET SCI	Narasimhan R, 2013, V49, P27, J SUPPLY CHAIN MANAG	Craighead C, 2009, V27, P405, J OPER MANAG
Koufteros X, 2007, V25, P847, J OPER MANAG	Soosay C, 2008, V13, P160, SUPPLY CHAIN MANAG	Salvador F, 2013, V49, P87, J SUPPLY CHAIN MANAG	Wynstra F, 2010, V27, P625, J PROD INNOVAT MANAG
Bhaskaran S, 2009, V55, P1152, MANAGE SCI	Jean R, 2012, V43, P1003, DECISION SCI	Wong C, 2013, V146, P566, INT J PROD ECON	Koufteros X, 2012, V48, P93, J SUPPLY CHAIN MANAG
Choi T, 2006, V24, P637, J OPER MANAG	Panayides P, 2009, V122, P35, INT J PROD ECON	Oke A, 2013, V49, P43, J SUPPLY CHAIN MANAG	Caridi, M, 2012, V136, P207, INT J PROD ECON
Ettlie J, 2006, V37, P117, DECISION SCI	Wang L, 2011, V134, P114, INT J PROD ECON	He Y, 2014, V147, P260, INT J PROD ECON	Modi, S, 2010, V46, P81, J SUPPLY CHAIN MANAG
Wagner S, 2014, V32, P65, J OPER MANAG	Chong A, 2011, V111, P410, IND MANAGE DATA SYST	Kim B, 2000, V123, P568, EUR J OPER RES	
Billington C, 2013, V22, P1464, PROD OPER MANAG	Seo Y-J., 2014, V19, SUPPLY CHAIN MANAGEM		
Bellamy M, 2014, V32, P357, J OPER MANAG	Blome C, 2013, V49, P59, J SUPPLY CHAIN MANAG		
McIvor R, 2004, V32, P179, OMEGA-INT J MANAGE S	Roy S, 2010, V63, P1356, J BUS RES		
Lau A, 2007, V107, P1036, IND MANAGE DATA SYST	Cao M, 2010, V128, P358, INT J PROD ECON		
Jayaram J, 2013, V51, P1958, INT J PROD RES	Fawcett S, 2012, V55, P163, BUS HORIZONS		
	Hilletofth P, 2011, V111, P184, IND MANAGE DATA SYST		
	Pero M, 2010, V15, P115, SUPPLY CHAIN MANAG		

4 The Main Topics in the Literature and the Characteristics of the Clusters

In this section the main characteristics of the clusters are discussed. However, it is important to highlight some general features of the literature on this topic. Regarding the methodology used, there is a predominance of quantitative empirical studies and concerning the nature of the samples, there was a predominance of the use of information from industrial companies.

When it comes to the theoretical perspective, the analysis of the papers showed that there was no dominant theory on the relationship between innovation and supply chains as more than 30 different theories were mentioned. The resource-based view was the theory with the largest number of articles, followed by the knowledge-based view and transaction cost economics. Moreover, there is a recent trend regarding the use of the resource-based view of the firm, which was heavily cited in recent publications. Another important factor is that there are a considerable number of papers that do not mention their theoretical basis.

4.1 Cluster 1: Supply Network Structural Characteristics

The 11 papers that are part of this cluster study, in general, the structural characteristics of the supply chain network, with special focus on the supply base. The supply base is understood as the "portion of a supply network that is actively managed by a buying company" (Choi and Krause 2006, p. 637).

The supply network of a firm, and specially the supply base, has been viewed as an important source of innovation—in addition to the operational benefits of managing it effectively—and its structural characteristics have a great influence on a firm's innovation outputs (Bellamy et al. 2014). The supply network provides critical conduits for knowledge and information flows and the structural characteristics define the way in which firms manage knowledge and information sharing (or integration) with their partners (Billington and Davidson 2013; Jayaram and Pathak 2013; Bellamy et al. 2014).

Information and knowledge integration is an effective strategy to achieve superior innovation or new product development performance, and the context of new product development is important and promising for knowledge integration (Jayaram and Pathak 2013). As the capability share knowledge and information between firms, mainly as a result of the growth of the Internet, often makes it easier for companies to access external resources than to develop them internally (Billington and Davidson 2013), open innovation is addressed by Billington and Davidson (2013) as a network structure that can facilitate the overall relationship between firms, especially the sharing of information, knowledge and decision making and, therefore, collaboration in research and development of new products and processes. In addition, Ettlie and Pavlou (2006) argue that information and knowledge sharing makes the development of technology-based new products possible.

Although firms use many mechanisms to help preserve and stimulate the creation of knowledge, it is still difficult for many firms to transfer internal knowledge to actors that are external to the firm and vice versa. Accordingly, it is the responsibility of the firms to find the right partners and build what Jayaram and Pathak (2013) call 'enterprise-wide knowledge architectures'. Thus, "to achieve product co-development with suppliers and customers, managers should identify, assess and qualify competent partners as a major supply base" (Lau et al. 2007, p. 1054). The importance of the supplier selection for integrating them in the new products process, considering "not only the capabilities, but also the culture of the supplier, which will have an impact on the buying firm's ability to interact with the supplier effectively" has to be emphasised (Petersen et al. 2005). Therefore, Lau et al. (2007) discuss three types of co-development: supplier co-development (SC); customer co-development (CC); and internal co-development (IC). The type of co-development determines the main partner(s) in the innovation process.

Regarding the level of involvement of the supply chain partners, Petersen et al. (2005) suggest three basic forms of supplier involvement in product development: white-box, grey-box and black-box approaches. In summary, in the white-box approach, the suppliers are consulted about new product development and the integration is informal. In the grey-box model, the supplier and the customer work alongside each other and the supplier provides expertise, suggestions and other inputs to the product development effort but typically will not assume sole responsibility for developing parts, let alone modules, for the final product (Koufteros et al. 2007). Finally, a black-box approach implies that each company will concentrate on certain tasks and components. In this case, the supplier can be "trusted" to develop parts and components.

Besides the level of involvement, it is also important to discuss when the partners will participate in the innovation or new product development process. Several authors (McIvor and Humphreys 2004; Petersen et al. 2005; Lau et al. 2007), highlight the role of early supplier, and client, involvement in the design process as a central attribute for the success of the co-development of new products.

The management of the supply network also can be seen as a cost sharing mechanism and a way of optimizing the research and development process. Bhaskaran and Krishnan (2009) propose a model which includes the interfirm interaction, the co-development process, technological uncertainty, the information structure and decision sequence. Depending on the type of project, the investment and revenue are shared. Wagner and Bode (2014) discuss the important differences between process and product innovation sharing, and the role of supplier-relationship-specific investments and safeguards for the investments for supplier innovation sharing.

4.2 Cluster 2: Supply Chain Trust and Collaborative Advantage

The 13 articles in Cluster 2 focus on the relationships, as opposed to the structural characteristics. The two most important features for the authors in this cluster are trust between partners (Panayides and Lun 2009; Wang et al. 2011; Fawcett et al. 2012; Blome et al. 2013; Jean et al. 2014) and building alliances (Roy et al. 2004; Soosay et al. 2008; Blome et al. 2013).

For Fawcett et al. (2012, p. 163), "trust is at the heart of a collaborative innovation capability". The objective of the relationships is to gain collaborative advantage, which is defined as "strategic benefits gained over competitors in the market place through supply chain partnering and partner enabled knowledge creation, and it relates to the desired synergistic outcome of collaborative activity that could not have been achieved by any firm acting alone".

Trust between supply chain partners can be seen as a catalyst for collaborative innovation (Fawcett et al. 2012). It is important to search for supply chain partners with distinctive complementary capabilities and create unique collaborative relationships with them to generate unparalleled process and product innovation (Fawcett et al. 2012). In this context, trust is an essential element of relational architecture and "without a foundation of trust, collaborative alliances can neither be built nor sustained" (Fawcett et al. 2012, p. 164). Fawcett et al. (2012) identified four stages of trust: limited trust, transactional trust, relational trust, and collaborative trust. In the last stage, relationships entail a common belief leading parties to view supply chain partners' capacity and capabilities as an extension of their own business. Soosay et al. (2008) also describe trust as one of the most important characteristics to reinforce collaboration and, as a consequence, improve innovation performance.

Trust allows supply chain partners to build collaborative relationships (Roy et al. 2004; Fawcett et al. 2008; Cao and Zhang 2010; Hilletofth and Eriksson 2011). In the supply chain context, building collaborative relationships can help firms share risks, access complementary resources, reduce transaction costs and enhance productivity, and, therefore, enhance profit performance and competitive advantage over time (Cao and Zhang 2010; Chong et al. 2011). According to Cao and Zhang (2010), by collaborating, supply chain partners can work as if they were part of a single enterprise and such collaboration can increase joint competitive advantage. Collaborating with supply chain partners can involve activities such as sharing information, synchronizing decisions, sharing complementary resources, and aligning incentives with partners' costs and risks (Cao and Zhang 2010).

Roy et al. (2004) propose a framework in which the link between interactions and innovation generation is moderated by several factors, which can be grouped as internal or external. In the set of internal and dyadic buyer/supplier relationship factors, they highlight IT adoption, commitment and trust. The authors focused on the upstream supply chain relationships. Roy and Sivakumar (2010) studied innovation generation considering upstream and downstream relationships. In this study, the authors highlight the importance of complexity and globalization as moderator effects for the relationship between interaction and innovation generation. Chong et al. (2011) emphasizes that through strategic supplier partnerships, organizations can work closely with suppliers who can share responsibility for the success of products, in a relationship characterised by trust.

In addition to the partnership with suppliers and clients in the new product development process, it is important to coordinate the different functions inside the company. Hilletofth and Eriksson (2011) defend the involvement of members of the main functions of the company in the design stage of new products and single out the role of the supply chain in the success of the products and the improvement of performance. The model presupposes a strong view on the demand side and a consumer-oriented perspective.

Finally, Jean et al. (2012) discuss the role of power-dependence and study the supplier dependence on the buyer as a moderator of the effects of supplier market knowledge acquisition, relationship learning, systems collaboration, and technological uncertainty on supplier innovation generation. The authors claim to provide "a strong theoretical and empirical foundation for understanding how suppliers can augment their innovation capabilities by working with their customers in cross-border exchange relationships, and thus improve performance outcomes" (Jean et al. 2012, p. 1030).

4.3 Cluster 3: Supplier and Customer Long Term Integration

The six papers which compose Cluster 3 highlight the importance of supplier and customer long term integration. Topics such as partnership, strategic alignment and strategic relationships are discussed by the authors. According to Wong et al. (2013, p. 567), "external integration involves the strategic alignment of business processes, information sharing and joint collaboration with suppliers and customers" and helps firms to establish mutual understanding and gain information through network relationships.

Strategic relationships with supply chain partners, are defined by Oke et al. (2013, p. 44) "in terms of the extent to which the relationship is enduring and on a long-term basis". Considering the risks involved in the innovation process, suppliers are more likely to align with customers for innovation if there is a long-term relationship in place (Oke et al. 2013). In addition to the importance of building long-term relationships with partners, the authors highlight the need to create strategic collaboration with the most important partners, which creates mutual benefits. For Lee et al. (2014) integration with other supply chain actors presupposes partnership, which is characterized by a long-term commitment between the collaborators. The authors emphasize that integration in the context of NPD has different forms, internal or external to the firm boundaries, such as cross-functional team integration, intra-process or concurrent integration, resource integration, supply chain or external integration. For the authors, supplier integration has a positive effect on customer integration and they recommend that managers adopt the practice of supplier integration first. According to Salvador and Villena (2013), integrating suppliers into NPD projects offers manufacturers the potential for substantial improvements in the new product being designed.

Kim (2000) approaches coordination of the innovation process as a way to manage supplier innovation, considering that the innovation generated by the client company can lead to reduced costs for the supplier and, consequently, a reduction in the prices of their products. The coordination of innovation, for Kim (2000), is based on the long term relationship between client and supplier, which is characterised by trust and shared information and decision making.

Finally, for Narasimhan and Narayanan (2013) it is crucial that companies align their internal research and development strategies with the knowledge available in the supply chain in order to achieve better performance with regard to innovation (Narasimhan and Narayanan 2013). The authors define innovation as the process of generating changes in products, processes and services that result in the creation of value for the firm and its customers, through the knowledge generated by the company and/or its supply chain partners. Thus, the main reason to collaborate with other companies is to share and leverage resources unavailable internally.

4.4 Cluster 4: Emergent Topics

Cluster 4, which was the last cluster to emerge and is composed of five articles, is a set of papers that approach some trends in the topic. However, it is important to highlight that, considering that the analysis is based in the co-citation of papers, the newest studies in the area are not included in any cluster because they were not co-cited at the time the analysis was carried out.

The papers in this cluster mainly deal with topics related to strategy, such as knowledge management and supply chain knowledge, strategic supplier selection, supplier strategic focus on innovation, supply chain efficiency and product modularity. Supply chain strategy, knowledge, and action are key antecedents to firm performance (Craighead et al. 2009; Wynstra et al. 2010). Supply chain knowledge, in turn, can be understood in terms of three constructs: learning progression, use of existing knowledge, and organizational memory (Craighead et al. 2009). Companies "need to fit a supply chain's innovation–cost strategy to knowledge elements in a way that enhances action and creates superior firm performance" (Craighead et al. 2009, p. 418).

Efficiency is a core concept for operations management that influences firms' success in a general way and a central facet of supply chain management is the efficient flow of materials within the organization and across the firm's boundaries (Modi and Mabert 2010). Modi and Mabert (2010) study the relationship between efficient supply chain management and innovation and conclude that over time a firm's supply chain performance and supply chain stability positively influence the volume of its innovations.

Supply chain efficiency is also related to supplier selection. As firms become more dependent on their suppliers, the capabilities of those suppliers serve as key resources in the development of the buyer's own capabilities and performance (Koufteros et al. 2012). Strategic supplier selection has a positive effect on firm performance, including innovation performance (Wynstra et al. 2010; Koufteros et al. 2012). Moreover, supplier product development activity is directly affected by the supplier's position in the supply chain, by an explicit strategic focus on innovation and by commitment to customer development (Wynstra et al. 2010). The selection and the position in the supply chain will affect supplier innovation and, consequently, the customer innovation process and performance.

Finally, the product characteristics also influence the way that suppliers and clients participate in the innovation process (Caridi et al. 2012). The level of modularity, for example, is significantly related to new product performance (Caridi et al. 2012). Thus, identifying and qualifying the appropriate partners as a supply base for module design and production enhances the firm's capability to modularize products successfully, by leveraging the technological resources from the supply base.

5 Conclusions

Innovation is a complex process which is becoming more and more important for companies as markets become more competitive. This chapter has described and discussed how the relationship between innovation and supply chain management is addressed in the literature, identifying the intellectual structure of the topic. The analysis of the literature shows that the topic has aroused the interest of researchers from different parts of the world in recent years and that the theme has the potential for continuous growth. The dispersed character of the publications that are sources of information and the theoretical perspectives used also reinforce the broader character of the theme.

Different ways of addressing the topic were found in different journals and in different contexts. The importance of strong collaboration among supply chain partners for innovation performance is clear, even though that collaboration is seen and discussed in different ways in the literature.

After a bibliometric analysis of 114 studies, the intellectual base of the field was identified, composed by 35 studies. From this intellectual base, four main clusters were identified: the papers which are part of Cluster 1 focus on the structural characteristics of the supply chain network. Cluster 2 is predominately characterized by the study of supply chain trust and collaborative advantage. Authors in Cluster 3 highlight the importance of supplier and customer long term integration. And Cluster 4 is composed of a set of papers that explore some new trends on the topic.

However, in addition to identifying particular features of each cluster, it is also possible to find great similarities between the four groups of studies. As common characteristics, we can highlight the willingness to collaborate and the importance of communication between firms.

This chapter contributes to theory by identifying the different approaches that address the relationship between innovation and supply chains in the literature, and it contributes to practice by providing some ideas to stimulate this relationship and improve performance.

As a recommendation for future research, we highlight the emergence of new topics which can be explored in the future, such as the importance of new technologies for the relationship between innovation and supply chain, the study of the fit between innovation capabilities and strategies and supply chain strategies, and the effects of supply chains on the different types of innovation (for example, product or process, radical or incremental).

Finally, as a limitation, the study is based on the analysis of published papers available in the ISI Web of Science database. Accordingly, themes which are in vogue at the moment, such as new technologies—the Internet of Things, virtual reality, autonomous vehicles and drones—and their importance for supply chains and innovation, were not considered in this study. Moreover, considering that the cluster analysis is based on the co-citation of the papers, the newest studies in the area are not included in any cluster because they were yet to be co-cited at the time the analysis was conducted.

Acknowledgements The authors acknowledge the FCT—Fundação para a Ciência e a Tecnologia (Portuguese Foundation for Science and Technology) for the financial support of this work by means of a PhD scholarship with co-participation of the European Social Fund.

Appendix: Studies About the Relationship Between Supply Chain and Innovation

Paper	Year	Journal
Bruce, M; Moger ST	1999	Technology Analysis & Strategic Management
Kim, B	**2000**	**European Journal of Operational Research**
Sobrero, M; Roberts EB	2002	Research Policy
Roy S; Sivakumar K; Wilkinson IF	**2004**	**Journal of the Academy of Marketing Science**
McIvor R, Humphreys P	**2004**	**Omega**
Kim B, Oh H	2005	Supply Chain Management: An International Journal
Petersen KJ, Handfield RB, Ragatz GL	**2005**	**Journal of Operations Management**
Ettlie JE, Pavlou PA	**2006**	**Decision Sciences**
Choi TY, Krause DR	**2006**	**Journal of Operations Management**

(continued)

Paper	Year	Journal
Lau AKW, Yam RCM, Tang EPY	**2007**	**Industrial Management & Data Systems**
Koufteros XA, Cheng TCE, Lai KH	**2007**	**Journal of Operations Management**
Soosay CA, Hyland PW, Ferrer M	**2008**	**Supply Chain Management: An International Journal**
Tether BS, Tajar A	2008	Research Policy
Lee J, Veloso FM	2008	Journal of Product Innovation Management
Craighead CW, Hult GTM, Ketchen Jr DJ	**2009**	**Journal of Operations Management**
Bhaskaran RS, Krishnan V	**2009**	**Management Science**
Bakhshi H, McVittie E	2009	Innovation: Management, Policy & Practice
Panayides PM, Lun YHV	**2009**	**International Journal of Production Economics**
Modi SB, Mabert VA	**2010**	**Journal of Supply Chain Management**
Pero M, Abdelkafi N, Sianesi A, Blecker T	**2010**	**Supply Chain Management: An International Journal**
Wynstra F, von Corswant F, Wetzels M	**2010**	**Journal of Product Innovation Management**
Lin YC, Wang YC, Yu CH	2010	International Journal of Production Economics
Roy S, Sivakumar K	**2010**	**Journal of Business Research**
Cao M, Zhang Q	**2010**	**International Journal of Production Economics**
Hilletofth P, Eriksson D	**2011**	**Industrial Management & Data Systems**
Chong AYL, Chan FTS, Ooi KB, Sim JJ	2011	Industrial Management & Data Systems
Lau AKW	2011	Industrial Management & Data Systems
Johnsen TE	2011	International Journal of Operations & Production Management
Zolghadri M, Amrani A, Zouggar S, Girard P	2011	International Journal of Computer Integrated Manufacturing
Wang LW, Yeung JHY, Zhang M	**2011**	**International Journal of Production Economics**
Bouncken RB	2011	Engineering Management Journal
Hernández-Espallardo M, Sánchez-Pérez M, Segovia-López C	2011	Technovation
Lee KH, Kim JW	2011	Business Strategy and the Environment
Koufteros X, Vickery SK, Droge C	**2012**	**Journal of Supply Chain Management**
Berghman L, Matthyssens P, Vandenbempt K	2012	Industrial Marketing Management

(continued)

Paper	Year	Journal
Fawcett SE, Jones SL, Fawcett AM	**2012**	**Business Horizons**
Caridi M, Pero M, Sianesi A	**2012**	**International Journal of Production Economics**
Hsieh KN, Tidd J	2012	Technovation
Jean RJ, Kim D, Sinkovics RR	**2012**	**Decision Sciences**
Langenberg KU, Seifert RW, Tranchez JS	2012	International Journal of Production Economics
Hazen BT, Overstreet RE, Cegielski CG	2012	The International Journal of Logistics Management
Machikita T, Ueki Y	2012	Asian Journal of Technology Innovation
Bendoly E, Bharadwaj A	2012	Production and Operations Management
Salvador F, Villena VH	**2013**	**Journal of Supply Chain Management**
Kuhne B, Gellynck X, Weaver RD	2013	Supply Chain Management: An International Journal
Golgeci I, Ponomarov SY	2013	Supply Chain Management: An International Journal
Tomlinson PR, Fai FM	2013	International Journal of Production Economics
Peitz M, Shin D	2013	Journal of Economic Behavior & Organization
Fitjar RD, Rodriguez-Pose A	2013	Research Policy
Cabigiosu A, Zirpoli F, Camuffo A	2013	Research Policy
Jayaram J, Pathak S	**2013**	**International Journal of Production Research**
Tracey M, Neuhaus R	2013	Journal of Purchasing & Supply Management
Peng DX, Verghese A, Shah R, Schroeder RG	2013	Journal of Supply Chain Management
Narasimhan R, Narayanan S	**2013**	**Journal of Supply Chain Management**
Oke A, Prajogo DI, Jayaram J	**2013**	**Journal of Supply Chain Management**
Blome C, Schoenherr T, Kaesser M	**2013**	**Journal of Supply Chain Management**
Vickery SK, Koufteros X, Droge C	2013	IEEE Transactions on Engineering Management
Wong CWY, Wong CY, Boon-itt S	**2013**	**International Journal of Production Economics**
Billington C, Davidson R	**2013**	**Production and Operations Management**
Fox GL, Smith J, Cronin Jr JJ, Brusco M	2013	International Journal of Operations & Production Management
Ganotakis P, Hsieh WL, Love JH	2013	Production Planning & Control

(continued)

Paper	Year	Journal
Germani M, Mandolini M, Mengoni M, Peruzzini M	2013	International Journal of Computer Integrated Manufacturing
Cheng JH, Chen MC, Huang CM	2014	Supply Chain Management: An International Journal
He YQ, Lai KK, Sun HY, Chen Y	**2014**	**International Journal of Production Economics**
Jean RJ, Sinkovics RR, Hiebaum TP	2014	Journal of Product Innovation Management
Wagner SM, Bode C	**2014**	**Journal of Operations Management**
Yeniyurt S, Henke Jr. JW, Yalcinkaya G	2014	Journal of the Academy of Marketing Science
Ge ZH, Hu QY, Xia YS	2014	Production and Operations Management
Schoenherr T, Griffith DA, Chandra A	2014	International Journal of Production Research
Tan YC, Ndubisi NO	2014	Journal of Business & Industrial Marketing
Seo YJ, Dinwoodie J, Kwak DW	**2014**	**Supply Chain Management: An International Journal**
Storer M, Hyland P, Ferrer M, Santa R, Griffiths A	2014	The International Journal of Logistics Management
Wu GD	2014	International Journal of Simulation Model
Jafarian M, Bashiri M	2014	Applied Mathematical Modelling
Saenz MJ, Revilla M, Knoppen D	2014	Journal of Supply Chain Management
von Massow M, Canbolat M	2014	International Journal of Production Research
Gualandris J, Kalchschmidt M	2014	Journal of Purchasing & Supply Management
Ma XF, Kaldenbach M, Katzy B	2014	Technology Analysis & Strategic Management
Pulles NJ, Veldmann J, Schielle H, Sierksma H	2014	Journal of Supply Chain Management
Hernández JE, Lyons AC, Mula J, Poler R, Ismail H	2014	Production Planning & Control: The Management of Operations
Chong AYL, Zhou L	2014	International Journal of Production Economics
Bellamy MA, Ghosh S, Hora M	**2014**	**Journal of Operations Management**
Liao SH, Kuo FI	2014	International Journal of Production Economics
Singh PJ, Power D	2014	International Journal of Production Research
Lee VH, Ooi KB, Chong AYL, Seow C	2014	Expert Systems with Applications

(continued)

Paper	Year	Journal
Lefebvre VM, Raggi M, Viaggi D, Sia-Ljungström C, Minarelli F, Kühne B, Gellynck X	2014	Creativity and Innovation Management
Manasakis C, Pretrakis E, Zikos V	2014	Southern Economic Journal
Piening EP, Salge TO	2015	Journal of Product Innovation Management
Ren S, Eisingerich AB, Tsai H	2015	Journal of Business Research
Wang J, Shin H	2015	Production and Operations Management
Golgeci I, Ponomarov SY	2015	Technology Analysis & Strategic Management
Arsenyan J, Büyüközkan G, Feyzioglu O	2015	Expert Systems with Applications
Zhang, H. P.	2015	International Journal of Simulation Model
Steven Jifan Ren, Caihong Hu, E.W.T. Ngai and Mingjian Zhou	2015	Production Planning and Control
Stefanie Herrmann, Helen Rogers, Marina Gebhard and Evi Hartmann	2015	Production Planning and Control
Janice E. Carrillo, Cheryl Druehl	2015	Decision Sciences
Maarten Sjoerdsma, Arjan J.van Weele	2015	Journal of Purchasing & Supply Management
Eman S. Nasr, Marc D. Kilgour, Hamid Noori	2015	European Journal of Operational Research
Erica Mazzola, Manfredi Bruccoleri, Giovanni Perrone	2015	Journal of Purchasing & Supply Management
David Elvers and Chie Hoon Song	2016	Journal of Business & Industrial Marketing
Wu, S. B; Gu, X; Wu, G. D. and Zhou, Q	2016	International Journal of Simulation Model
Seong No Yoon, DonHee Lee, Marc Schniederjans	2016	Technological Forecasting & Social Change
Ricarda B. Bouncken, Boris D. Pluschke, Robin Pesch, Sascha Kraus	2016	Review of Management Science
Rosanna Fornasiero, Andrea Zangiacomi, Valentina Franchini, João Bastos, Americo Azevedo and Andrea Vinelli	2016	Production Planning & Control
Min Zhang, Xiande Zhao, Chris Voss, Guilong Zhu	2016	International Journal of Production Economics
Olov H.D. Isakssona, Markus Simeth, Ralf W. Seifert	2016	Research Policy
Graciela Corral de Zubielqui, Janice Jones and Larissa Statsenko	2016	Entrepreneurship Research Journal
I. Robert Chiang and S. Jinhui Wu	2016	IEEE Transactions on Engineering Management
Divesh Ojha, Jeff Shockley, Chandan Achary	2016	International Journal of Production Economics

(continued)

Paper	Year	Journal
María Isabel Roldán Bravo, Antonia Ruiz Moreno and Francisco Javier Llorens-Montes	2016	Supply Chain Management: An International Journal
Giovanna Lo Nigro	2016	Journal of Business Research
Edward C. S. Ku, Wu-Chung Wu, Yan Ju Chen	2016	Information Systems and e-Business Management
Ricardo Zimmermann, Luís M.D. Ferreira, António C. Moreira	2016	Supply Chain Management: An International Journal
Tingting Yana, Arash Azadegan	2017	International Journal of Production Economics

The papers that are part of the intellectual base of the topic are presented in bold

References

Arlbjorn JS, Paulraj A (2013) Special topic forum on innovation in business networks from a supply chain perspective: current status and opportunities for future research. J Supply Chain Manag 49:3–11

Bellamy MA, Ghosh S, Hora M (2014) The influence of supply network structure on firm innovation. J Oper Manag 32:357–373

Berghman L, Matthyssens P, Vandenbempt K (2012) Value innovation, deliberate learning mechanisms and information from supply chain partners. Ind Mark Manag 41:27–39

Bhaskaran SR, Krishnan V (2009) Effort, revenue, and cost sharing mechanisms for collaborative new product development. Manag Sci 55:1152–1169

Billington C, Davidson R (2013) Leveraging open innovation using intermediary networks. Prod Oper Manag 22:1464–1477

Blome C, Schoenherr T, Kaesser M (2013) Ambidextrous governance in supply chains: the impact on innovation and cost performance. J Supply Chain Manag 49:59–80

Cao M, Zhang Q (2010) Supply chain collaborative advantage: a firm's perspective. Int J Prod Econ 128:358–367

Caridi M, Pero M, Sianesi A (2012) Linking product modularity and innovativeness to supply chain management in the Italian furniture industry. Int J Prod Econ 136:207–217

Charvet FF, Cooper MC, Gardner JT (2008) The intellectual structure of supply chain management: a bibliometric approach. J Bus Logist 29:47

Choi TY, Krause DR (2006) The supply base and its complexity: implications for transaction costs, risks, responsiveness, and innovation. J Oper Manag 24:637–652

Chong AYL, Chan FTS, Ooi KB, Sim JJ (2011) Can Malaysian firms improve organizational/ innovation performance via SCM? Ind Manag Data Syst 111:410–431

Craighead CW, Hult GTM, Ketchen DJ Jr (2009) The effects of innovation–cost strategy, knowledge, and action in the supply chain on firm performance. J Oper Manag 27(16)

Denyer D, Tranfield T (2009) Producing a systematic review. In: Buchanan DA, Bryman A (eds) The Sage handbook of organizational research methods. Sage, London

Ettlie JE, Pavlou PA (2006) Technology-based new product development partnerships. Decis Sci 37:117–147

Fahimnia B, Tang CS, Davarzani H, Sarkis J (2015) Quantitative models for managing supply chain risks: a review. Eur J Oper Res 247:1–15

Fawcett SE, Magnan GM, McCarter MW (2008) Benefits, barriers, and bridges to effective supply chain management. Supply Chain Manag Int J 13:35–48

Fawcett SE, Jones SL, Fawcett AM (2012) Supply chain trust: the catalyst for collaborative innovation. Bus Horiz 55:163–178
Gerdsri N, Kongthon A, Vatananan RS (2013) Mapping the knowledge evolution and professional network in the field of technology roadmapping: a bibliometric analysis. Technol Anal Strateg Manag 25:403–422
Golgeci I, Ponomarov SY (2013) Does firm innovativeness enable effective responses to supply chain disruptions? An empirical study. Supply Chain Manag Int J 18:604–617
Hilletofth P, Eriksson D (2011) Coordinating new product development with supply chain management. Ind Manag Data Syst 111:264–281
Jayaram J, Pathak S (2013) A holistic view of knowledge integration in collaborative supply chains. Int J Prod Res 51:1958–1972
Jean R-JB, Kim D, Sinkovics RR (2012) Drivers and performance outcomes of supplier innovation generation in customer-supplier relationships: the role of power-dependence. Decis Sci 43:1003–1038
Jean RJ, Sinkovics RR, Hiebaum TP (2014) The effects of supplier involvement and knowledge protection on product innovation in customer-supplier relationships: a study of global automotive suppliers in China. J Prod Innov Manag 31:98–113
Kim B (2000) Coordinating an innovation in supply chain management. Eur J Oper Res 123:568–584
Koufteros XA, Cheng TCE, Lai KH (2007) "Black-box" and "gray-box" supplier integration in product development: Antecedents, consequences and the moderating role of firm size. J Oper Manag 25:847–870
Koufteros X, Vickery SK, Droege C (2012) The effects of strategic supplier selection on buyer competitive performance in matched domains: does supplier integration mediate the relationships? J Supply Chain Manag 48:93–115
Lau AKW, Yam RCM, Tang EPY (2007) Supply chain product co-development, product modularity and product performance – empirical evidence from Hong Kong manufacturers. Ind Manag Data Syst 107:1036–1065
Lee V-H, Ooi K-B, Chong AY-L, Seow C (2014) Creating technological innovation via green supply chain management: an empirical analysis. Expert Syst Appl 41:6983–6994
McIvor R, Humphreys P (2004) Early supplier involvement in the design process: lessons from the electronics industry. Omega Int J Manag Sci 32:179–199
Modi SB, Mabert VA (2010) Exploring the relationship between efficient supply chain management and firm innovation: an archival search and analysis. J Supply Chain Manag 46:81–94
Narasimhan R, Narayanan S (2013) Perspectives on supply network-enabled innovations. J Supply Chain Manag 49:27–42
Oke A, Prajogo DI, Jayaram J (2013) Strengthening the innovation chain: the role of internal innovation climate and strategic relationship with supply chain partners. J Supply Chain Manag 49:43–58
Panayides PM, Lun YHV (2009) The impact of trust on innovativeness and supply chain performance. Int J Prod Econ 122:35–46
Petersen KJ, Handfield RB, Ragatz GL (2005) Supplier integration into new product development: coordinating product, process and supply chain design. J Oper Manag 23:371–388
Roy S, Sivakumar K (2010) Innovation generation in upstream and downstream business relationships. J Bus Res 63:1356–1363
Roy S, Sivakumar K, Wilkinson IF (2004) Innovation generation in supply chain relationships: a conceptual model and research propositions. J Acad Mark Sci 32:61–79
Salvador F, Villena VH (2013) Supplier integration and NPD outcomes: conditional moderation effects of modular design competence. J Supply Chain Manag 49:87–113
Silva LF, Moreira AC (2017) Collaborative new product development and the supplier/client relationship: cases form the furniture industry. In: Garcia Alcaraz JL, Alor-Hernández G, Maldonado Macias AAAA, Sanchez-Ramírez C (eds) New perspectives on applied industrial tools and techniques. Springer International Publishing, Heidelberg, pp 175–195

Soosay CA, Hyland PW, Ferrer M (2008) Supply chain collaboration: capabilities for continuous innovation. Supply Chain Manag Int J 13:160–169

Strozzi F, Colicchia C, Creazza A, Noè C (2017) Literature review on the 'Smart Factory' concept using bibliometric tools. Int J Prod Res 20

Wagner SM, Bode C (2014) Supplier relationship-specific investments and the role of safeguards for supplier innovation sharing. J Oper Manag 32:65–78

Wang LW, Yeung JHY, Zhang M (2011) The impact of trust and contract on innovation performance: the moderating role of environmental uncertainty. Int J Prod Econ 134:114–122

Wong CWY, Wong CY, Boon-itt S (2013) The combined effects of internal and external supply chain integration on product innovation. Int J Prod Econ 146:566–574

Wynstra F, von Corswant F, Wetzels M (2010) In chains? An empirical study of antecedents of supplier product development activity in the automotive industry. J Prod Innov Manag 27:625–639

Zimmermann R, Ferreira L, Moreira AC (2016) The influence of supply chain on the innovation process: a systematic literature review. Supply Chain Manag Int J 21:289–304

Zupic I, Cater T (2015) Bibliometric methods in management and organization. Organ Res Methods 18:429–472

Ricardo A. Zimmermann is a research fellow at the Department of Economics, Management, Industrial Engineering, and Tourism in the University of Aveiro, Portugal, and is a member of the Research Unit on Governance, Competitiveness and Public Policies (GOVCOPP). His research interests have been focused on supply chain management, innovation management, and strategic management. Ricardo has work experience in companies in Brazil and Portugal in areas such as strategic management, quality management, risk management, project management, budget and costs management, and corporate governance. He has also experience as a consultant and as an assessor in quality awards.

Luís Miguel D. F. Ferreira is currently Assistant Professor of Logistics and Supply Chain Management at the Department of Mechanical Engineering, University of Coimbra, Portugal. He obtained a bachelor's degree in mechanical engineering from the University of Coimbra. He received his master's degree and PhD from Instituto Superior Técnico—University of Lisbon, Portugal. His research interests include topics related to supply chain management, supply chain risk management, sustainable supply chain management, and international purchasing. His research has been published in *Supply Chain Management: An International Journal, Production Planning and Control*, among others. He has also been deeply involved in consultancy projects with public institutions and private companies. He is member of CEMMPRE research unit.

António Carrizo Moreira obtained a bachelor's degree in electrical engineering and a master's degree in management, both from the University of Porto, Portugal. He received his PhD in management from UMIST—University of Manchester Institute of Science and Technology, England. He has a solid international background in industry leveraged by working for a multinational company in Germany as well as in Portugal. He has also been involved in consultancy projects and in research activities. He is assistant professor at the Department of Economics, Management, Industrial Engineering, and Tourism, University of Aveiro, Portugal, where he headed the bachelor and master degrees in management for 5 years. He is member of GOVCOPP research unit.

Part II
The Importance of Supplier-Client Relationships

Coordination of New Product Development and Supply Chain Management

Per Hilletofth, Ewout Reitsma, and David Eriksson

Abstract New product development (NPD) and supply chain management (SCM) enable companies to respond to new demands in a responsive manner. The scarcity of research addressing the coordination of NPD and SCM is notable. The purpose of this research is to identify and examine linkages between NPD and SCM through a case study that includes a Swedish furniture wholesaler. Several linkages that stress the need of using an integrative NPD process where the design functions are aligned with other main functions of the company were identified. For example, it was observed that a strong focus on the demand side (NPD) has induced high demands on the supply side (SCM) of the case company. Therefore, the NPD process to a larger extend needs to incorporate main supply functions and other sales-related functions that support the commercialization of the product. This promises to create a consumer-oriented business, especially needed in markets where products have short life cycles and where having a short time to market is crucial. Within future research, it will be interesting to expand this research to companies that operate in different markets and/or have different objectives and to provide an inclusive description of the consumer-oriented business model.

1 Introduction

A company's ability to compete and survive in the market is connected to its ability to innovate, which concerns the ability to design, manufacture, and deliver new products and services to the market (Ellram and Stanley 2008; Marsillac and Roh 2014; Sansone et al. 2017). In order to realize this, new product development (NPD) must result in innovative consumer-desired products and supply chain management (SCM) must provide consumer-desired supply chain solutions, and this needs to be done in collaboration (Hilletofth 2010). SCM also needs to establish the right balance between responsiveness and efficiency for various types of markets, consumers and

P. Hilletofth (✉) · E. Reitsma · D. Eriksson
Department of Industrial Engineering and Management, School of Engineering, Jönköping University, Jönköping, Sweden
e-mail: prof.p.hilletofth@gmail.com; ewout_reitsma@outlook.com

A. C. Moreira et al. (eds.), *Innovation and Supply Chain Management*, Contributions to Management Science, https://doi.org/10.1007/978-3-319-74304-2_2

products. It is vital to be first to market with new products at a reasonable cost (Petersen et al. 2005) and still achieve the right balance. Thus, NPD is considered a key strategic activity in many companies and a short time-to-market (TTM) is considered as critical to long-term success.

Companies can reduce the TTM by working more integrated and with parallel activities (Morash et al. 1996; Smith and Reinertsen 1998; Zacharia and Mentzer 2007). The majority of research in this field has been on the integration within design and between design, manufacturing, and marketing (e.g. Drejer 2002; Nobelius 2004; Olausson 2009; Turkulainen 2008; Vandevelde and van Dierdonck 2003). Even though SCM (i.e. managing the supply, manufacturing, and distribution of the product) affects a new product's TTM (Wynstra et al. 2003), research that focuses on NPD and SCM coordination is lacking (Carillo and Franza 2006; Hilletofth and Eriksson 2011; Hilletofth et al. 2010; Van Hoek and Chapman 2007). Not coordinating NPD and SCM could limit the ability to deliver products in an innovative and responsive way, which becomes increasingly important in mature and highly competitive markets (Christopher et al. 2004). Thus, the coordination of NPD and SCM on a macro level promises to create a consumer-oriented company that understands how consumer-desired products are developed efficiently (NPD) and which consumer-desired supply chain solutions (SCM) should be offered. Accordingly, it is important to conduct research that aims to better understand why and how NPD and SCM should be coordinated.

The purpose of this research is to identify and examine linkages between NPD and SCM through a case study, which is centered on a furniture wholesaler (hereafter called FurnitureCo for anonymity) and includes in total 29 companies. Empirical data has been collected since 2009, mainly from in-depth and semi-structured interviews with key persons representing senior and middle management in the case companies.

The remainder of this chapter is structured as follows: To begin with, a literature review on NPD and critical success factors (CSFs) for NPD are presented in Sect. 2. After that, the research methodology is further elaborated on in Sect. 3. Thereafter, the empirical findings are presented and discussed in Sects. 4 and 5. Finally, the research is concluded in Sect. 6.

2 Literature Review

2.1 New Product Development

There are multiple ways to manage the organizational aspect of NPD. In many cases, responsibility for new product ideas is assigned to product managers, high-level management committees, cross-functional teams or a specific product development department (Kotler et al. 2009; Sethi et al. 2001). Wheelwright and Clark (1992) argue that companies should form an aggregate project plan in order to ensure that the collective set of development projects will accomplish the development objectives as well as build the organizational capabilities needed for successful NPD. This should ensure that the development resources are used for the appropriate types and mix of projects.

Many researchers advocate that NPD needs to be structured into a stage-gate process (e.g. Cooper 1990; Ulrich and Eppinger 2012) and it has been shown that best practice firms have implemented these processes to a greater extent than other firms (Griffin 1997). The stage-gate processes discussed in the literature differ in terms of the number and titles of stages and gates, which is also true for stage-gate processes used in companies (Phillips et al. 1999). Cooper and Kleinschmidt (1986) describe the stage-gate process as dividing NPD into a predetermined set of stages, each consisting of prescribed, multifunctional and parallel activities with gates that are quality control checkpoints. Kotler et al. (2009) further argue that the process should be project oriented, target a specific market segment, and that the project leader should be responsible for reaching a set of known objectives at each gate. In addition, senior managers should review the project at the gates and decide if the project should continue, stop, hold or recycle. In most cases, companies work with several parallel projects in different stages of their own process. However, due to project terminations at different gates, there are usually more projects in the early stages compared to the late stages (Kotler et al. 2009). The stage-gate approach visualizes the NPD processes for people in the company and clarifies responsibilities for the different internal stakeholders (Cooper 1998). The stage-gate approach also facilitates a better balance between innovation and creativity, and business management.

Different researchers advocate for different number of stages (e.g. Barcley 1992; Cooper and Kleinschmidt 1986; Karkkainen et al. 2001; Kotler et al. 2009) and some researchers argue that stage-gate processes are too linear and rigid to handle innovative projects in a more competitive and global world (e.g. Cooper 2014; Sommer et al. 2015). This criticism is one of the reasons why new stage-gate processes, with different number of stages, have been proposed over time. For example, Cooper (2014) proposes a more adaptive and accelerated stage-gate process that uses methods from software development. Sommer et al. (2015) further developed this process into an agile/stage-gate hybrid process. Ulrich and Eppinger (2012) have a similar approach, but they have a clearer focus on the content of the process and design engineering work.

The process developed by Ulrich and Eppinger (1995, 2012) consists of six phases, including planning, concept development, system-level design, detail design, testing and refinement, and production ramp-up. In the process, typical responsibilities and activities of the key business functions during each phase of development are described. Within the model, Ulrich and Eppinger (2012) address eight functions, including marketing, design, manufacturing, research, finance, legal, service and sales. Supply chain is not mentioned as a function or area. However, some typical supply chain activities are discussed within the function manufacturing (e.g. identification of suppliers for key components). Ulrich and Eppinger (2012) discuss that the NPD process will differ depending on the company's context and the specific project. Furthermore, they argue that the generic NPD process is most likely used for market-pull products and that the NPD process should vary based on seven types of products, including technology-push products, platform products, process-intensive products, customized products, high-risk products, quick-build products, and complex systems.

2.2 *Critical Success Factors for New Product Development*

New product introductions are failing at a disturbing rate, underlining that NPD is a complex task. The study of Ogama and Pillar (2006) shows that this rate is as high as 50%, and potentially as high as 95% in the United States and 90% in Europe. CSFs for NPD derived from the literature can be divided into different categories based on four types of characteristics, including market, strategy, product, and process characteristics (Table 1).

There are four CSFs for NPD within the category market characteristics, including the market potential, product life cycle length, competitor's aggressiveness, and the competitive response intensity (e.g. Cheng and Shiu 2008; Cooper et al. 2004; Henard and Szymanski 2001; Kotler et al. 2009). The market potential equivalents the total opportunity of a company to make business. The business opportunity's length is limited by the lifetime of the offered product and its width is reduced by competitor's aggressiveness. Both the width and breadth of the business opportunity are reduced by the competitive response intensity.

There are five CSFs for NPD within the category strategy characteristics, including marketing and technological synergy, order of entry (timing), dedicated human and R&D resources, fit with organizational culture, and brand power (e.g. Carillo and Franza 2006; Cheng and Shiu 2008; Cooper and Kleinschmidt 1986; Dowling and Helm 2006; Gerwin and Barrowman 2002; Hamm and Symonds 2006; Henard and Szymanski 2001). Market synergy means congruency between existing marketing skills of the firm and marketing skills required to serve the market. Technological synergy refers to congruency between a company's existing technological skills and the technological skills needed to develop desirable products and efficient processes. Most companies focus on brand power, as it is connected to the added value of the brand name to the product. Therefore, branding has become a powerful tool to attract and retain consumers.

There are five CSFs for NPD within the category product characteristics, including product advantage (unique/superior product), product meets consumer needs, product price, product technological sophistication, and product innovativeness (e.g. Cheng and Shiu 2008; Cooper et al. 2004; Droge et al. 2008; Hamm and Symonds 2006; Henard and Szymanski 2001; Kotler et al. 2009; Van Kleef et al. 2005). In essence, the offered product's value as perceived by consumers determines what product characteristics are important for success, not the offered product's cost.

There are 14 CSFs for NPD within the category process characteristics, which can be divided in two main CSFs: a strategic and holistic process view, and a structured approach to NPD. This implies that the internal and external processes of the company need to be formulated based on the requirements of consumers, ultimately

Table 1 CSFs for new product development

Category	CSF	References
Market characteristics	Competitive response intensity	Cheng and Shiu (2008), Cooper et al. (2004), Henard and Szymanski (2001), Jain (2001), and Kotler et al. (2009)
	Market potential	
	Product life cycle length	
	Competitors' aggressiveness	
Strategy characteristics	Market and technological synergy	Carillo and Franza (2006), Cheng and Shiu (2008), Cooper et al. (2004), Cooper and Kleinschmidt (1986), Dowling and Helm (2006), Gerwin and Barrowman (2002), Hamm and Symonds (2006), Henard and Szymanski (2001), Karlsson and Åhlström (1996), Kotler et al. (2009), and Lummus and Vokurka (1999)
	Order of entry (timing)	
	Dedicated human and R&D resources	
	Fit with the organization's culture	
	Brand power	
Product characteristics	Product advantage (unique/ superior product)	Cheng and Shiu (2008), Cooper et al. (2004), Droge et al. (2008), Hamm and Symonds (2006), Henard and Szymanski (2001), Jain (2001), Kotler et al. (2009), and Van Kleef et al. (2005)
	Products meets consumer needs	
	Product price	
	Product technological sophistication	
	Product innovativeness	
Process characteristics	Strategic and holistic view	Barczak et al. (2009), Carillo and Franza (2006), Cheng and Shiu (2008), Ciappei and Simoni (2005), Cooper (1990), Cooper and Kleinschmidt (1986), Cooper et al. (2004), Droge et al. (2008), Gerwin and Barrowman (2002), Gupta and Wilemon (1990), Hamm and Symonds (2006), Henard and Szymanski (2001), Holger (2002), Iansiti (1995), Jain (2001), Karlsson and Åhlström (1996), Kess et al. (2010), Kotler et al. (2009), Karkkainen et al. (2001), Lummus and Vokurka (1999), Schmidt et al. (2009), Park et al. 2010, Swink et al. (1996), Van Kleef et al. (2005), and Wheelwright and Clark (1992)
	Structured approach	
	Market/consumer oriented	
	Project-oriented	
	Team-based (instead of group of experts)	
	Cross-functional (concurrent design)	
	Segmentation-based	
	Market intelligence driven	
	Proficiency of the NPD activities	
	Technological proficiency	
	Reduced lead-time (responsiveness	
	Cooperation with suppliers and customers	
	Information technology support	
	Senior management involvement	

aiming to achieve the overreaching goals. Furthermore, the processes affecting NPD success require support from information technology and involvement from senior management (e.g. Barczak et al. 2009; Cheng and Shiu 2008; Droge et al. 2008; Gerwin and Barrowman 2002; Hamm and Symonds 2006; Henard and Szymanski 2001; Schmidt et al. 2009).

3 Research Methodology

The purpose of this research is to identify and examine linkages between NPD and SCM. Being a novel area, where context-dependent knowledge is desired, a case study was considered to be the most appropriate method (Yin 2009). The case study is centered on FurnitureCo, but also includes an additional 28 companies that work or compete with FurnitureCo. FurnitureCo is a furniture wholesaler and the additional companies include retailers, wholesalers, agents, and transportation companies. This longitudinal study was initiated in 2009 and includes secondary data dating back to 1999. Through this approach, it has been possible to gain a deep understanding of both the phenomenon and context being studied, allowing for both a good construct and a good story (Dyer and Wilkins 1991; Eisenhardt 1989) through an abductive approach (Dubois and Gadde 2002; Eriksson 2015; Kovacs and Spens 2005).

Data was initially gathered through interviews with senior and middle management at FurnitureCo. Voice recording and note taking were used to collect data. Throughout the research, respondents at FurnitureCo have continuously reviewed data and findings, increasing the trustworthiness of the research (Lincoln and Guba 1985). In addition to interviews, protocols from internal meetings, reports from NPD projects and observations from working at FurnitureCo were used in the research. FurnitureCo also provided free access to internal sales and logistics systems, from which sales data, inventory levels, and purchase data was collected. Additional data includes retailers, analysis of sales data from 16 retailers and interviews with 12; suppliers in China, factory visits and interviews; competitors, review of 10 years of financial statements and interviews with seven senior managers; and an agent company based in China, travelled with four people from the company in China for 5 days.

The data analysis has primarily been centered on model building. The models have been used to structure and understand the business model and include supply chain flows (materials and information), NPD processes, pricing, marketing and sales activities. This approach has allowed the researchers to continually validate the findings with both the initial informant, but also with other informants, active in several companies, reducing the likelihood of bias through, for example, self-promotion.

In order to improve the quality of the research, the criteria for trustworthiness that are proposed by Lincoln and Guba (1985) has been followed. These include credibility, transferability, dependability and confirmability (see also Eriksson 2015; Hulthén 2002). Credibility was increased through the long and deep engagement with the empirical setting, the use of multiple informants and sources of data, and by allowing informants to review data and findings. Transferability is increased by a thick description and dependability is increased by giving insights to the process of inquiry. Confirmability has been increased through the research process itself, systematic combining (Dubois and Gadde 2002), which emphasizes the consistency between data, findings, framework, and theory.

4 Empirical Findings

This case study is centered on FurnitureCo, a furniture wholesaler whose headquarters and central distribution center are located in Sweden. Independent manufacturers in China produce FurnitureCo's furniture and are also responsible for the sourcing of materials. In order to ensure reliable supply, FurnitureCo collaborates with these manufacturers. FurnitureCo works with a supporting team in China, so as to overcome issues inherent with the geographical and cultural distances. The Chinese manufactured products and items are transported in full containers to the port of Gothenburg and then either delivered to FurnitureCo's distribution center or directly to retailers (around 1% of the volume).

These products and items can be delivered as knockdown products, items ready for assembly, or as complete products. The products are delivered to a retailer whenever a consumer places an order in a store. When the product is received by the retailer, the consumer will either be notified that their order can be collected or have their order home delivered. Fast delivery is vital, as retailers notice that there is an explicit consumer need for fast delivery when an order is placed. Only one retailer has chosen to collect the orders at FurnitureCo's warehouse instead of having them delivered, which is facilitated by FurnitureCo's flexible business model that stimulates collaboration. Retailers allocate resources for show room furniture and need to make sure that the display furniture is well thought out. Wholesalers tend to focus on low cost production, as many consumers are price sensitive. However, there is demand for furniture in all price ranges and FurnitureCo focuses on consumers in the premium segment. According to retailers, FurnitureCo is at level three out of five on the premium scale, where the fourth level is the highest level reached by most furniture wholesalers in Sweden. Examples of level two furniture wholesalers are IKEA and other wholesalers that offer non-branded furniture. Level one reflects low quality furniture sold in stores that focus on low price. The described premium scale does not indicate the actual quality of the furniture, only the furniture's premium as perceived by consumers.

A strong brand name enables furniture wholesalers to attract and retain consumers and the brand in the furniture industry is usually connected with the retailer. For instance, in Sweden consumers perceive that they buy a retailer furniture, for example 'Mio' sofa, an 'Europamöbler' table, or a 'Svenska hem' bed. In reality, consumers buy furniture that is sourced from common distributors. This indicates that the furniture industry differs greatly from the apparel, computer and home

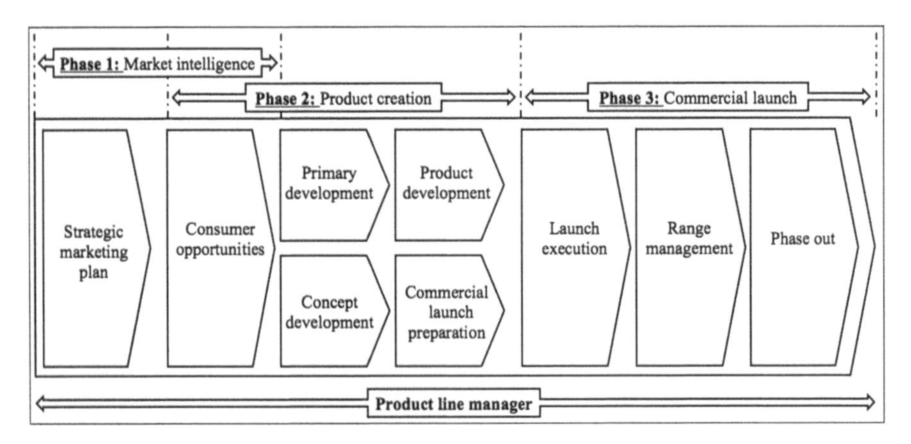

Fig. 1 FurnitureCo's new product development process

electronics industry, where consumers often ask for the brand of the manufacturer and not the retailer.

In the furniture industry, the product life cycle is not related to the innovativeness of the product and within all price segments, there are products, colors, and materials that go in and out of fashion. High-end furniture may be perceived as modern for a short time period, but some of the most modern and most fashionable products available on the market have remained the same for more than 30 years. Low cost furniture may either have a lifecycle of several years or of the time needed to sell one shipment.

FurnitureCo has adopted a strategy that aims to add value by gaining insights into the explicit and implicit needs of consumers. This embodies a transformation from being cost and volume orientated to striving to provide superior value through being consumer-oriented and becoming innovative. Furthermore, FurnitureCo has defined NPD as the major business process that aims to develop products that are innovative, add perceived consumer value, and can have a premium price. FurnitureCo states that SCM is a function that has to support the NPD process and how it relates to SCM is further described below in parallel with FurnitureCo's business performance records.

FurnitureCo's approach, which strives to develop consumer-oriented products, is described within the NPD process and is adopted for all target markets. The purpose of this process is to manage the product lifecycle by incorporating all areas of creating and selling products. During the NPD process, the CEO has the overall responsibility and managers, purchasing/logistics, and marketing are responsible for assigned areas. As of today, these assigned areas are not clearly defined. FurnitureCo's NPD process consists of three main phases, including market intelligence, product creation, and commercial launch (Fig. 1).

Within the first phase of FurnitureCo's NPD process, which is consumer opportunities, the aim is to collect consumer insight. FurnitureCo needs to go beyond what consumers know about themselves and are able to express. Consumer insight, segmentation, and product development facilitates the consumer perceived value. The quality of the product is believed to be quotient of the consumer perceived value and the consumer expected value.

Consumer insights can be collected by photo-documenting the homes of potential and actual consumers in everyday situations, ultimately enabling FurnitureCo to identify business opportunities they were unaware of. The collection of these insights can be performed by external personnel supervised by FurnitureCo. Since FurnitureCo adopted the NPD process six years ago, one NPD project is managed each year. The NPD process resulted in the identification of new business opportunities and the number of consumer available products tripled.

Within the product creation phase, the aim is to address well-understood consumer needs by defining and developing consumer relevant and innovative products. Similar to described above, the initial step of this phase is consumer opportunities and a room or furniture type is targeted based on an assessment of the product platform. Afterwards, a number of squares in the product platform are targeted during primary development. Designers are restricted to only make decisions relating to their targeted square by using a list of pre-chosen materials (materials matrix). The manager of the NPD process is responsible for the materials matrix, which only include materials that are sourced securely and ensure coherence and flexibility between different furniture collections.

During the NPD process, FurnitureCo's employees might come up with ideas that have not gone through the phases mentioned above. An example of consumer opportunities that are identified outside the prescribed process can be found in FurnitureCo, where its bestselling furniture collections includes products co-developed between the manager of purchasing/logistics and the manufacturer. Still, the materials matrix and the product platform form are leading in primary development of the furniture. FurnitureCo's consumer opportunity identification process is summarized in Fig. 2 and FurnitureCo's founder states that the process generates many residual ideas that may be used in the future, depending on internal capabilities and prioritization.

FurnitureCo invests a large amount of time during the early stages of the NPD process and external personnel (mainly students) invests most of the time, while the project is guided by internal personnel. This enables FurnitureCo to keep economic investment to a minimum during the opportunity identification process. This approach ensures successful NPD, ultimately causing FurnitureCo's new products to be well received among retailers. This can be exemplified by what one retailer told FurnitureCo's CEO: 'You are really good at making new furniture. They have great features and offer a lot of flexibility to the end consumer'.

FurnitureCo did notice that a high focus on NPD in combination with decreased sales has had a negative effect on supply chain performance. Due to complex market setting, it is difficult to determine what the impact of NPD was on business performance. However, the CEO is aware of the bad fit between the business model and processes and argues that future new products launches must be well-thought-out. This shows that even if is product has all the potential to be successful in the market, the supply chain capabilities might not support the introduction of new

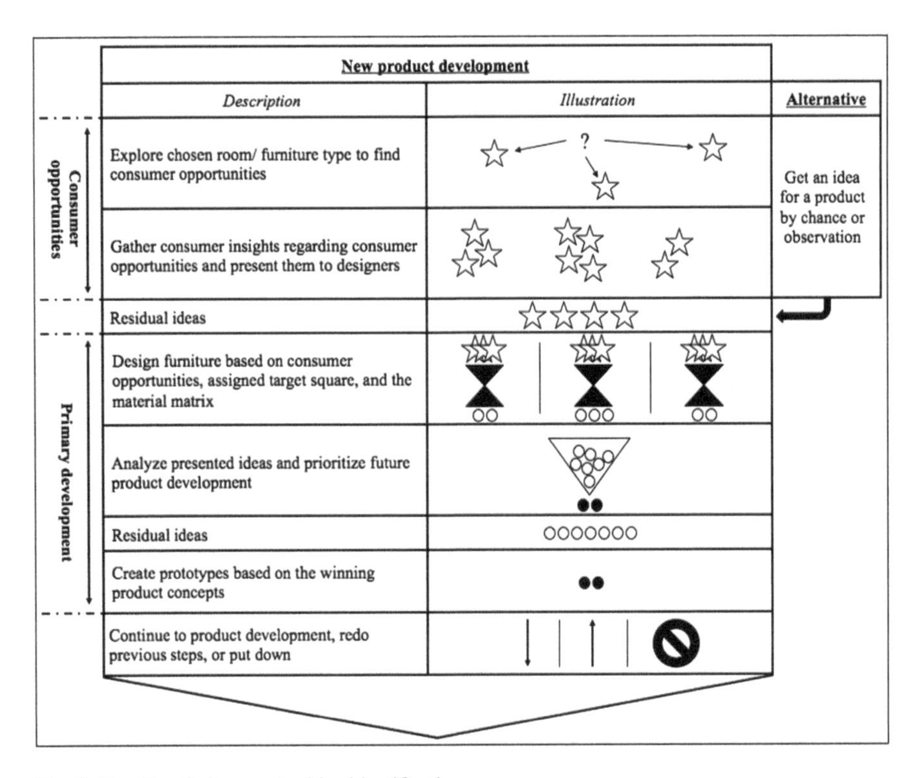

Fig. 2 FurnitureCo's opportunities identification process

products. Therefore, there is a need to balance value creation and value delivery. This is confirmed by FurnitureCo's manager of purchasing/logistics, who states that he struggles with capital that is tied up in inventory and increasing complexity due to the high number of products and items relative to turnover. The high number of inbound products and items in relation to the amount of consumer available products is reduced by using a modular design. Within the marketing-size, managers believe that scale benefits in the supply chain can be realized through focusing on demand creation and higher volumes. As of today, managers agree that FurnitureCo needs to focus on increasing sales.

FurnitureCo decided to implement modular design and a new forecasting system to support the NPD process, whilst not changing the sourcing structure. The company used to source products from Malaysia and China, whereas now they only rely on production in China. The first implication of this change is that the lead-time from placed order to delivery at FurnitureCo is about 16 weeks, which has put requirements on the accuracy of the forecasting. Chinese manufacturers have been able to negotiate contracts from companies originating from several countries. The US market has been preferred by these manufacturers over the Scandinavian and Northern European markets for two reasons. Firstly, since US companies have been able to place larger order, it is easier for manufacturers to reduce set-up times and to simplify production planning. Secondly, US companies have a demand for furniture that has a distressed finish, while the Scandinavian market requests a pristine (often white) finish. Therefore, Scandinavian furniture requires a cleaner manufacturing environment, which is hard to achieve in factories that are tailored to supply a distressed finish. Furthermore, Chinese manufacturers lost labor capacity after the Chinese New Year and when floods affected the factories in the fall of 2010.

The products developed by FurnitureCo are highly innovative and offer solutions for problems that consumers have not been able to articulate. These products are sold to consumers by retailers who have an important role when it comes to communicating the benefits of FurnitureCo's products. This means that it is vital to coordinate NPD with this part of SCM. A retailer who sold the most expensive dining table offered by FurnitureCo without showing it to the consumer was asked by FurnitureCo's CEO how she was able to do it. She casually replied: 'I like your furniture'. In this case, the retailer conveyed the value of the innovative NPD process to the consumer. However, it is not a priority at FurnitureCo to make retailers love and understand their furniture, as sales representatives are assessed on their ability to get display items sold to retailers. In essence, FurnitureCo is a testament to the need of coordinating NPD and SCM on a macro level. NPD has a huge impact on supply chain performance and collaboration in the supply chain seems to determine how well the benefits of innovative products are conveyed to the consumer.

5 Discussion

There are six CSFs for NPD that were identified in both the case study and the literature (Table 2). The first CSF for NPD that was identified in both the case study and the literature (e.g. Barczak et al. 2009; Cheng and Shiu 2008; Cooper et al. 2004; Droge et al. 2008; Kotler et al. 2009; Karkkainen et al. 2001; Schmidt et al. 2009) is 'a holistic view from strategy to commercialization'. This indicates that multiple functions such as marketing and sales, product developments, R&D, sourcing, manufacturing, and distribution need to be incorporated in the NPD process. With this business model, a supply chain representative needs to be involved, as different supply chain competencies (sourcing, manufacturing, and distribution) have to be incorporated in the NPD process. By having this focus, feedback can be provided from a logistics point of view during the different development stages (Table 2). Furthermore, having a supply chain representative in the NPD process provides an opportunity to coordinate NPD and SCM.

The second CSF for NPD that was identified in both the case study and the literature (e.g. Barczak et al. 2009; Cheng and Shiu 2008; Cooper et al. 2004; Droge et al. 2008; Hamm and Symonds 2006; Jain 2001; Kotler et al. 2009; Van Kleef et al. 2005; Schmidt et al. 2009) is 'development of products based on needs identified through market intelligence (consumer-oriented)'. This implies that the NPD process should be driven by consumer needs and requirements identified through market intelligence, rather than by technology improvements alone. This can be realized by

Table 2 CSFs for NPD and their linkage to SCM

CSF	Linkage to SCM
A holistic view from strategy to commercialization	Different supply chain competences have to be involved in the NPD process to provide feedback from a logistics point of view. This also creates an opportunity to address NPD and SCM in parallel at an early stage
Development of products based on needs identified through market intelligence (consumer-oriented)	The provided supply chain solutions need to be developed based on consumer demand. When companies gather information concerning the needs of new products, they also need to collect information regarding service needs in order to develop the most appropriate supply chain solutions
Development of products based on a segmentation model	The requirements of consumers may differ when it comes to lead-times, service levels, and preferred supply chain solutions. Therefore, several supply chain solutions are required in order to become successful in the market and all operations should be directed based on the same overall segmentation model
Development of new and innovative products in accordance with consumer preferences	Innovation should not be restricted to products, meaning that other areas such as supply chain solutions should also be included. This can be achieved in the NPD process through involvement of supply chain representatives and by exchanging information between NPD and SCM
Developing products rapidly and moving them quickly and efficiently to the market	TTM is not solely affected by NPD, but also by SCM (e.g. sourcing, manufacturing, and distribution). This implies that supply chain representatives should be involved early during NPD to reduce TTM
Incorporating all the activities supporting the commercialization (integrative NPD approach)	SCM and NPD need to be strategically coordinated in order to ensure that new products are introduced to market successfully, that the product assortment is updated according to product life cycles, and that obsolete products are properly phased out

having a profound understanding of consumers and their requirements, which is similar to what is needed within SCM, as the provided supply chain solutions also need to be developed based on consumer needs and requirements. This indicates that, when companies gather information concerning the need for new products, they also should gather information regarding consumer service needs in order to develop the most appropriate supply chain solutions.

The third CSF for NPD that was identified in both the case study and the literature (e.g. Cooper et al. 2004; Hilletofth et al. 2010; Kotler et al. 2009) is ‘development of products based on a segmentation model’. This implies that the NPD process needs to be based on a segmentation model and has to include a set of market segments, since preferences of consumers may vary greatly. Therefore, products should be developed to meet needs that have been identified within specific market segments. This relates to SCM, as consumers' preferences may also differ when it comes to lead-times, service levers, and supply chain solutions (e.g. procurement and distribution alternatives). This implies that several solutions are required to become successful in the market and that all operations should be directed based on the same segmentation model in order to create a consumer-oriented company. Thus, the same overall segmentation model should be applied in NPD and SCM, which requires coordination.

The fourth CSF for NPD that was identified in both the case study and the literature (e.g. Barczak et al. 2009; Cheng and Shiu 2008; Droge et al. 2008; Hamm and Symonds 2006; Jain 2001; Kotler et al. 2009) is 'development of new and innovative products in accordance with consumer preferences'. This implies is that new (or modified old) products need to be innovative and developed continuously in accordance with consumer preferences in order to be attractive on the market. However, companies offer more than a product; they offer a value package that includes the physical product and related services such as delivery and assembly. Thus, a company's supply chain solutions need to be innovative as well. This can be realized by involving supply chain representatives and by establishing an exchange of information between NPD and SCM, which requires coordination.

The fifth CSF for NPD that was identified in both the case study and the literature (e.g. Carillo and Franza 2006; Hamm and Symonds 2006; Jain 2001; Kotler et al. 2009; Van Hoek and Chapman 2007) is 'developing products rapidly and moving them quickly and efficiently to the market'. This implies that the right products need to be developed rapidly and moved efficiently to the market due to shortening product life cycles and rapid product obsolescence. This indicates the importance of SCM, as TTM is not solely determined in the NPD process. For instance, the ramp-up of coursing, manufacturing, and distribution are activities of SCM that affect a new product's TTM. Thus, supply chain representatives should be involved early in the NPD process to shorten TTM.

The sixth CSF for NPD that was identified in both the case study and the literature (e.g. Carillo and Franza 2006; Kotler et al. 2009; Sharifi et al. 2006; Van Hoek and Chapman 2007) is 'incorporating all the activities supporting the commercialization (integrative NPD approach)'. This implies that NPD needs to assist the ramp-up of various supply chain activities such as sourcing, manufacturing, distribution, and marketing and sales. These activities need to be addressed in parallel and requires an integrative NPD process. This indicates that SCM needs to be coordinated with NPD in order to successfully introduce new products to the market, to ensure that the product assortment is updated according to their life cycles, and to ensure that obsolete products are phased out properly.

6 Conclusion

Within this research, several linkages between NPD and SCM have been identified. These linkages stress the need of using an integrative NPD process where the design functions are aligned with other main functions of a company. Integration and alignment can be achieved by involving member from other main functions in the NPD process (Portioli-Staudacher et al. 2003; Sharifi et al. 2006; Van Hoek and Chapman 2006). These members should be involved as soon as possible and provide feedback from their function's perspective during the NPD process. The topic of using an integrative NPD process has been addressed in concurrent design, where concepts such as 'design for manufacturing' and 'design for supply chain' are discussed (Appelqvist et al. 2004; Ellram et al. 2007; Perks et al. 2005; Sharifi et al. 2006; Sharifi and Pawar 2002). These concepts imply that NPD processes are aligned and integrated with other main functions within the company.

Even though the NPD process is supposed to be cross-functional, usually only marketing, product development, and R&D are involved in this process. This indicates that the NPD process to a larger extend need to incorporate main supply functions and other sales-related functions that support the commercialization of the product. Apart from the NPD process enabling the efficient flow of new products, it assists the support ramp-up of various supply chain activities. Thus, companies need to be aware of the impact the NPD process has on supply chain performance and success. The overarching goal is to create a consumer-oriented company that understands how consumer-desired products (e.g. innovative, customized, and affordable) are developed efficiently (NPD) and which consumer-desired supply chain solutions should be offered (SCM). This goal can be reached by systematically coordinating NPD and SCM on a macro level, not by just working with these domains individually.

This consumer-oriented business model is especially beneficial in markets where products have short life cycles and when new products launches are negatively affected by a long TTM. In this type of market, companies need to produce products that are desired by consumers and launch these new products to market quickly and effectively. Since the furniture industry is fashion driven, many furniture wholesalers operate in above described environment. These companies may encounter problems due to a high emphasis on either value creation processes on the demand-side of the company (NPD) or on the value delivery processes on the supply-side of the company (SCM).

Case company FurnitureCo started to implement the consumer-oriented business model by developing unique and value adding products and bringing them quickly and effectively to the market. This distinguishes them from competitors, making it possible to avoid competing with companies such as IKEA based on price and volume. IKEA has a small assortment when it comes to a specific item group (e.g. kitchen table) and focuses on offering the lowest price possible. In contrast, FurnitureCo cannot compete on price and instead focuses on producing consumer desired products. An example of this value differentiation pursued by FurnitureCo is that they offer customization options within their assortment, where IKEA does not. On the other hand, IKEA has more item groups than FurnitureCo and therefore has a significantly broader assortment. IKEA has also a different degree of control in their supply chain, since they own their stores. This means that they do not have to collaborate with retailers and can solely focus on the strategic, tactical, and operational alignment of distributors.

By researching furniture wholesaler FurnitureCo's consumer-oriented business model, a number of practical implications were identified. It was observed that a strong focus on the demand-side (NPD) has induced high demands on the supply-side of the company (SCM), emphasizing the importance of coordinating NPD and SCM. This finding is not supported by literature (e.g. Esper et al. 2010; Hilletofth et al. 2009; Juttner et al. 2007) and is probably generalizable to companies that differ in size, country of origin, and in business environment. Furthermore, this research identified the social aspects of globalization and the inherent consequences of increased geographical and cultural distances. FurnitureCo does not employ one full-time employee for NPD, warehousing employs two people, while the manufacturing in China employs a large amount of people. In a market where consumer demand is increasing, so do the requirements of the demand-supply chain. When offering premium products that differentiate based on added value instead of price, it may be financially beneficial to move production or assembly closer to the consumer market, which results in increased domestic employment.

Within future research, it will be interesting to further investigate the coordination between NPD and SCM in companies that operate in different markets and/or have different objectives. Also, there are several processes and aspects within FurnitureCo that need to be investigated to provide an inclusive description of the consumer-focuses business model, including (but not limited to) the effects on sourcing, distribution, and information systems. In short, the research needs to be conducted in more case companies and need to include more processes and aspects. This will contribute to the overall understanding of the consumer-oriented business model and its effects. Another direction for further research could be the investigation of the opportunity to move manufacturing closer to the consumer market in the increased cost is absorbed by higher contribution margins. Probably higher flexibility and better supply chain responsiveness is realized, as short distances between production and consumption will decrease transportation lead-times. However, more research needs to be conducted to find out if having a consumer-oriented business model and offering superior consumer value may justify manufacturing in high cost countries.

References

Appelqvist P, Lehtonen JM, Kokkonen J (2004) Modelling in product and supply chain design: literature survey and case study. J Manuf Technol Manag 15(7):675–686

Barcley I (1992) The new product process: part 2. Improving the process of new product development. R&D Manag 22(4):307–317

Barczak G, Griffin A, Kahn K (2009) Perspective: trends and drivers of success in NPD practices: results of the 2003 PDMA best practices study. J Prod Innov Manag 26(1):3–23
Carillo JE, Franza RM (2006) Investing in product development and production capabilities: the crucial linkage between time-to-market and ramp-up time. Eur J Oper Res 171(2):536–556
Cheng C, Shiu E (2008) Critical success factors of new product development in Taiwan's electronics industry. Asia Pac J Mark Logist 20(2):174–189
Christopher MC, Lowson R, Peck H (2004) Creating agile supply chains in the fashion industry. Int J Retail Distrib Manag 32(8):367–376
Ciappei C, Simoni C (2005) Drivers of new product success in the Italian sport shoe cluster of Montebelluna. J Fash Mark Manag 9(1):20–42
Cooper RG (1990) Stage-gate systems: a new tool for managing new products. Bus Horiz 33 (3):44–54
Cooper RG (1998) Product leadership: creating and launching superior new products. Perseus Books, New York, NY
Cooper RG (2014) What's next? After stage gate. Res Technol Manag 57(1):20–31
Cooper RG, Kleinschmidt EJ (1986) An investigation into the new product process: steps, deficiencies and impact. J Prod Innov Manag 3(2):71–85
Cooper RG, Edgett S, Kleinschmidt E (2004) Benchmarking best NPD practice. Res Technol Manag 47(6):43–55
Dowling M, Helm R (2006) Product development success through cooperation: a study of entrepreneurial firms. Technovation 26(4):483–488
Drejer A (2002) Integrating product and technology development. Int J Technol Manag 24 (2/3):124–142
Droge C, Calantone R, Haramancioglu N (2008) Characterizing the role of design in new product development: an empirically derived taxonomy. J Prod Innov Manag 22(2):111–127
Dubois A, Gadde LE (2002) Systematic combining: an abductive approach to case research. J Bus Res 55(7):553–560
Dyer WG, Wilkins A (1991) Better stories, not better constructs, to generate better theory: a rejoinder to Eisenhardt. Acad Manage Rev 16(3):613–619
Eisenhardt KM (1989) Building theories from case study research. Acad Manage Rev 14(4):532–550
Ellram LM, Stanley LL (2008) Integrating strategic cost management with a 3DCE environment: strategies, practices, and benefits. J Purch Supply Manag 14(3):180–191
Ellram LM, Tate WL, Carter CR (2007) Product-process-supply chain: an integrative approach to three-dimensional concurrent engineering. Int J Phys Distrib Logist Manag 37(4):305–330
Eriksson D (2015) Lessons on knowledge creation in supply chain management. Eur Bus Rev 27 (4):346–368
Esper T, Ellinger A, Stank T, Flint D, Moon M (2010) Demand and supply integration: a conceptual framework of value creation through knowledge management. J Acad Mark Sci 38(5):5–18
Gerwin D, Barrowman NJ (2002) An evaluation of research on integrated product development. Manag Sci 48(7):938–953
Griffin A (1997) The effect of project and process characteristics on product development cycle time. J Mark Res 34(1):24–35
Gupta AK, Wilemon D (1990) Improving R&D/Marketing relations: R&D's perspective. R&D Manag 20(4):277–290
Hamm S, Symonds W (2006) Mistakes made on the road to innovation. Business Week IN. Inside Innovation, November, pp 27–31
Henard D, Szymanski D (2001) Why some new products are more successful than others. J Mark Res 38(3):362–375
Hilletofth P (2010) Demand-supply chain management. Doctoral thesis, Chalmers University of Technology, Gothenburg, Sweden
Hilletofth P, Eriksson D (2011) Coordinating new product development with supply chain management. Ind Manag Data Syst 111(2):264–281

Hilletofth P, Ericsson D, Christopher M (2009) Demand chain management: a Swedish industrial case study. Ind Manag Data Syst 109(9):1179–1196
Hilletofth P, Ericsson D, Lumsden K (2010) Coordinating new product development and supply chain management. Int J Value Chain Manag 4(1/2):170–192
Holger E (2002) Success factors of new product development: a review of the empirical literature. Int J Manag Rev 4(1):1–40
Hulthén K (2002) Variety in distribution networks: a transvection analysis. Doctorate thesis, Chalmers University of Technology, Gothenburg, Sweden
Iansiti M (1995) Science-based product development: an empirical study of the mainframe computer industry. Prod Oper Manag 4(4):335–339
Jain D (2001) Managing new product development for strategic competitive advantage. In: Lacobucci D (ed) Kellog marketing. Wiley, New York, NY
Juttner U, Christopher MC, Baker S (2007) Demand chain management: integrating marketing and supply chain management. Ind Mark Manag 36(3):377–392
Karkkainen H, Pippo P, Tuominen M (2001) Ten tools for customer-driven product development in industrial companies. Int J Prod Econ 69(2):161–176
Karlsson C, Åhlström P (1996) The difficult path to lean product development. J Prod Innov Manag 13(4):283–295
Kess P, Law KMY, Kanchana R, Phusavat K (2010) Critical factors for an effective business value chain. Ind Manag Data Syst 110(1):63–77
Kotler P, Keller KL, Brady M, Goodman M, Hansen T (2009) Marketing management. Person Education Limited, Harlow
Kovacs G, Spens K (2005) Abductive reasoning in logistics research. Int J Phys Distrib Logist Manag 35(2):132–144
Lincoln YS, Guba EG (1985) Naturalistic inquiry. Sage, Newbury Park, CA
Lummus R, Vokurka R (1999) Defining supply chain management: a historical perspective and practical guidelines. Ind Manag Data Syst 99(1):11–17
Marsillac E, Roh JJ (2014) Connecting product design, process and supply chain decisions to strengthen global supply chain capabilities. Int J Prod Econ 147:317–329
Morash EA, Dröge C, Vickery S (1996) Boundary spanning interfaces between logistics, production, development. Int J Phys Distrib Logist Manag 26(8):43–62
Nobelius D (2004) Linking product development to applied research: transfer experiences from an automotive company. Technovation 24(4):321–334
Ogama S, Pillar FT (2006) Reducing the risk of new product development. MIT Sloan Manag Rev 47(2):65–71
Olausson D (2009) Facing interface challenges in complex product development. Doctoral thesis, Linköping University, Linköping
Park J, Shin K, Chang T-W, Park J (2010) An integrative framework for supplier relationship management. Ind Manag Data Syst 110(4):495–515
Perks H, Cooper RG, Jones C (2005) Characterizing the role of design in new product development: an empirically derived taxonomy. J Prod Innov Manag 22(2):111–127
Petersen KJ, Handfield RB, Ragatz GL (2005) Supplier integration into new product development: coordinating product, process and supply chain design. J Oper Manag 23(3/4):371–388
Phillips R, Neailey K, Broughton T (1999) A comparative study of six stage-gate approaches to product development. Integr Manuf Syst 10(5):289–297
Portioli-Staudacher A, Van Landeghem H, Mappelli M, Redaelli C (2003) Implementation of concurrent engineering: a survey in Italy and Belgium. Robot Comput Integr Manuf 19 (3):225–238
Sansone S, Hilletofth P, Eriksson D (2017) Critical operations capabilities for competitive manufacturing: a systematic review. Ind Manag Data Syst 117(5):801–837
Schmidt J, Sarangee K, Montoya M (2009) Exploring new product development project review practices. J Prod Innov Manag 26(5):520–535

Sethi R, Smith DC, Park CW (2001) Cross-functional product development team, creativity and the innovativeness of new consumer products. J Mark Res 38(1):73–85
Sharifi S, Pawar KS (2002) Virtually co-located teams sharing teaming experiences after the event? Int J Oper Prod Manag 22(6):656–679
Sharifi H, Ismail HS, Reid I (2006) Achieving agility in supply chain through simultaneous 'design of' and 'design for' the supply chain. J Manuf Technol 17(8):1078–1098
Smith PG, Reinertsen DG (1998) Developing products in half the time: new rules, new tools. Nostrand Reinhold, New York, NY
Sommer AF, Hedegaard C, Dukovska-Popovska I, Steger-Jensen K (2015) Improved product development performance through agile/stage-gate hybrids. Res Technol Manag 58(1):34–44
Swink M, Sandvig J, Mabert V (1996) Customizing concurrent engineering processes: five case studies. J Prod Innov Manag 13(3):229–244
Turkulainen V (2008) Managing cross-functional interdependencies: the contingent value of integration. Doctoral thesis series 2008/8, Helsinki University of Technology, Finland
Ulrich KT, Eppinger SD (1995) Product design and development. McGraw-Hill, New York, NY
Ulrich K, Eppinger S (2012) Product design and development. McGraw-Hill, New York, NY
van Hoek R, Chapman P (2006) From tinkering around the edge to enhancing revenue growth: supply chain-new product development. Supply Chain Manag Int J 11(5):385–389
Van Hoek R, Chapman P (2007) How to move supply chain beyond cleaning up after new product development. Supply Chain Manag Int J 12(4):239–244
Van Kleef E, Trijp H, Luning P (2005) Consumer research in the early stages of new product development: a critical review of methods and techniques. Food Qual Prefer 16(3):181–201
Vandevelde A, van Dierdonck R (2003) Managing the design-manufacturing interface. Int J Oper Prod Manag 23(11):1326–1348
Wheelwright S, Clark K (1992) Creating project plans to focus product development. Harv Bus Rev 70(2):70–82
Wynstra F, Weggeman M, Van Weele A (2003) Exploring purchasing integration in product development. Ind Mark Manag 32(1):69–83
Yin RK (2009) Case study research: design and methods. Sage, London
Zacharia ZG, Mentzer JT (2007) The role of logistics in new product development. J Bus Logist 28 (1):83–110

Per Hilletofth (PhD) is a Professor of Operations and Supply Chain Management at Jönköping University in Sweden. His research focuses on demand–supply integration, operations strategy, supply chain relocation, product development, and decision support. He has editorial assignments in several international journals.

Ewout Reitsma (MSc) is a PhD student at Jönköping University in Sweden. His research agenda consists of various research subjects including new product development, supply chain management, demand–supply integration, operations strategy, and information systems.

David Eriksson (PhD) is a researcher and lecturer at Jönköping University in Sweden. His research agenda consists of various research subjects including corporate social responsibility, methodology, new product development, supply chain management, sustainability, and waste management. He has editorial assignments in several international journals.

An Investigation of Contextual Influences on Innovation in Complex Projects

Lone Kavin and Ram Narasimhan

Abstract There is paucity of literature on supplier-enabled-innovation in complex project-contexts. Based on literature from repetitive-manufacturing-contexts, this conceptual chapter identifies *innovation-fostering-practices* and develops a conceptual-framework relating them to innovation-performance. The framework suggests that new knowledge is the basis of innovation and leveraging knowledge from the supply-network is a key element along with absorptive-capacity and R&D-investment in creating new knowledge. New knowledge, however, must be exploited to create innovative new products and successfully commercialized. Suppliers can play an important role in ensuring successful exploitation of new knowledge. We posit that innovation-fostering-practices mediate the exploitation of new knowledge into superior innovation-performance. Thus, the proposed conceptual-framework incorporates the exploration and exploitation-aspects of innovation. Since contextual-differences can play a major role in the efficacy of these innovation-practices, our conceptual-framework might not fit complex project-environments in its entirety. To better understand the contextual-influences in complex projects, we evaluate the applicability of theoretical arguments from repetitive-manufacturing literature and our conceptual-framework to complex project-environments. The chapter utilizes a qualitative study to carry out this assessment. The results, while pointing to the usefulness of our framework and the innovation-fostering-practices, highlight the influence of the contextual-factors in complex projects. We develop practically useful conclusions for leveraging the supply-base for enhancing innovation in complex projects.

L. Kavin
Southern Denmark University, Odense, Denmark

R. Narasimhan (✉)
Michigan State University, East Lansing, MI, USA
e-mail: narasimh@broad.msu.edu

A. C. Moreira et al. (eds.), *Innovation and Supply Chain Management*, Contributions to Management Science, https://doi.org/10.1007/978-3-319-74304-2_3

1 Introduction

Complex projects (CP) are typically large projects initiated based on an innovation-pull from a client demanding a customized-solution (Olhager 2010). This requires development of a unique product based on specialized production-processes, which often takes place off-site based on temporary relationships. The supply-networks of CPs are often designed by the client and characterized by multiple suppliers involved at different times (Bygballe and Jahrem 2009). As configuration of the supply-network often involves competitive tendering to complete every new project at lowest possible cost, value-creating activities related to performance, flexibility and innovativeness are complicated as no standardized or legacy business-processes exist among the actors (Bygballe and Jahrem 2009). Due to temporary organization and high complexity of each project, requiring many specialized, but interdependent suppliers, knowledge-sharing to create innovative new products does not take place to a high degree (Vrijhoef and Koskela 2000). This is often caused by lack of trust and power-imbalances between actors. In CPs, management of the supply-network is often handled by an engineering-procurement-and-construction (EPC) firm based on formal contractual relations (Bygballe and Jahrem 2009). However, despite need for innovation in CPs, management generally focuses primarily on costs of discrete deliveries. In general, inability to foster innovation in CPs is one of the main reasons why CPs suffer from both low productivity and rising costs (Vrijhoef and Koskela 2000).

In CPs, suppliers can play an important role in ensuring successful exploitation of new knowledge. Being reliant on suppliers in the supply-network, the EPC-firm plays a crucial role in fostering innovation through its management of the different suppliers and deliveries throughout projects. The literature is lacking a framework that addresses how to foster innovation in CPs. While there is paucity of literature on supplier-enabled innovation in CP-contexts, much has been written on innovation in repetitive-manufacturing-contexts. We posit that context will affect how innovation-fostering-practices mediate the exploitation of new knowledge into superior innovation-performance. It is useful to understand how innovation-fostering-practices from repetitive-manufacturing-contexts can be modified or rendered effective in CPs. Such understanding can help the EPC-firm manage the supply-network and direct efforts to areas that are fruitful for innovation-success. Hence, there is a need to explore how innovation-fostering-practices from a repetitive-manufacturing-context can be applied in CP-contexts. Therefore, the objective of this conceptual chapter is to develop a theoretical-framework to identify *innovation-fostering-practices* based on innovation-literature and to assess the applicability of the framework to CPs, delineate relevant contingencies and develop propositions for future research.

The reminder of the chapter is structured as follows: First, we define innovation and develop a conceptual framework based on literature in repetitive-repetitive-manufacturing-contexts. We explicate innovation-factors and innovation-fostering-practices in repetitive-manufacturing-firms. Next, we discuss the distinguishing characteristics of repetitive-manufacturing and CP-contexts. After that, we discuss the methodology used to collect case-data in CP-contexts. Finally, we use the case-data to contrast CP- and repetitive-manufacturing-environments, and analyze the applicability of our framework to CP-contexts. We contribute to the literature by offering propositions of theoretical interest.

2 Review of Innovation in Repetitive-Manufacturing Networks

In this section, we review the innovation-literature from repetitive-manufacturing briefly to identify factors that underpin innovation-success. We also discuss "innovation-fostering-practices" and their role in enhancing innovation-success in repetitive-manufacturing-networks. Finally, we specify the different contexts of supplier-enabled-innovation in repetitive repetitive-manufacturing and CPs.

2.1 *Antecedents of Innovation*

Innovation is generally understood as development and exploitation of new knowledge (Tidd and Bessant 2009: 16). The characteristics of supplier-enabled-innovation are accordingly defined by Narasimhan and Narayanan (2013: 28) as "*the process of making changes to products, processes and services that result in new value creation to the organization and its customers by leveraging knowledge efforts of the firm and (or) that of its supply-network partners.*"

Within the realm of this and similar definitions on the relationship between supply-network and the innovation-process, four factors have been shown to be particularly relevant in the exploration-stage of the innovation-process. We examine how they contribute to innovation-performance.

First, *supply-network* influences innovation-performance through *open* and *closed* innovation-strategies pursued by firms. Open innovation suggests that interactions with suppliers are high allowing utilization of supplier-knowledge to generate new ideas. Interaction with suppliers can be co-development, joint-venturing and/or sourcing, where focus is on problem-solving (Chesbrough 2006). The firm might involve suppliers in multiple stages of the development-process (Song and Di Benedetto 2008). The ability of the firm to enrich its knowledge-base through integration of suppliers and external knowledge-sourcing during the R&D-phase is important in this "outside-in" innovation-process (Enkel et al. 2009: 312). Open innovation-processes can also be "inside-out" where the firm exploits its ideas externally in different markets by selling its intellectual-property. However, to successfully innovate, it is important to have efficient in-house-R&D (Narasimhan and Narayanan 2013). In closed innovation, R&D is confined within the focal-firm to control and manage ideas internally (Chesbrough 2006). Simultaneous use of open and closed innovation-approaches can improve innovation-performance (Narasimhan and Narayanan 2013).

Second, *new knowledge* improves innovation-performance by combining different explicit and tacit knowledge-sets (Kothandaraman and Wilson 2001). Exploiting the capabilities in a supply-network entails effective integration of the knowledge within the network and that of the focal-firm. Thus, knowledge-sharing-mechanisms with suppliers and common knowledge-bases are important to create new knowledge (Cho and Lee 2013). Exploiting knowledge within a supply-network requires investments in technologies, processes and people across suppliers by focusing on supplier-development efforts towards specific R&D-efforts and aligning these with the focal-firm's R&D-activities (Mahapatra et al. 2010). Two-way-exchange of knowledge via purposive communication, transparent processes, teaching "the right thinking", and encouragement of teaming have been shown to be effective (Noordhoff et al. 2011).

Third, literature suggests that *R&D-strategy* determines how a firm involves the supply-network (Tidd and Trewhella 1997). If a *first-move-strategy* is pursued, R&D develops proprietary product-technologies in-house. In a *follower*-strategy, R&D is in collaboration with suppliers to achieve low-cost imitation. Today, firms seldom rely on only internal knowledge and ideas or on immediate suppliers, but on a larger network of potential suppliers. However, to benefit from the supply-network, the focal-firm must achieve strategic-alignment across the network (Handfield et al. 2015). The firm must be aware of which technologies it needs and plan to acquire them. Further, R&D-strategy must be supported by financial and knowledge-resources to leverage the supply-network for innovation (Chiesa 2001). Thus, both R&D-strategies require a firm to manage relationships effectively to increase the likelihood of innovation-success. The strategic importance of relationships with suppliers for innovation has been noted by several studies (Handfield et al. 2015).

Fourth, *absorptive-capacity* affects innovation-performance. Absorptive-capacity is a firm's "ability to recognize value of new external information, assimilate it, and apply it to commercial-ends" (Cohen and Levinthal 1990). Firms with superior absorptive-capacity are better at transforming "exploration" successes into "exploitation"-successes. In order to exploit absorptive-capacity and use it when an opportunity is recognized, retained knowledge must be integrated with additional knowledge (Smith et al. 2005). Knowledge-integration is thus an important aspect of managing and fostering innovation. In repetitive-manufacturing, the focal-firm plays a central role in knowledge-integration for innovation. It must exploit network-knowledge to supplement internal knowledge to enhance absorptive-capacity and pursue innovation-opportunities.

These four factors constitute important antecedents of innovation in repetitive-manufacturing-networks.

2.2 Innovation-Fostering-Practices

"Innovation-fostering-practices" is a construct based on integration of many adjacent concepts of innovation in the context of repetitive-manufacturing/supply-networks.

Innovation-fostering-practices can be defined as practices that improve performance of the innovation-process pursued by a focal-firm. Main reason to collaborate with other actors in the supply-network is to gain access to resources, particularly knowledge (Zimmermann et al. 2016). Innovation-fostering-practices consist of many complementary and related concepts that facilitate the innovative process (Golgeci and Ponomarov 2013), and engage the supply-network in achieving innovation-success (Isabel et al. 2016).

The practices suggested by literature fall into three categories: *general management, knowledge-management and organizational-fostering-practices*. Collectively, we refer to these as "*innovation-fostering-practices*". This categorization provides a basis for "context-dependent understanding and learning" (Flyvbjerg 2006) about how to influence innovation in CP-contexts.

2.2.1 General Management-Practices

Firms generally use a "stage-gate"-approach throughout the innovation-life-cycle (Cooper 2000). Before initiating a new stage, a decision is made as to whether the innovation-effort should continue or terminate (Cooper 2000). In the stage-gate-approach, the "degree of openness" at each stage to supply-network-partners with different knowledge-capabilities is important. To improve innovation, firms adopt different degrees of openness in terms of outside-in and inside-out practices (Gassmann and Enkel 2004). Tradeoffs are made between internal appropriability (control), and developing capabilities (flexibility) in managing the degree of openness (Narasimhan and Narayanan 2013). It is feasible to use an open innovation-approach in ideation (exploration phase) and closed innovation-practices in later stages.

The twin objectives of flexibility and control influence management-processes and the firm's ability to innovate. Flexible management-practices help the firm in generating and acquiring new knowledge and exploiting opportunities (Narasimhan and Narayanan 2013). In flexible situations, sourcing-practices are often modified though different contract-structures. However, literature suggests that some firms prefer strict control over flexibility to have productive and appropriable R&D-investments (Rizzi et al. 2014). Need for control might prevent firms from involving the supply-network at any stage of the innovation-process. In sum, general management-practices pertain to the degree of control to be exercised, flexibility, contract-structures and resource-commitments that are made for innovation.

2.2.2 Knowledge-Management-Practices

Knowledge-management to capture, share and use knowledge from suppliers is salient in innovation (Maier 2007). Relationships in supply-network influence how firms interact dynamically to create tacit and explicit-knowledge in complex ways (Snowden 2002). The "triggers" for innovation occur from former projects, or acquisition of new capabilities. Innovation can also occur via a "probe-and-learn" experimental-approach to acquire new knowledge (Bessant and Von Stamm 2007). Identification and codification of existing knowledge is important for using data and information, contextualizing and giving them meaning, and operational usefulness (Cohen and Levinthal 1990). Codified-knowledge is easy to share whereas relational-knowledge is difficult to share outside the location where it is developed. It is therefore critical to convert tacit-knowledge into explicit-knowledge that can be shared across supply-network-members (Koskinen 2000).

Another aspect in utilizing network-knowledge is how knowledge is stored and retrieved (Nonaka and Takeuchi 1995). The main aspect is codification of tacit-knowledge and providing incentives to contribute to, retrieving, sharing and reusing relevant knowledge. It is information-sharing that leads to new knowledge across the supply-network (Tranfield et al. 2006). Some of the mechanisms to help knowledge-sharing are: "*organizational-translators*," who express the interests of one actor in terms of the firm's perspective; "*knowledge-brokers*"; and "*boundary-objects and practices*" (Carlile 2002) such as shared documents like quality manuals, prototypes, and common database.

These aspects of knowledge-management-practices facilitate the integration of the knowledge in supply-networks into a firm's innovation-initiative.

2.2.3 Organizational-Fostering-Practices

Organizational-fostering-practices include networking, establishing a creative climate, organizational-structure and others. *Networking* that impacts how organizations pursue interaction with suppliers and integration of internal knowledge-resources is a common way of fostering innovation, as it gives access to resources that can be used in implementation of new ideas (Kchaich Ep Chedli 2014). Three major resources that can be exchanged though networking are: information, "motivation" and material resources (Jenssen and Koenig 2002). A large network will provide a rich base of heterogeneous knowledge to use in innovation (Greve 1995). Networking that increases heterogeneous-knowledge has been shown to have a greater impact on the discovery of new opportunities (Mcevily and Zaheer 1999) and on creativity due to recombination of opportunities (Alves et al. 2007). The architecture of a supply-network can influence access to heterogeneous-information and knowledge. In supply-networks with a high intensity of interaction, information will disseminate faster than in a network with low intensity (Sligo and Massey 2007). Further, a clearly articulated and shared sense-of-purpose is needed across the network to foster innovation (Musiolik et al. 2012). Networking extends "peripheral-visions" (Brown 2004) for foster idea-generation and considering new options (Chesbrough 2006). In leveraging knowledge of its supply-network the firm needs to *emphasize "probe-and-learn"-practices* that extend learning across boundaries and into networks (Mariotti and Delbridge 2012).

Fostering a *creative climate* in the organization is essential for utilizing knowledge in the network and improving innovation-performance (Isaksen and Tidd 2006). This is done by giving employees freedom to generate new ideas, experiment with existing ideas to come up with new products in-house and in cooperation with the supply-network (Wei et al. 2013). Studies have shown that top-management must allocate appropriate resources to unleash the creative potential of the firm and its network (Kelley 2001). Training employees in basic innovation-skills and the use of innovation-tools is essential for a creative climate. Innovation-fostering-practices that result in a creative climate include encouraging experimentation and *risk-taking* without fear of failure throughout the innovation-life-cycle can improve internal R&D and innovation-performance through multiple path-ways (Muller and Hutchins 2012). *Creative climate* is developed by openness to new ideas, willingness to experiment, challenging new ideas and solutions, willingness to take risks and rewarding risk-taking-behavior and through structural and *infrastructural support* to both individuals and project-teams (Muller and Hutchins 2012). Regular cross-functional meetings to discuss innovation-projects and to promote leadership at the individual level are also helpful (Hersey et al. 2001).

Organization-structure influences innovation-performance through new organizational-forms and team-structures to increase knowledge-creation, dissemination and utilization of knowledge. Structure impacts how organizations learn and integrate knowledge both internally and with suppliers (Mintzberg 1983). In order to increase innovation-performance organizational-structure must be accompanied by channels for unorthodox ideas to flow and create capacity to deal with "off-message"-signals (Prahalad 2004). This can be done by regular solicitation of input on the firm's website, and involving representatives from complementary industries in ideation-sessions. Further, *"innovation-boards"* may exist at several levels to screen, prioritize, and fund innovation-projects (Muller and Hutchins 2012). The different innovation-fostering-practices are illustrated in Fig. 1.

Based on the foregoing review of ideas from repetitive-manufacturing and supply chain management (SCM)-literature, it can be argued that innovation is principally based on creation of *new knowledge*, which is dependent on and influenced by the integration *supply-network, R&D* and the *absorptive-capacity* of a firm. The conversion of new knowledge into successful innovation is *moderated* by management-practices, knowledge-management-practices and organizational-fostering-practices that define the environment in which innovation occurs. Our discussions thus far

Fig. 1 Innovation fostering practices that underpin innovation performance

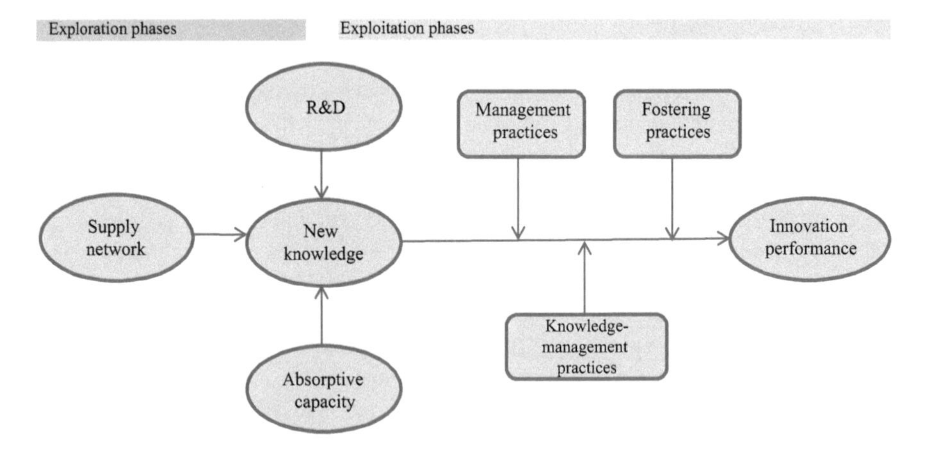

Fig. 2 Conceptual model of knew knowledge and innovation performance

suggest the conceptual-model shown in Fig. 2 that captures the innovation-factors and innovation-fostering-practices discussed in this section.

2.3 *Distinction Between Repetitive-Manufacturing and CP-Contexts*

Generally, literature suggests that the SCM-framework consists of three major and closely related components: supply-chain-network-structure, business-processes, and management-components, as suggested by Cooper et al. (1997). The supply-chain-network-structure concerns membership-aspects, horizontal-structure, vertical-structures and horizontal-position of a firm, and different kinds of process-links between the supply-network-partners (Lambert et al. 1998). Business-processes are activities that produce a specific output of value to customers (Davenport 1993). As indicated above, drivers for process-integration vary from process-link to process-link and over time, so allocating scarce resources among the different process-links across the supply-network is crucial (Lambert et al. 1998). The management-component is the managerial-variables by which business-processes are integrated and managed across the supply-network (Lambert et al. 1998). To specify the different contexts of supplier-enabled innovation in repetitive-manufacturing and CPs, the three components of Cooper's SCM-framework will be used.

2.3.1 Repetitive-Manufacturing-Context

In this section, the concept of SCM in a repetitive-manufacturing-context is succinctly described. Within a repetitive-manufacturing-context innovation and development of new products takes place based on a perceived understanding of customer-requirements and needs (Wortmann et al. 1997) and that the new product-development/engineering-process takes place under uncertainty where no activities are based on actual customer-orders (Rudberg and Wikner 2004). Furthermore, the innovation-process is typically initiated based on a push in the form of an overall assessment of future needs which is broken-down to input-plans based on net-requirements and pushed forward through the upstream supply-network (Olhager 2010). Because the value-creating-activities generally become frozen far-down the up-stream supply-network, much uncertainty related to demand exists, which means that innovation-fostering-practices will be based on speculation related to the innovative-outcome (Rudberg and Wikner 2004).

Initiation and development of new product is typically triggered by a focal-firm, with only a few network-actors involved (Wowak et al. 2015). The business-processes will be concerned with achieving economies-of-scale by developing a standard-design to be produced in high-volume. This can be realized as the activities within the supply-network typically are carried out by permanent, vertical-relationships along the individual chains supplying several development-projects based on long-term-relationships (Bygballe et al. 2013). In addition, value-creating-activities will be related to delivery, cost, and quality (Olhager 2010). This means that business-processes are characterized by specialization and modularization of components to speed-up the production-process. Indeed, this gives the supply-network-members leeway related to design and innovation, often realizing large cost-savings as they experience lack of restrictions in production (Kam-Sing Wong 2014) and allowing for mass-customization (Rudberg and Wikner 2004).

The management-focus is on the vertical-connections between the focal-firm and tier-one suppliers (Bygballe et al. 2016). Due to modularized product-architectures, it is possible to reuse modules in new product-development, so innovation often takes place within a shorter time-scale and at lower level of technological-risk (Koh et al. 2015). Furthermore, production is often aimed at consumer-goods, where customers have lower level of technological-knowledge (Cheng and Kam 2008). Indeed, focus is on integrating supply-chains, where often the manufacturer has power to manage/coordinate the other actors. The relation between partners in direct-exchange, where one partner produces outputs that serve as inputs for the actor in the next step of the process, is the key management-task. This is due to the actors being dependent on vertical-connections in the supply-network to deliver customer-value (Bygballe et al. 2013). Management will, in addition, be preoccupied with integrating disparate-systems as basis for mass-production through long-term-obligations and goals to achieve efficiency across the entire supply-network (Cao and Zhang 2011). Co-development, joint-venture, and sourcing are commonly used practices (Farahani et al. 2014). Furthermore, management of a relatively low level of risk and rights of intellectual-properti are needed (Bygballe and Jahrem 2009).

2.3.2 Complex Project-Contexts

In this section, we describe the concept of SCM in CP-contexts. Fostering innovation within CP-contexts where both the design and engineering-activities are driven by customer-orders is different from the repetitive-manufacturing-context in important ways. Innovation and development of new products take place under high security as the requirements are explicitly described by the client removing demand-uncertainty (Rudberg and Wikner 2004). Instead of coping with demand-uncertainty, interaction with new suppliers and commitment to a project become the primary-concerns (Bygballe et al. 2013). The innovation-process is thus typically initiated based on a "pull" where a client, external to the supply-network, places an actual order (Olhager 2010). Activities will be concerned with identifying and utilizing knowledge and capabilities to fulfill a client's needs.

Triggered by an order, a temporary supply-network is configured based on clients' choices, with formal contracts determining responsibilities and authorities throughout the entire process. The supply-chain-network-structure is often designed by the client and characterized by multiple supply-network-actors involved at different times. This affects the configuration of the supply-network as the client often relies on competitive-bidding to accomplish every new project at the lowest possible cost, resulting in distinctly different supply-chains for each project (Bygballe et al. 2013). Thus, a new supply-network is set up each time a client initiates a new product-development-project.

The business-processes will focus on achieving efficiency in delivering a customized-solution to a specific client based on the requirements (Bygballe et al. 2013). Here, the value-creating-activities are related to performance, flexibility, and innovativeness (Rudberg and Wikner 2004). This is complicated due to the temporary organization created from a project-coalition of different firms engaged in successful completion of delivering the customized solution related to discrete time, financial, and technical-goals (Vrijhoef and Koskela 2000). Integrating business-processes, and creating an efficient flow by managing reciprocal-interdependencies among suppliers, is therefore an important feature (Bygballe and Jahrem 2009).

The element of interaction and particularly, of alignment of internal business-processes with knowledge available in the supply-network are not only seen as critical preconditions for specialization but also as imperatives for superior-performance (Narasimhan and Narayanan 2013). However, due to discontinuous-demand and technical and financial uniqueness of each project, supply-chain-integration is especially challenging (Eriksson 2015). This is also caused by complex demand-requirements and production-conditions (e.g., offsite-production) along with complexity of each project requiring a high number of specialized but interdependent suppliers and their activities (Gil 2009).

The combination of high complexity and customization coupled with long duration usually requires inter-organizational integration to enhance coordination, flexibility, compliance, joint problem-solving, and knowledge-exchange across the supply-network (Eriksson 2015). However, low transaction-frequency and uniqueness of each project make supply-chain-integration and partnering challenging. Eriksson (2015), stated that supply-network-activities in CP-contexts contain sets of interdependent supply-network-dimensions related to the strength, scope, duration, and depth of projects, which will affect integration of activities and collaboration within the supply-network. Management must therefore focus on systematically mobilizing, coordinating, and adjusting various actors' contributions to achieve the project's goal and fulfill a client's objectives. This job is often handled by an EPC-firm based on formal contractual relations. This key management-task will have to address the many dyadic-relationships between the participants in the project-coalition and the general contractor–sub-contractor relationships (Bygballe et al. 2013). Many sub-contractors are often small firms, so achieving a balance between objectives, project-description, and organizational-arrangements, in addition to mobilizing and adjusting the contributions of participants in the project-coalition, is of great importance (Bygballe et al. 2013). The management-component is thus related to integration of various systems and managing high level of technological-risk by involving customers, different suppliers, and stakeholders throughout the business-processes. Furthermore, there is a need to continuously integrate different perspectives through an open innovation-approach throughout a longer time-horizon (Caldwell and Howard 2014).

Key characteristics of SCM within the two contexts are summarized in Table 1. As can be seen the two contexts differ markedly creating the need to better understand how literature-based understanding of innovation in repetitive-manufacturing-context can be utilized in CP-contexts.

3 Data Collection and Methodology

Methodologically, this chapter utilizes a qualitative-approach to develop a conceptual-framework, building on current literature on innovation and inductive-reasoning. Three strategies are utilized for data-collection in this research. First, literature on innovation was used to develop a framework for comparative-analysis of innovation in repetitive-manufacturing and CP-contexts. To assess the applicability of the innovation-framework, supply-network-level data were collected by means of semi-structured interviews with CEOs and Heads of departments in the Danish offshore-wind-industry as key-informants, to investigate the phenomenon of interest (Dubé and Paré 2003). The objective of the interviews was to develop a contextual-understanding of the factors pertaining to innovation in offshore-wind-industry, a CP-environment. We focused on supplier-relationships, with the utility-company as the focal-firm coordinating the overall innovation.

In offshore-wind-industry, innovation needs to encompass both product and process-dimensions. As a subsidized-industry, offshore-wind must innovate to decrease costs throughout the life-cycle of the CP. It offers a sharp contrast to the repetitive-manufacturing-industry (see Table 1) with high-dependence of the focal-firm on its supply-network.

Table 1 Comparing characteristics of supply chain management contexts

Characteristics	Repetitive manufacturing context	Complex project context
Product-related	Make-to-stock/make-to-order/assembly-to-order	Engineering-to-order
Engineering-related	Engineering-to-stock	Engineering-to-order
Initiation of innovation	Push	Pull
Purpose of innovation	Achieve competitive success	Deliver a customized solution related to discrete time, financial, and technical goals
Market demand	Predictable based on forecasts	Described by a client
Forecast mechanism	Algorithmic	Consultative
Customer drivers	Cost	Availability
Type of product	Standard	Customer-specific
Type of production system	Productivity-related	Flexibility-related
Production system	Volumetric	Non-volumetric
Production process	Industrialized	Specialized
Degree of product standardization	High	Low
Degree of offsite production	Low	High
Productivity	High	Low
Costs	Relatively low	Relatively high
Product variety	Low	High
Product life cycle	Relatively short	Relatively long
Order-winning manufacturing output	Delivery, cost, and quality	Performance, flexibility, and innovativeness
Information enrichment	Highly desirable	Obligatory
Manufacturing carried out by	Permanent vertical relationships along the individual chain supplying several projects	Temporary organizations made of a project coalition of different firms engaged in the successful completion of the project
Configuration of supply network	Long-term relationships between supply network actors	Temporary supply networks
Project configuration method	One actor takes the role of the integrator	Client's choice of procurement strategy and forms of contracts which determine responsibilities and authorities in the entire process
Focus	Speculation	Commitment

(continued)

Table 1 (continued)

Characteristics	Repetitive manufacturing context	Complex project context
Supply chain focus	Integration of supply chains	Dyadic relations between participants in the project coalition; general contractor–subcontractor relationships
Key management task	The relation between partners in direct exchange, where one partner produces outputs that serve as inputs for the actor in the next step of the process	Achieving balance between objectives, project description, and organizational arrangements; mobilizing and adjusting contributions of participants in the project coalition
Handling reciprocal interdependencies	A focal firm (often the manufacturer) has the power to manage/coordinate the other actors	Systemic mobilizing, coordinating, and adjusting the various actors' contributions to achieve the project's goals and fulfill the client's objectives
Focus of interdependencies	Vertical connections in the supply network	Many, often small, firms acting as subcontractors

Adapted from Jonsson and Rudberg (2014), Arlbjørn and Haug (2010: 183) and Bygballe et al. (2013)

Given our objective to develop contextually-based understanding of innovation, a single case-study of a network in a CP-environment is appropriate. Flyvbjerg (2006), in discussing "five misunderstandings" about case-studies, addresses theory-development based on the rich-information from a single, exemplar case-study. He observes: *"a case-study is the detailed examination of a single example of a class of phenomena"*.

We chose to carry out the case-study with a Danish utility-company as: (1) it operates in a CP-environment, (2) it was interested in participating in the research-project, and (3) we had unfettered access to senior-management in the utility-company and to key-informants in several firms in its supply-network.

Our selection of the EPC-firm and its supplier-network was guided by following factors: the EPC-firm focuses on supply-network-enabled-innovation in its offshore-wind-division. Its suppliers possess distinct-capabilities needed to build an offshore-wind-park. These include foundations, tower, nacelle and blades, assembly and installation, electrical-infrastructure and power-transmission. We followed "selective sampling" in selecting sub-contractors for the study (Neergaard 2007: 39–40). We requested the top-management across all levels of the offshore-wind-power supply-network to participate. The informants were first contacted by phone to explain the objective of the research. Subsequently, an email was sent with more information about the study emphasizing the academic nature of the research to ensure maximum cooperation from sub-contractors. An appointment was made to conduct a semi-structured interview with CEO or Heads of Department after they agreed to participate in the study. Due to confidentially-reasons, names of companies are not divulged. The interviews utilized an interview-guide and lasted an average of 2 h each. Most were conducted on-site except for a few phone-interviews due to distances involved.

Literature recommends a priori-selection of relevant variables and constructs from previous studies to develop the initial protocol in qualitative theory-building-research (Eisenhardt 1989). The initial interview-guide was based on literature and extant theoretical-frameworks on contextual-contingencies and innovation-practices. The interview-protocol encompassed two main-themes: (a) process and critical characteristics of projects in offshore-wind-industry and dependencies across supply-network and (b) factors affecting innovation in offshore-wind-industry. Our principal aim was to understand the *unique characteristics of the supply-network* in CPs. In order to validate and confirm interview-findings, interview-data were transcribed, verified with key-informants and industry-experts, and used in subsequent analysis (Yin 2014: 45).

We followed the guidelines in literature by incorporating our understanding of the contextual-contingencies from initial interviews in subsequent interviews to capture a rich-base of information for analysis (Yin 2014). For example, understanding the uniqueness of contracting with suppliers during the initial stage of data-collection enabled us to probe for the motivations that suppliers have for participating in offshore-wind-industry even if monetary-payoffs are not sufficient to warrant such participation.

The data-collection-approach used in this study is appropriate given our interest in investigating and obtaining an in-depth-understanding of innovation in a dynamic and CP-context (Yin 2014: 24).

The interview-data were analyzed through data-categorizing and recombination (Miles et al. 2014) to allow for new aspects to emerge that were initially not under investigation. By using this approach, we could identify salient aspects relevant to the research-questions and assess how innovation-fostering-practices in the CP-environment of offshore-wind-industry differ from a repetitive-manufacturing-context.

Studies that use comprehensive information from a single company and its supply-network can be criticized for potential lack of generalizability of findings. While this may be a valid criticism of our study, developing theoretical understanding of a new context using the phenomenological-approach requires establishment of the *existence* of a phenomenon and the *principal mechanisms* at work there in. In this, exploratory-phase, a comprehensive single case-study has its place (Thomas 2011: 138–140). Through comprehensive data-collection, inclusion of relevant literature, data-triangulation, and careful analysis of data based on literature, we have developed a set of findings that are logically consistent. This approach enhances the validity of our findings (Flyvbjerg 2006).

3.1 Validity of Case Data

Maxwell (1992) asserts that validity in qualitative research has five aspects: *descriptive-validity, interpretive-validity, theoretical-validity, generalizability* and *evaluative-validity*. Descriptive-validity refers to "factual accuracy" in capturing the data collected. Descriptive-validity was ensured through careful recording, transcription and verification of interview-data with key-respondents to ensure factual accuracy. Interpretive-validity is concerned with the *meaning of data* in the context being studied. Construction of meaning from interview-data is necessarily ideational. Maxwell (1992) observes: "*…understanding is most central to interpretive research*". In interpreting the interview-data, we stayed close to what the respondents conveyed. Theoretical-validity pertains to relationships that are posited among the salient concepts that comprise the theory. Theoretical-validity is ensured by validity of the concepts and the postulated relationships. We ensured theoretical-validity by relying on current-literature to identify the principal-constructs relating to innovation and by utilizing inductive reasoning to posit relationships. In qualitative-studies generalizability must do with development of a theory that "makes sense" and "shows how the same processes in different situations can lead to different results" (Maxwell 1992). In our study, we have sought to infer conclusions that were consistent with data from specific CP-contexts. They may have applicability in similar CP-contexts. We don't make a stronger claim than that. Evaluative-validity is not particularly relevant in qualitative research. Our qualitative-study meets these criteria for validity.

4 Assessment of the Framework in a Complex Project

The framework in Fig. 2 is based on inductive-reasoning from innovation-literature on repetitive-manufacturing-contexts. We discuss applicability of this framework in CP-contexts within the Danish offshore-wind-industry next, emphasizing the innovation-fostering-practices.

4.1 Innovation Antecedents in the Danish Offshore-Wind-Industry

When the governmental Department of Energy initiates an offshore-wind-project by putting it on tender a utility-provider (EPC) will bid and win the rights to commission and install the wind-power-farm. It thus becomes the focal-firm of a new supply-network. The actual supply-network will be determined by which specialised sub-suppliers (wind-turbine-generator-manufacturer, foundation-provider, assembly-and-installation-provider, provider-of-the-electrical-infrastructure and power-transmission-provider) win the next tender-round on the specified requirements by the EPC.

Specialized suppliers are continuously researching and developing new knowledge within their respective areas to be able to win an upcoming tender. The EPCs are also engaged in R&D to put up a realistic tender emphasising supplier-enabled-innovation. When they put-up a tender, they require three different technological-solutions along with a price for each solutions to be delivered: The key-informant at the EPC-firm stated: "*to promote innovation, we require the bidders to submit at least three different ways of solving their part of the final solution ... so that in reality it becomes a portfolio of offers from each supplier".*

Based on the supplier-enabled-innovations of the winning suppliers, the EPC uses its absorptive-capacity to recognize the value of each supplier's innovation-efforts and assimilate it to the wider project based on former knowledge and experiences. It is an "outside-in" open innovation-process based on co-development where the EPC combines different explicit and tacit-knowledge-sets within the supply-network to increase its absorptive-capacity and create new knowledge through a two-way-exchange of knowledge.

4.2 Management-Practices in the Danish Offshore-Wind-Industry

In the case of the offshore-wind-industry a "stage-gate"-approach is used. The innovation-life-cycle starts at each potential supplier. Before initiating the first-stage, the EPC decides wheatear the innovation-effort should continue. In the initial-phases, with each of the potential suppliers, the innovation-approach is closed and controlled due to intellectual-property-concerns related to the chance of winning the tender on the next offshore-wind-turbine-project. Subsequent phases of the innovation-process are "open", where the individual suppliers collaborate with the EPC on innovation. The EPC tries to use flexible management-practices to generate and acquire new knowledge and knowledge-exploiting opportunities. Sourcing-practices are modified though different contract-structures. When it comes to commercialization of the final wind-farm and generated power the EPC uses a closed-approach to utilize the value created. Thus, the EPC utilizes both flexibility and control-oriented approaches to manage the innovation-process.

Externally oriented and flexible management-practices across the supply-network, the different phases and across projects make it more likely for the EPC to capture and use ideas and technologies affecting innovation. For instance, the EPC uses different contract-structures to control the diverse sub-deliverables throughout projects. The contracts are however project-specific and linked to time, cost, quality and innovation-objectives of the single project, not obligations across projects, which affects innovation-performance to a higher degree.

4.3 Knowledge-Management-Practices in the Danish Offshore-Wind-Industry

To capture, share and use knowledge from suppliers in the innovation-process, the EPC interacts dynamically with industry domestically and internationally to create tacit and explicit-knowledge. Most innovation-projects within supplier-firms are triggered by former projects or acquisition of new capabilities. Each of them tries to identify and codify their existing knowledge and use their data and information in operationally useful ways. However, they don't share their codified-knowledge; only relational-knowledge is shared if different suppliers collaborate on several projects.

Knowledge on each offshore-wind-industry-project is stored at the separate suppliers. Due to different project-managers on each project and complex retrieving-systems, data and new knowledge are seldom used in new projects or across the supply-network. By using a "probe-and-learn"-strategy, the EPC could promote network-based learning by extending learning across the supply-network. The portfolio-approach coupled with a "probe-and-learn"-strategy would differ from the typical approach of engaging suppliers in innovation in repetitive-manufacturing. However, in CP-contexts, knowledge-management-practices must enable technological-change through efficient information-flows. This is also necessitated by the length of the project, complexity and uncertainty related to the innovation-outcome. In contrast to repetitive-manufacturing, the "*uncertainties of the market are not important*" in CP-contexts as innovation-needs are defined by the requirements of the client.

The wind-industry-organizations are trying to promote knowledge-sharing among firms by expressing the common-interests in reducing the-cost-of-producing-energy across the industry to be competitive compared to the cost-of-energy from fossil-fuels. Besides the organizations emphasize and create shared documents in terms of innovation-progress within the industry, standardizing efforts and by promoting common network-meetings and workshops to integrate the innovation-initiatives within the supply-network.

4.4 Innovation-Fostering-Practices in the Danish Offshore-Wind-Industry

In this section, the organizational-aspects of how the EPC can foster innovation in the CP-context of the Danish offshore-wind-industry are discussed.

4.4.1 Networking

Whereas firms can choose between closed and open approaches to innovation in repetitive-manufacturing, innovation in projects *must be open at any stage of the R&D-process,* defining a sharp contrast. In a project, the EPC-firm must choose suppliers according to the project-plan and criteria stipulated by the client, regardless of prior relationships. An executive from the EPC-firm stated: "*We must drive invitations to tender where all (suppliers) must be equally judged*". Networking, in projects requires more extensive efforts to select and integrate suppliers with different knowledge-sets and is done at the start of each project. Typically, in repetitive-manufacturing, networking is with suppliers with whom the focal-firm has had prior relationships in a stable network.

Intensity of network-interactions is greater in repetitive-manufacturing due to previous experiences and interactions. In contrast, the intensity of interactions in a project-environment is likely to be lower. The CEO of the assembly-and-installation-provider observed: *"Networking is good to create some approaches, but it is not in the network it [innovation] happens"*, suggesting that intensity of interactions might be considerably lower, governed primarily by contractual-agreements.

The *heterogeneity* of information, motivation and resources is higher in a project-network, where the suppliers belong to different industries with diverse knowledge and capabilities. Therefore, the EPC-firm cannot rely on past interactions as do focal-firms in repetitive-manufacturing-networks, but must promote purposive-networking to a greater degree in each CP, taking into consideration unique project-requirements and different set of network-actors. This is also necessitated by lower network-intensity and lower level of trust among suppliers due to the absence of past interactions. The data suggest that knowledge gained through networking can, however, improve learning and innovation-performance in both contexts. The case-study data suggest that the EPC-firm engages in networking-activities in different networks supporting all sub-deliveries. In contrast, in repetitive-manufacturing the focal-firm typically interacts with tier-one suppliers only who, in turn, manage sub-suppliers and lower-tier-suppliers. The objectives of the EPC-firm are to utilize both explicit and tacit-knowledge within not only its own network or industry, but also in *peripheral or complementary industries*. The CEO of the electrical-infrastructure-firm observed: *"...Danish Research Consortium for wind-power was created, with representatives from major universities, industrial-companies, and even small industries along with us...,"* which underscores the need for integrating different industrial and academic networks. Networking across industries is likely to be more due to the high complexity of the integrated-solution sought in CP-contexts such as the design, installation, operation and maintenance of off-shore-wind-parks. Networking as an innovation-fostering-practice is useful in both contexts with important differences induced by context.

4.4.2 Organizational-Structures

Organizational-structure impacts how organizations learn and integrate internally and externally with suppliers. Whereas a divisional-form or a professional-bureaucracy (Mintzberg 1983) might be appropriate to foster innovation in a repetitive-manufacturing-context, the CP-contexts requires organizational-structures that can accommodate high uncertainty and complexity. Due to the need for *system-wide-integration* of innovations in projects, there is a greater need to integrate the knowledge and technical-skills of the sub-suppliers and to *foster collaboration among diverse suppliers*—large versus small, domestic versus international, same versus different industry and different levels of knowledge-asymmetry. This implies that primarily, it is the EPC-firm that is responsible for promoting collaboration across the supply-network and in multiple-tiers of the supply-network. This contrasts with repetitive-manufacturing-firms where there is greater reliance on tier-one suppliers to promote collaboration across the network. Therefore, the organizational-structure, must be able to cope with high level of uncertainty induced by information-asymmetry associated with different technologies and constraints of being "locked-in" by decisions taken in earlier phases of the project, without stifling creativity (Roberts 1991).

In CP-contexts, it is necessary to occasionally change the organizational-structure to respond to the changing and uncertain nature of the project. For example, the CEO of assembly-and-installation-firm told us this was the case with the Anholt site where it was not possible to fasten anything to the sea-bed. To install turbines and the electrical-infrastructure, it was necessary to use knowledge and experiences within the network based on ad-hoc teams and external support-functions illustrating adaptations and fluidity of organizational-structures.

Absorptive-capacity pertaining to organizational-structures must be higher in CP-contexts to recognize and react to contingencies, and exploit new information by internalizing and synthesizing it with knowledge resting in the network. Thus, organizational-structure must be fluid to adapt to these varied requirements without adversely impacting the innovation-outcome; the degree of adaptation required in terms of organizational-structures is much lower in a repetitive-manufacturing-context due to the relative stability of technologies, relationships and supply-network.

4.4.3 Encouraging Risk-Taking

Risk-taking is essential for promoting a creative environment, which was identified as an innovation-fostering-practice from the literature. Due to more unfamiliar firms interacting in a CP, the *perception-of-risk* is higher than in a repetitive-manufacturing-context. Employees might be reluctant to pursue new ideas than in a repetitive-manufacturing-context where they often know the collaboration-partners and might feel more secure based on prior experiences with them (Ross and Athanassoulis 2010). In addition, in mass-produced, commercialized-products it is easier to estimate the probability of technical or commercial-success based on prior-knowledge of products and markets. The employees might find it easier to suggest new products or additional features to a current-product for markets they know, than coming up with integrated-solutions to an unfamiliar-client, which characterizes the CP-environment. Another risk-element arises when goals and values of a supplier are different from the other suppliers or the EPC-firm. It can be especially challenging due to intellectual-property-concerns. Albeit difficult, encouraging risk-taking is more critical in fostering innovation in CP-contexts. As the organizational-climate and encouraging risk-taking might differ across the diverse firms in CP-contexts, a well-articulated and shared sense-of-purpose might be of even greater importance than in a repetitive-manufacturing-context. To encourage risk-taking, the EPC-firm creates a range within their specifications. Within this range, the EPC-firm incentivizes the suppliers to come-up with a portfolio of technical-solutions, which promotes risk-taking. In offshore-wind-industry, where the EPC-firm might not have chosen the collaborators, or have prior-knowledge of all suppliers, practices are required to *promote-trust* and *build relations* that enable clear and regular communication, providing inputs for problem-solving and shared innovation. Although trust-building is equally important in the repetitive-manufacturing-context, in CPs the EPC-firm cannot rely on furthering loyalty and trust based on repeated past interactions. Instead, it must rely to a greater extent on relation-building-practices such as justice, joint problem-solving, transparency in interactions and openness in sharing information. Building relations counters different kinds of risk based on different cultures, knowledge-assets and cooperative-practices (Xiwei et al. 2010).

The fostering-practices, therefore, must *encourage risk-taking to a higher degree in CP-contexts* to engage and motivate disparate-suppliers for creation of new value in various networks.

4.4.4 Creating Incentives

Incentives to innovate are typically different in CPs. The literature from repetitive-manufacturing-contexts overemphasizes the competitive-component and misses relational-perspectives and knowledge-integration, which are significant in a project-environment (Hayes and Walsham 2003). Innovations in a repetitive-manufacturing-context can be initiated by a focal-firm and potential partners whenever they come-up with a new idea they decide to pursue. In CP-contexts, however, innovation-efforts begin when the client recognizes a need, regardless of how it fits into the potential suppliers' plans, current commitments and capacity to engage with the client. In the offshore-wind-industry, the actors don't know when a project will be initiated or what the criteria will be for bidding. Thus, suppliers with the most comprehensive knowledge and experience might not be able to participate in the project due to prior commitments. The informant from the transmission-firm stated: *"It is often the case there might only be one or two suppliers left that we can contract with"*. In some cases the supplier agrees to participate, but might not agree to the delivery-times.

When initiated by a government, the formal-requirements might be so burdensome that the income for a small or medium-sized supplier with specific project-related knowledge will be exceeded by the cost of adhering to the procedures and formal-requirements. The business-development-manager at the foundation-supplier observed: "*...This means that we incur administrative-costs, before the project even starts, during the project and in the end of the project. These projects or activities that we participate in, they are relatively so small that it hardly pays for us...*" The incentive to participate in projects is *related to learning-outcomes* and experiences that might be useful in in commercial-projects in the future. To foster innovation and to ensure participation by the suppliers in the project, management must create incentives and opportunities for the suppliers to *learn* by providing timely feedback and exchanging information. The informant from one of the suppliers observed:

"...within oil-and-gas they have a big center to foster learning; something similar would benefit the offshore-wind-industry..." underscoring the importance of learning. In sharp contrast to repetitive-manufacturing, where incentives are profits and growth, in CPs incentives pertain to *network-based-learning*.

4.4.5 Infra-structural-Practices

A clearly articulated and shared sense-of-purpose is noted as an essential innovation-fostering-practice in literature. This is applicable to CP-contexts where diverse firms collaborate on a solution based on an integrated set of technologies and knowledge-inputs. The Head of energy-analysis at the transmission-firm stated *"I think these long-term plans with broad political-agreement around objectives provide a stable environment and better opportunities for innovation*", underscoring the importance of a shared sense-of-purpose. What is needed in CP-contexts is the spontaneity and flexibility to engage in open innovation-approaches when an opportunity or a need arises. The infra-structural-practices in CP-contexts, therefore, require an increased management focus on resource-acquisition and achieving external support to provide cohesion and morale (Quinn and Rohrbaugh 1983).

The stage-gate-model to improve innovation-performance does not apply to the CP-context. Practices to foster stability and control must be augmented with practices to foster collaboration and valuing the different relational and knowledge-resources of the suppliers. Conflicts will occur when people with different backgrounds have to work together. Current literature from repetitive-manufacturing does not adequately recognize this innovation-challenge. For example, an employee from a university was involved in co-development of a new material for a supplier. However, he left the team. Meanwhile the university developed a competing product, which happens to be similar to one of the supplier's, and sold it to the supplier's closest competitor. This has hampered the supplier's confidence in being open and sharing information with anyone. To foster innovation-success in CPs the potential for inter-personal or inter-firm conflicts must be anticipated and effective steps to protect intellectual-property must be taken. In contrast, in repetitive-manufacturing the repetitiveness of interactions and relative stability of the supply-network coupled with relational-governance minimizes opportunism to a degree.

4.4.6 Communication

Literature emphasizes communication as a mechanism to share knowledge and increase innovative-performance. In contrast to a repetitive-manufacturing-context where a stable set of actors collaborate on innovation, innovation in CP-contexts are complicated as the suppliers don't have the same level of trust and knowledge of each-other or similar culture or ways of communicating. Consequently, the norm of confidentiality doesn't exist within CP-contexts. Therefore the concepts of open communication and transparency cannot be readily applied in CP-contexts. To facilitate communication in CP-contexts, practices must focus on building trust and confidentiality. Trust enhances the strength of supplier-relationships, establishes partnering-roles and increases the willingness to cooperate and share information across the supplier-network (Pinto et al. 2009). Investment in stable inter-organizational communication-systems, which is often seen in repetitive-manufacturing-contexts, might not be a solution in CP-contexts. For suppliers in a CP with intellectual-property to protect, a stable IT-system will be less useful for exchange of information because of potential loss of intellectual-property to other project-participants. Fostering-practice to increase communication must instead include techniques to securely connect data, documents, web-pages and aggregate information from different suppliers based on the specific needs of the project.

It is important to ensure transparent processes and simultaneously let the different suppliers work together in an intelligent, creative and efficient way. The dynamics of a project-environment further make it important that the suppliers are able to easily create their own ad-hoc teams that can contribute with the right expertise, knowledge and value needed at the right time. By implementing web-based, team-collaboration-platforms to temporarily store, organize and share all information related to a specific project it is possible to encourage communication in CP-contexts where the suppliers can retrieve the knowledge they need or take advantages of other suppliers' experience whenever they need it.

5 Conclusions and Implications

In this section, we summarize the conceptual-framework of fostering-practices and define opportunities to apply these in CP-contexts.

Figure 2 depicts how firms can integrate suppliers in the innovation-process to leverage knowledge within supply-networks. The conceptual-model is suggestive of the following propositions in CP-contexts. We state these succinctly for reasons of brevity. The phrase "contextually-relevant" in the following propositions is intended to capture important contextual-differences discussed in previous sections.

P1: *With respect to the conceptual-framework, innovation-performance is positively moderated by contextually relevant management-practices surrounding innovation. The strength of moderation will be higher in CPs.*

P2: *Coordination in CPs will demonstrate greater ambidexterity in that flexibility and control will be emphasized to a greater degree.*

P3: *Innovation-performance is positively moderated by knowledge-management-practices. The strength of moderation will be higher in CPs.*

P4: *Innovation in CPs will favor an open innovation-approach.*

P5: *The intensity of relational governance-approaches will be higher in CPs to support innovation-objectives.*

P6: *Network-based organizational-learning will be higher in CPs compared to the repetitive-manufacturing-context.*

The *innovation-fostering-practices can be applied in CP-contexts to different degrees*. However, due to integration of many different sub-systems in CP-contexts, the closed innovation might not work well in CP-contexts. Instead, firms must apply an *open approach to innovation with externally-oriented and flexible management-practices*. Further, fostering of innovation in CP-contexts requires much more extensive *networking to develop integrated-solutions*. In addition, a more formal collaboration-approach with *use of contracts* as well as relational-governance-practices is needed to manage all transactions with suppliers and sub-suppliers. Contractual-governance might not dominate in this respect. Despite that, CP-contexts require more flexible and organic organizational-structures. This requires a *well-developed absorptive-capacity to adapt to changes in the environment in CP-contexts*. In CP-contexts, encouraging risk-taking will demand a higher level of management-commitment, building on a clearly articulated and shared sense-of-purpose. However, incentives in the CP-context are quite different than in a repetitive-manufacturing-context. This presents an increased challenge to encourage commitment and delivering on time. Management must embrace spontaneity and flexibility to utilize different relational and knowledge-resources of the suppliers in the network. Two-way-exchange of knowledge must be emphasized in CP-contexts through transparent processes and the ability to retrieve knowledge within the network whenever it is needed via effective communication.

Based on these findings, we conclude that generally the innovation-fostering-practices from repetitive-manufacturing-context can be applied to CP-contexts. However, *the moderating-factors play a much more critical role for improving the innovation-performance* in CP-contexts. Understanding the contextual-differences between projects and repetitive-manufacturing can allow firms in a CP-environment to suitably emphasize the innovation-fostering-practices.

Our conclusions contribute to innovation-literature in several ways:

1. By delineating the important distinctions of the CP-environment, they identify different roles for the fostering-practices.
2. By explicating the differences from a repetitive-manufacturing-environment, we pave the way for a contingent theory of innovation-performance.
3. We promote theory-development by the joint-examination of fostering-practices, knowledge-management-practices and organizational-practices vis-à-vis supply-networks.

5.1 Managerial Implications and Future Research

As apparent from the above discussion, the impact of the fostering-practices is much more extensive in CP-contexts. This might be due to the temporary nature of the project, focusing primarily on ensuring each supplier's accomplishment of their individual parts of the project, and the EPC-firm's focus on fulfilling the overall objectives of the clients and future efficiency of its own business. In contrast, innovation in a repetitive-manufacturing-context is initiated in larger vertically-integrated supply-chains, where each network supplies several firms with similar components and have a long-term relational-orientation. The implication for managing innovation in CP-contexts, therefore, is to have an increased focus on relationship-management as the short-term commitment to the project is relatively unimportant to the longer-term interests of survival and growth of each supplier's own business (Winch 2002: 335). Compared with innovation in a repetitive-manufacturing-context, management in CP-contexts must have an increased awareness on creating incentives for suppliers to contribute. To have a satisfactory innovative-outcome, suppliers must buy-into project-objectives rather than pursuing their own objectives at the expense of the project's objectives (Winch 2002). It would therefore be interesting to map the networking-activities and content of information-exchange in CP-contexts to explore what kind of knowledge is sought and shared by different suppliers in order to improve learning-outcomes and experiences. If suppliers find it more attractive to be part of a project-team, and if the project had a significant impact on their learning which could promote future growth and survival of their individual business, it might be possible to further increase the innovative-output of CPs.

In mapping and investigating the networking-activities, research showing how formal and informal organizational-structures are organized throughout the different innovation-phases in CP-contexts would be helpful in adjusting the managerial-practices. It can be conjectured that the organizational-structure in the initial-phases of a project needs to be more flexible in order to capture all the knowledge-inputs, whereas more formal-structures might be appropriate to manage the underlying processes and make it easier to coordinate the activities in the later phases. As the modes of organization affect the fostering-practices related to power, communication and trust in CP-contexts, understanding the different formal and flexible practices needed in different phases would help achieve risk-taking and effective communication within CP-contexts.

References

Alves J, Marques MJ, Saur I, Marques P (2007) Creativity and innovation through multidisciplinary and multisectoral cooperation. Creat Innov Manag 16(1):27–34

Arlbjørn JS, Haug A (2010) Business process optimization. Aarhus, Academica

Bessant J, Von Stamm B (2007) Twelve search strategies which might save your organization. AMI Executive Briefing, London

Brown J (2004) Minding and mining the periphery. Long Range Plann 37(1):143–151

Bygballe LE, Jahrem M (2009) Balancing value creating logics in construction. Constr Manag Econ 27(7):695–704

Bygballe LE, Håkansson H, Jahrem M (2013) A critical discussion of models for conceptualizing the economic logic of construction. Constr Manag Econ 31(2):104–118

Bygballe LE, Sward AR, Vaagaasar AL (2016) Coordinating in construction projects and the emergence of synchronized readiness. Int J Proj Manag 34:1479–1492

Caldwell N, Howard M (2014) Contracting for complex performance in markets of few buyers and sellers: the case of military procurement. Int J Oper Prod Manag 34(2):270–294
Cao M, Zhang Q (2011) Supply chain collaboration: impact on collaborative advantage and firm performance. J Oper Manag 29(3):163–180
Carlile PR (2002) A pragmatic view of knowledge and boundaries: boundary objects in new product development. MIS Q 13(4):442–455
Cheng SK, Kam BH (2008) A conceptual framework for analysing risk in supply networks. J Enterp Inf Manag 22(4):345–360
Chesbrough H (2006) Open business models: how to thrive in the new innovation landscape. Harvard Business School Press, Boston
Chiesa V (2001) R&D strategy and organisation: managing technical change in dynamic contexts. Imperial College Press, London
Cho DW, Lee YH (2013) The value of information sharing in a supply chain with a seasonal demand process. Comput Ind Eng 65(1):97–108
Cohen WM, Levinthal DA (1990) Absorptive-capacity: a new perspective on learning an innovation. Adm Sci Q 35(1):128–152
Cooper MC, Lambert DM, Pagh JD (1997) Supply chain management: more than a new name for logistics. Int J Logist Manag 8(1):1–14
Cooper RG (2000) Doing it right: winning with new products. Ivey Bus J 64(6):54–60
Davenport T (1993) Process innovation: reengineering work through information technology. Harvard Business School Press, Boston
Dubé L, Paré G (2003) Rigor in information systems positivist case research: current practices. MIS Q 27(4):597–635
Eisenhardt KM (1989) Building theories from case study research. Manage Rev 14(4):532–550
Enkel E, Gassmann O, Chesbrough H (2009) Open R&D and open innovation: exploring the phenomenon. R&D Manag 39(4):311–316
Eriksson PE (2015) Partnering in engineering projects: four dimensions of supply chain integration. J Purch Supply Manag 21(1):38–50
Farahani RZ, Rezapour S, Drezner T, Fallah S (2014) Competitive supply chain network design: an overview of classifications, models, solution techniques and applications. Omega 45(1):92–118
Flyvbjerg B (2006) Five misunderstandings about case-study research. Qual Inq 12(2):219–245
Gassmann O, Enkel E (2004) Towards a theory of open innovation: three core process archetypes. R&D Manag Conf 6(1):1–18
Gil N (2009) Developing cooperative project client-supplier relationships: how much to expect from relational contracts? Calif Manage Rev 51(2):144–169
Golgeci I, Ponomarov SY (2013) Does firm innovativeness enable effective responses to supply chain disruptions? An empirical study. Supply Chain Manag 18(6):604–617
Greve A (1995) Networks and entrepreneurship – an analysis of social relations, occupational background, and use of contacts during the establishment process. Scand J Manag 11(1):1–24.
Handfield RB, Cousins PD, Lawson B, Petersen KJ (2015) How can supply management really improve performance? A knowledge-based model of alignment capabilities. J Supply Chain Manag 51(3):3–17
Hayes M, Walsham G (2003) Knowledge sharing and ICTs: a relational perspective. In: Easterby-Smith M, Lyles MA (eds) The Blackwell handbook of organizational learning and knowledge management. Blackwell, Malden, pp 54–77
Hersey P, Blanchard KH, Johnson DE (2001) Management of organizational behavior leading human resources. Prentice Hall, New Jersey
Isabel M, Bravo R, Ruiz A, Francisco L-M (2016) Supply network-enabled innovations: an analysis based on dependence and complementarity of capabilities. Supply Chain Manag Int J 21 (5):627–641
Isaksen S, Tidd J (2006) Meeting the innovation challenge: leadership for transformation and growth. Wiley, Chichester

Jenssen JI, Koenig HF (2002) The effect of social networks on resource access and business start-ups. Soc Netw 10(8):1039–1046
Jonsson H, Rudberg M (2014) Classification of production systems for industrialized building: a production strategy perspective. Constr Manag Econ 32(1–2):53–69
Kam-Sing Wong S (2014) Impacts of environmental turbulence on entrepreneurial orientation and new product success. Eur J Innov Manag 17(2):229–249
Kchaich Ep Chedli M (2014) Obtained resources through individual networking inside the organization, creativity of the supervisor and innovation. Econ Manag Finan Markets 9(4):376–394
Kelley T (2001) The art of innovation: lessons in crea tivity from IDEO, America's leading design firm. Crown, New York
Koh ECY, Forg A, Kreimeyer M, Lienkamp M (2015) Using engineering change forecast to prioritise component modularisation. Res Eng Design 26(4):337–353
Koskinen KU (2000) Tacit knowledge as a promoter of project success. Eur J Purch Supply Manag 6(1):41–47
Kothandaraman P, Wilson DT (2001) The future of competition: value-creating networks. Ind Mark Manag 30(4):379–389
Lambert DM, Cooper MC, Pagh JD (1998) Supply chain management: implementation issues and research opportunities. Int J Logist Manag 9(2):1–20
Mahapatra SK, Narasimhan R, Barbieri P (2010) Strategic interdependence, governance effectiveness and supplier performance: a dyadic case study investigation and theory development. J Oper Manag 28(6):537–552
Maier R (2007) Knowledge management systems: information and communication technologies for knowledge management. Springer, Berlin
Mariotti F, Delbridge R (2012) Overcoming network overload and redundancy in latent ties overcoming network overload and redundancy in interorganizational networks: the roles of potential and latent ties. Organ Sci 23(2):511–528
Maxwell J (1992) Understanding and validity in qualitative research. Harv Educ Rev 62 (3):279–301
Mcevily B, Zaheer A (1999) Bridging ties: a source of firm heterogeneity in competitive capabilities. Strateg Manag J 20:1133–1156
Miles MB, Huberman AM, Saldaña J (2014) Qualitative data analysis – a methods sourcebook. Sage, Thousand Oaks
Mintzberg HT (1983) Structure in fives: designing effective organizations. Prentice Hall, Englewood Cliffs
Muller A, Hutchins N (2012) Open innovation helps Whirlpool Corporation discover new market opportunities. Strateg Leadersh 40(4):36–42
Musiolik J, Markard J, Hekkert M (2012) Networks and network resources in technological innovation systems: towards a conceptual framework for system building. Technol Forecast Soc Chang 79(6):1032–1048
Narasimhan R, Narayanan S (2013) Perspectives on supply network-enabled innovations. J Supply Chain Manag 49(4):27–42
Neergaard H (2007) Udvælgelse af cases i kvalitative undersøgelser. Samfundslitteratur, Frederiksberg
Nonaka I, Takeuchi H (1995) The knowledge creating company: how Japanese companies create the dynamics of innovation. Oxford University Press, New York
Noordhoff CS, Kyriakopoulos K, Moorman C, Pauwels P, Dellaert BG (2011) The bright side and dark side of embedded ties in business-to-business innovation. J Mark 75(5):34–52
Olhager J (2010) The role of the customer order decoupling point in production and supply chain management. Comput Ind 61(9):863–868
Pinto JK, Slevin DP, English B (2009) Trust in projects: an empirical assessment of owner/ contractor relationships. Int J Proj Manag 27(6):638–648
Prahalad CK (2004) The blinders of dominant logic. Long Range Plann 37(2):171–179
Quinn RE, Rohrbaugh J (1983) A spatial model of effectiveness criteria: towards a competing values approach to organizational analysis. Manag Sci 29(3):363–378

Rizzi F, Frey M, Testa F, Appolloni A (2014) Environmental value chain in green SME networks: the threat of the Abilene paradox. J Clean Prod 85(1):265–275
Roberts EB (1991) Entrepreneurs in high technology: lessons from MIT and beyond. Oxford University Press, New York
Ross A, Athanassoulis N (2010) The social nature of engineering and its implications for risk taking. Sci Eng Ethics 16(1):147–168
Rudberg M, Wikner J (2004) Mass customization in terms of the customer order decoupling point. Prod Plan Control 15(4):445–458
Sligo FX, Massey C (2007) Risk, trust and knowledge networks in farmers' learning. J Rural Stud 23(2):170–182
Smith K, Collins C, Clark K (2005) Existing knowledge, knowledge creation capability, and the rate of new product introduction in high-technology firms. Acad Manag J 48(2):346–357
Snowden D (2002) Complex acts of knowing: paradox and descriptive self-awareness. J Knowl Manag 6(2):100–111
Song M, Di Benedetto CA (2008) Supplier's involvement and success of radical new product development in new ventures. J Oper Manag 26(1):1–22
Thomas G (2011) How to do your case study – a guide for students & researchers. Sage, London
Tidd J, Bessant J (2009) Managing innovation: integration technological, market and organizational change. Wiley, West Sussex
Tidd J, Trewhella MJ (1997) Organizational and technological antecedents for knowledge acquisition and learning. R&D Manag 27(4):359–375
Tranfield D, Young M, Partington D, Bessant J, Sapsed J (2006) Knowledge management routines for innovation projects: developing a hierachical process model. From knowledge management to strategic competence. Imperial College Press, London, pp 126–149
Vrijhoef R, Koskela L (2000) The four roles of supply chain management in construction. Eur J Purch Supply Manag 6(3–4):169–178
Wei YS, O'Neill H, Lee RP, Zhou N (2013) The impact of innovative culture on individual employees: the moderating role of market information sharing. J Prod Innov Manag 30 (5):1027–1041
Winch GM (2002) Managing construction projects. Blackwell, Oxford
Wortmann JC, Mustlag DR, Timmermans PJM (1997) Customer-driven manufacturing. Chapman & Hall, London
Wowak KD, Craighead CW, Ketchen DJ, Hult GTM (2015) Toward a 'theoretical toolbox' for the supplier-enabled fuzzy front end of the new product development process. J Supply Chain Manag 52(1):66–81
Xiwei W, Stößlein M, Kan W (2010) Designing knowledge chain networks in China – a proposal for a risk management system using linguistic decision making. Technol Forecast Soc Chang 77 (6):902–915
Yin RK (2014) Case study research design and methods. Sage, London
Zimmermann R, Ferreira LMDF, Moreira CA (2016) The influence of supply chain on the innovation-process: a systematic literature review. Supply Chain Manag 21(3):289–304

Lone Kavin is a postdoc at the Department of Entrepreneurship and Relationship Management at the University of Southern Denmark in Kolding. She holds a PhD within supply chain management focusing on supply network enabled innovation within a non-repetitive manufacturing context. Her research is concerned with innovation in supply networks with a special emphasis on the offshore wind energy sector.

Ram Narasimhan is a University Distinguished Professor Emeritus and John H. McConnell Endowed Professor Emeritus at Michigan State University, USA. He is currently a researcher with the ReCoE project at the University of Southern Denmark. He has published nearly 140 articles on SCM, strategic sourcing, and buyer–supplier relationships in peer-reviewed journals. He is the recipient of "Distinguished OM Scholar Award" from the Academy of Management for lifetime scholarly contributions.

Necessary Governing Practices for the Success (and Failure) of Client-Supplier Innovation Cooperation

Romaric Servajean-Hilst

Abstract This chapter aims to empirically identify governance practices that are critical for the success of client-supplier innovation cooperation. To do so, we use Necessary Condition Analysis (NCA) to screen a large panel of contractual provisions and coordination practices that are theoretically recognized as influencing relationship performance. Based on survey data describing 160 client-supplier relationships on an innovation project, we empirically determine which of these practices are conducive to highest or lowest performing relationship performance. We identify 12 practices—including the necessity of considering a client/supplier as a key account, and regular involvement of the client's purchasing function—that are critical for creating a high-performing relationship, and 12 that lead to a low-performing relationship—that is, those that should be avoided. Our results provide deeper knowledge of the governance of client-supplier innovation cooperation, thanks to the paradigm change driven by the NCA approach. They also provide direct practical implications: practices to promote or to avoid in order to maximize successful innovation cooperation.

1 Introduction

In vertical innovation relationships, choosing the best form of governance is a crucial issue for both client and supplier. For more than 30 years now, the use of such innovation partnerships has been growing steadily. In parallel with the rise of the Open Innovation paradigm (Chesbrough et al. 2014), a wave of research and consulting whitepapers has accompanied this phenomenon. Although studies have highlighted successful companies and projects, few firms have been able to emulate these pioneers. Now, scholars are investigating the optimal forms of organization for cooperating in innovation—for traditional client-supplier cooperation as well as newer approaches. The literature recognizes that further work is required, specifically

R. Servajean-Hilst (✉)
i3-CRG, Ecole polytechnique, Palaiseau Cedex, France
e-mail: romaric.servajean-hilst@polytechnique.edu

A. C. Moreira et al. (eds.), *Innovation and Supply Chain Management*, Contributions to Management Science, https://doi.org/10.1007/978-3-319-74304-2_4

in the streams on Open Innovation (West and Bogers 2014) and Early Supplier Involvement in New Product Development (Johnsen 2009; Laursen and Andersen 2016; Säfsten et al. 2014).

In fact, managers, consultants, and researchers are still searching for the best governing practices they should use—as well as the worst ones they should avoid. To answer this unmet need, this research screens a wide range of contractual and coordination practices that are theoretically recognized as influencing the relationship performance of client-supplier cooperation in innovation. To do so, we use a methodological approach inspired by the pharmaceutical industry, where the first step in the drug-discovery pathway is to test large numbers of available compounds to see whether they produce an appropriate biochemical or cellular effect (Smith 2002).

In order to identify the best and worst governance practices, we systematically review data originating from a 2014 survey on the performance and governance of vertical innovation cooperation, using a new logic and method: Necessary Condition Analysis (Dul 2015). An illustration of NCA logic is as follows: to make an apple pie, having apples is a *necessary* condition, but it is not *sufficient*. To apply this logic, NCA uses graphical representations of the data distribution between X, the condition, and Y, the outcome. It is appropriate for identifying critical practices, since it represents a way to analyze causal inferences by finding the level of the independent variable—the condition—that *must* be present to attain a certain level of the dependent variable—the outcome. Without this level of independent variable, the targeted level of dependent variable will be absent (Dul 2015, p. 20).

This chapter is organized as follows. First, the literature background provides some definitions and a quick overview of previous work on supplier-customer collaboration in innovation that links governance practices with cooperation performance. We then present our causal inference development based on the previous literature, Transaction Cost Theory, and the Resource-Based View. The next section details the research method: first, the data collection and mobilized sample, and second, the application of NCA through the processes elaborated for the dichotomous and discrete analysis of our independent variables. The fourth section presents the raw positive results of the screening of governance practices with NCA. We conclude with a discussion of the results and their implications for the management of vertical innovation cooperation.

2 Literature Background

2.1 Definitions

For this chapter, we adopt the OECD's definition of **innovation** as "the implementation of a new or significantly improved product (good or service), or process, a new marketing method, or a new organizational method in business practices, workplace organization or external relations" (OECD and Statistical Office of the European

Communities 2005). A **vertical innovation cooperation** is a relationship between two firms where one is, or has been, a supplier to the other, and the two collaborate on a joint innovation project or program. The **governance** of this relationship refers to all structured and informal contractual and coordination practices that ensure and regulate the interactions within the relationship and the elements being exchanged, which may be: (1) products or services, (2) information, (3) financial, or (4) social (Håkansson and IMP Project Group 1982, p. 17). Here, "relationship governance" does not refer to the concept of governance modes proposed by transaction cost theory (Williamson 1975), but to the organization of the dyad. Finally, the **performance** of such a relationship denotes the degree to which the agreed objectives of a cooperation are achieved (Das and Teng 2000) and to the quality with which the dyad's resources are utilized (M. Le Dain et al. 2011)—though it also includes any "supernormal profit jointly generated" (Dyer and Singh 1998).

A **necessary condition** is a condition that must be present for a certain outcome to be achieved. If such a condition is absent, so is the outcome (Dul et al. 2010). For managers, it implies a condition that must be present in order to avoid failure (if the desired outcome is success). Nevertheless, the presence of the condition doesn't guarantee the outcome; it is necessary but not sufficient. A sufficient condition, on the other hand, induces the outcome in itself. Necessary Condition Analysis (NCA) is a recent data analysis technique successfully applied in both the social sciences and in operations management (Dul et al. 2010) and to vertical innovation cooperation management (van der Valk et al. 2016).

2.2 Literature Related to Vertical Innovation Cooperation

The performance and governance of relationships is a widely explored subject, and the link between governance and performance has often been emphasized. When this relationship includes joint innovation, it is mainly studied through three overlapping streams of strategy and relationship marketing research: (1) R&D alliances and technology partnerships (Doz 1996; Dussauge et al. 2000; Kauppila 2014); (2) Open Innovation (Chesbrough et al. 2014; West and Bogers 2014); and (3) the study of early supplier involvement in new product development (Clauss and Spieth 2016; Le Dain et al. 2011; Maniak and Midler 2008; Säfsten et al. 2014; van der Valk et al. 2016). All these streams share the same theoretical bases: Transaction Cost Theory and Resource-Based Views (Bogers 2012; Stanko and Calantone 2011).

The literature on R&D alliances focuses on how to pool technological resources between independent organizations. It is now being enriched by recent Open Innovation literature examining novel tools of cooperation (mainly based on digital innovations) and existing and emerging business models (Chesbrough et al. 2014). Early Supplier Involvement keeps the focus on vertical relationships and the best ways to coordinate them during new product development projects—which is part of what we define as innovation—and on how to share responsibilities between client and supplier in that context (Le Dain and Merminod 2014; Säfsten et al. 2014).

In these three literature streams, governance mechanisms are mainly distinguished based on whether they rely on contractual safeguards or on relational levers. Transactional or *contractual governance* is rooted in transaction cost theory (Williamson 1975), while *relational governance* is rooted in a resource-based view of interacting firms (Das and Teng 2000). Numerous studies validate these two approaches and demonstrate their complementarity (Bstieler and Hemmert 2015; Clauss and Spieth 2016; Poppo and Zenger 2002; van der Valk et al. 2016).

Other studies propose optimal governance sets for managing innovation cooperation (Chesbrough et al. 2014; Hausman and Johnston 2010; Lakemond et al. 2006). The recommended approach is to use sets of variables to measure governance, to ensure the reliability and validity of measures used for quantitative analysis (Churchill 1979; Hair et al. 2006). As a result, there are as yet relatively few studies on individual governance practices (Littler et al. 1995; Ragatz et al. 1997), and most works provide results based on aggregates of governance practices. However, such aggregates may have less practical relevance to managers.

Moreover, the unit of analysis of these studies is a focal innovation project. But the question of which governance to adopt remains open when this governance extends beyond the boundaries of the project (Johnsen 2009; Säfsten et al. 2014; Schiele 2012; Vanhaverbeke et al. 2014; West and Bogers 2014)—i.e., when the dyad is taken as the unit of analysis. While previous theories and academic works clarify the causal link between different governance modes and the performance of client-supplier cooperation in innovation, there is still room for, and empirical interest in, systematically screening governance practices and identifying the best—or worst—for relationship performance. Our research aims to fill this gap.

3 Causal Inference Development

Our propositions of causal inferences between governance practices and performance are theoretically founded in the previously cited literature and theories. To build these propositions, we separate the governance practices used in vertical innovation cooperation into two empirical types: (1) contractual provisions and (2) coordination practices. These practices and their theoretical link with relationship performance are described below.

3.1 Contractual Provisions

In a vertical innovation cooperation, contracts define the legal scope of the relationship. Contractual provisions are framed in a mutual agreement that defines common rules within the dyad. Each firm's expectations, rights, and obligations are written down in order to protect each party against the other's potential opportunistic behavior (MacNeil 1980; Williamson 1975) and to establish how each firm will share its inputs and outputs on the innovation project. First, *safeguard provisions* allow each party to impose its will on the other without its consent (*ibid.*) and maintain the effectiveness of the relationship in the event of opportunistic behaviors or disputes. Such provisions have a positive impact on relationship performance when their coercive effect is low. Nevertheless, as this effect increases, a point is reached where its impact becomes negative (Hausman and Johnston 2010)—specifically because too great a power imbalance in a relationship increases the risk of opportunistic behaviors (Provan and Skinner 1989).

Second, *sharing provisions* determine each firm's inputs into the project, their responsibilities within it, and the rules for sharing its outputs. By fixing both rules and objectives together, and by forming adaptive contractual provisions in some cases, both firms deepen their understanding of the project and improve the potential performance of the relationship. Usually, the firms agree whether responsibilities in terms of innovation design and further production will be supplier-driven, client-driven, or shared (Le Dain and Merminod 2014). The sharing of inputs and outputs is includes rules on sharing intellectual property. The most common principle for sharing provisions in complex collaboration contexts in order to achieve better performance is equity (Jap 2001). However, voluntarily imbalanced sharing can also elicit greater commitment from the favored firm (Ring and van de Ven 1994). When such imbalance is *not* voluntary, however, the effect on the relationship can be negative, since this also exacerbates opportunism.

Although safeguard and sharing provisions are aimed at ensuring the effectiveness of the relationship, they can also enhance its efficiency, as well as each party's commitment to the relationship. However, they also have the power to harm relationship performance, discouraging interfirm collaboration.

Causal Inference 1 The contractual provisions of a vertical innovation cooperation impact relationship performance.

3.2 Coordination Practices

The coordination practices used in vertical innovation cooperation are heterogeneous. They are the formal and informal practices, processes, tasks, tools, and routines through which interactions take place within and between client and supplier, and which are complementary to the contractual provisions.

Coordination practices can be divided into three categories. First, formal information sharing practice refers to the systems with which information and knowledge are shared within the dyad. Through tools, information systems, and processes, interactions are regulated and knowledge is shared and combined. These practices are put in place for the effectiveness of the relationship in an innovation context (Thomas 2013). They can also positively impact its efficiency by facilitating cross-learning (Nonaka and Takeuchi 1995; Sluyts et al. 2011). Nevertheless, the quality and importance of such practices strongly mediate these performance impacts, and coordination costs can also be a major drag on the success of vertical innovation cooperation (Sobrero and Roberts 2002; West and Bogers 2014).

A second coordination practice category is the involvement of each firm's business functions. Since a function has a dedicated role in the firm, its involvement in a joint innovation project reflects this role. The significance of its involvement reflects the relevance of its logics for the targeted innovation, from each firm's point of view. Since different functions can have different goals and interests at both the firm and dyadic levels (Doz 1987; Johnsen et al. 2008; Wheelwright and Clark 1992), their involvement can be complementary or contradictory. Further, the consistency within a firm regarding the other can influence relationship performance for better or worse: it can be source of either over-commitment or of information losses.

Third, control practices category denotes the formal and informal practices implemented to safeguard the interests of the dyad and of each cooperating firm. These practices complement contractual safeguard provisions, and are based on the policies, routines, and standards of both firms. They drive adjustment to the other coordination practices, and can help improve relationship efficiency (Dekker 2004; Doz 1996; Le Dain et al. 2011; Ring and van de Ven 1994). Control practices are mainly based on evaluating coordination practices, and are activated formally through adjustments to these practices, and also informally through influence strategies, either coercive or incentivizing, that one firm applies to the other (Hausman and Johnston 2010). Following the same inference as for safeguard provisions, these influences have the potential to encourage or discourage collaboration in relation to the level with which they are applied (*ibid.*).

Causal Inference 2 The coordination practices of a vertical innovation cooperation impact relationship performance.

4 Research Method

4.1 Data Collection

In order to empirically test a large number of governance practices that theoretically impact performance, we used data from a 2014 survey on vertical innovation cooperation, dyadic governance, and performance. This survey was specifically designed by the author with standard psychometric scales and survey instrument development techniques (Alreck and Settle 1985; Dillman 2007). The set of measurable variables mainly relies upon scales previously developed and validated in the three literature streams described above (see Table 1). The scales were adapted to the specific context of vertical innovation cooperation and completed with governance practices identified during preliminary qualitative interviews conducted with 35 managers working

Table 1 Examples of items used to identify governance practices

Governance mechanisms		Items		References
Contractual provisions	Safeguard provisions	The contract between the two companies includes the following provisions	Regular reporting in writing of all relevant transactions and exchanges between our two companies	Reuer and Ariño (2007)
			The written notice in the event of stoppage of the agreement by either party	Reuer and Ariño (2007)
	Sharing provisions	The generic rules for sharing the results of this project are	The contribution proportional to a company determines its share of results	Jap (2001)
		The sharing of responsibilities between your client and your firm is	The other company is responsible for the overall design of the targeted innovation. My company is responsible for its implementation	Le Dain and Merminod (2014)
Coordination	Information sharing mechanisms	We follow a well-defined process to manage our innovation cooperation		Sluyts et al. (2011)
		Product lifecycle management software (PLM) and computer-aided design (CAD) systems are implemented to manage information exchanges between our two companies in the innovation project		created
	Functions frequency of involvement	The frequency of contact of #### of my organization with this client	Top management	created
		The frequency of contact of the #### entities of my client with my firm	Purchasing	created
	Control mechanisms	In its general attitude	The other company focuses on the positive impact on our business of the success of our innovation cooperation	Hausman and Johnston (2010)
		The management of each company follows the progress and performance of their relationship through joint reviews		Sluyts et al. (2011)

on innovation cooperation. In addition, the proposed themes were ordered in order to reduce the impact of individual themes on others (Alreck and Settle 1985, p. 103).

The drafted survey instrument was discussed with scholars and practitioners with expertise in statistics or vertical innovation cooperation. Finally, it was pretested with five practitioners from different industries and different functions. The survey was sent to a cross-sectional sample of 4500 (not specifically innovative) supplier

Responding firms (suppliers)	
Aeronautics, naval and train construction	16
Agrifood	10
Automotive	30
Chemistry, plastics	17
Construction industry	2
Consulting, research and services	11
Cosmetics and pharmacy	6
Electronics, hardware and IT industry	36
Environment, energy, waste and water industries	9
Leisure and tourism industry	2
Mechanics	9
Metallurgy	9
Packaging	3

Client firms	
Aeronautics, naval and train construction	16
Agrifood, cosmetics and pharmacy	16
Automotive	30
Chemistry, plastics	17
Consulting, research and services	11
Electronics, hardware and IT industry	36
Environment and construction industries	11
Leisure and tourism industry	2
Mechanics, metallurgy and packaging	21

Fig. 1 Industry breakdown of supplier and client firms in our sample

firms in France by means of a self-administered internet-based survey accompanied by an introductory email presenting the intent of the study. After two rounds of follow-up reminders the response rate was 4% (179 answers).

We retained 160 completed questionnaires for this study, all representing contemporary vertical relationship and innovation projects. Figure 1 shows an industry breakdown of the responding supplier firms and their respective clients in our sample. Individual respondents were all managers working in top management (31%), sales (32%), and research & development (26%); 81% had been with their firm for over 3 years. They were asked to answer about a relationship with one of their clients, and about the most representative joint innovation project. In exchange for their participation, they were offered a summary of the results.

4.2 *Necessary Condition Analysis*

The logic and methodology of organizational determinants making necessary but not sufficient contributions to a desired outcome is a novel methodology that was recently reified by Dul (2015). NCA provides a more straightforward vision of the conditions needed to reach a certain outcome, making it better suited to the analysis of management mechanisms than variance-based analysis. In variance-based analysis, increasing levels of the conditions lead to a certain increase in the level of the outcome—on average, and not in each instance (van der Valk et al. 2016). NCA can be the foundation for stronger managerial implications, since it allows us to make recommendations such as, "If you don't create condition X, you can't reach result Y."

Following Dul's (2015) recommendations, we applied NCA to our dataset by considering successively that each of these measured governance practices is a condition (X_n) for the desired outcome (Y_m) that is highest relationship performance (Y_1) or lowest relationship performance (Y_3). X is the independent variable and Y the dependent variable. To do this, and to generate discriminant measures of

Table 2 Construction of the dependent variable: relationship performance

Classes	Nb	Effectiveness level (mean)	Efficiency level (mean)	Proactivity level of supplier (mean)	Proactivity level of client (mean)
Y_1 highest performing relationship	34	6.29	6	6.22	5.84
Y_2 mean performing relationship	82	5.59	5.62	5.30	4.60
Y_3 lowest performing relationship	44	4.82	5.18	4.17	3.17

Table 3 Algorithm for building contingency matrix for dichotomous independent variables

	Ym = 0	Ym = 1
Xn = 0	$\sum_{i=1}^{160} (Xn(i) - 1) * (Ym(i) - 1)$	$1 - \sum_{i=1}^{160} Xn(i) * Ym(i)$
Xn = 1	$1 - \sum_{i=1}^{160} (Xn(i) - 1) * (Ym(i) - 1)$	$\sum_{i=1}^{160} Xn(i) * Ym(i)$

performance, we first generated a performance scale with a hierarchical ascendant classification (Euclidian distance and Ward method) on the nine performance items measuring the effectiveness of the relationship, its efficiency, and the proactivity of each firm towards the other. Three classes were determined (Table 2). The liability of this scale was checked through the estimation of Cronbach's alpha, which was 0.781.

Then, for each case (X_n;Y_m), we established two processes to realize and analyze contingency matrices that were specific to (a) dichotomous and (b) discrete variables. For both, we also determined common inclusion and exclusion settings to reinforce the reliability of our tests by taking into account the effect size of the observed subjects relative to the sample (Goertz et al. 2013):

1. When the number of observations in a matrix's entry was less than 5% of the total observation of the total matrix, the value of this case was considered null.
2. When the number of observations in a considered matrix's area was less than twice the number of the number of observations considered null, the result of the test was not exploited, as the difference was not considered discriminating.

4.2.1 Analysis Process for Dichotomous Variables

The contingency matrices for dichotomous X variable are 2 × 2 square matrices constructed by applying the following formulas (Table 3) to our sample and the aforementioned settings. In the matrix, X_n are the conditions tested to reach the outcome Y_m. X_n (i) represents the value of X_n for the subject i in the sample comprising i = 160 subjects (dyads), Y_m (i) the value of Y_m for the same one.

$Y_1 = 1$ if relationship performance reaches the highest level, 0 if not or if it is absent, and where $Y_3 = 1$ if relationship performance reaches the lowest level and 0 if not or if it is absent. The analysis of the results of these matrices depends on whether the failure (Y_3) or the success (Y_1) of the relationship is considered. When looking for the conditions for a successful relationship, when the (0,1) entry is null and the (1,1) entry is non-null, the condition X_n is necessary. When looking for the conditions for a failing relationship Y_3, when the (0,0) entry is null and the (1,0) entry is non-null and very high, the condition X_n is necessary. And when looking for the absent conditions for failing relationship Y_3, when the (1,0) entry is null and the (0,0) entry is non-null, the absence of condition X_n is necessary.

4.2.2 Analysis Process for Discrete Variables

The contingency matrices for discrete X variables are constructed by distributing for each level of condition X_n the number of subjects of our sample presenting each level of relationship performance Y_m. As our discrete X variables are measured with five- or seven-level scales, the conditions that are measured are related to the level α of the variable X that is a condition for success or failure. Y_1 is the highest level of relationship performance (representing the successful relationship), Y_2 the intermediate level, and Y_3 the lowest level (representing the failing relationship).

When analyzing contingency matrices for discrete variables, the upper left empty zone indicates the existence of a necessary condition. It is underlined by a "ceiling line" that separates areas with observations from those without (Goertz et al. 2013). This line indicates which level of the X variable is necessary for a given level of the Y variable (van der Valk et al. 2016): the necessary condition for reaching the outcome Y_m is identified by the first level of X_α that is non-null.

5 Results

We summarize our positive results regarding the NCA of contractual provisions in Table 4 and Table 5.

We observe that in the contractual provisions, the provisions related to sharing of responsibilities are the only ones that can be necessary for a successful relationship. The necessary condition is completed when both firms participate in the conception of the innovation, and this participation from the supplier is significant. Regarding the contractual provisions conditioning the least successful relationship, the absence of a contract ($X_{GIPGouvContr\_pas}$) and the absence of a safeguard provision that allows the client to audit the supplier ($X_{ContrSauv\_Audit}$) appear to be two necessary conditions.

Table 4 Positive results of NCA on contractual provisions—discrete variables

$X_{PartagRespo}$	The client is responsible for entire conception of the innovation and my firm produces it	My firm provides feedbacks on conception choices and produces the innovation	My firm significantly participate to conception of a part of the innovation and produces it	My firm has the entire responsibility of conception and production of a part of the innovation	My firm has the entire responsibility for conception and production of the innovation
Y_1	3	1	18	6	5
Y_2	3	8	22	18	27
Y_3	3	3	18	10	10

Table 5 Positive results of NCA on contractual provisions—dichotomous variables and lowest performance

		$Y_3 = 0$	$Y_3 = 1$
$X_{GIPGouvContr\_pas}$	X = 0	0	1
	X = 1	44	115
$X_{ContrSauv\_Audit}$	X = 0	42	105
	X = 1	2	11

Table 6 Positive results of NCA on dichotomous coordination practices and highest performance

		$Y_1 = 0$	$Y_1 = 1$
$X_{CentralCoordM}$	X = 0	10	1
	X = 1	117	32

The positive results of the NCA of coordination practice measured with dichotomous variables are presented in Table 6 and Table 7.

We observe that having a central coordinator in the supplier firms ($X_{CentralCoordM}$) is a necessary condition for a successful relationship. Further, in Table 6, if the client is not considered strategic for the supplier ($X_{KAMpourM}$), and vice versa ($X_{KAMpourA}$), this is a necessary condition for an unsuccessful relationship. The same applies if there is no co-location of people from both firms ($X_{ColocchezM}$ and $X_{ColocchezA}$). Finally, no information system sharing, through an access to a part of the other firm's IS (X_{SI_acces}) and through a shared product lifecycle management system (X_{SI_PLM}), is a necessary condition for an unsuccessful relationship.

In Table 8, we present the positive results of NCA on discrete variables measuring coordination practices. We can observe that, to be a necessary condition for success, the frequency of contact from the general management of the supplier firm ($X_{ContactDGM}$) must be at least "sometimes": when it is "never" or "rarely," the highest performance is not significantly reached. Further, the frequency of contact from the operation function of the supplier ($X_{ContactOperatM}$) has to be at least "often" to reach success; from the quality function ($X_{ContactQlteM}$) it has to be "often" but not "always"; and from marketing ($X_{ContactMkgM}$) "sometimes" only—at any other

Table 7 Positive results of NCA on dichotomous coordination practices and lowest performance

		$Y_3 = 0$	$Y_3 = 1$
$X_{KAMpourM}$	X = 0	7	9
	X = 1	37	107
$X_{KAMpourA}$	X = 0	2	14
	X = 1	42	102
$X_{ColocchezM}$	X = 0	37	93
	X = 1	7	18
$X_{ColocchezA}$	X = 0	43	100
	X = 1	0	10
X_{SI_acces}	X = 0	35	80
	X = 1	7	28
X_{SI_PLM}	X = 0	28	67
	X = 1	3	38

frequency of contact from this function, this highest level of performance is not reached. Similarly, the frequency of contact from the finance function of the client firm ($X_{ContactFinA}$) needs to be at "rarely" or "sometimes," and no other level.

Below, we can also observe that another necessary condition for a successful relationship is an encouraging attitude from the client that is clearly perceived by the supplier ($X_{AttitudeAEncourag}$). When the supplier disagrees, or even just "moderately agree[s]" that its client "focuses on the positive impact on our business of the success of our innovation cooperation," the highest level of performance is never significantly reached. Three other necessary conditions for successful relationship are a certain number of reminders of the contract terms ($X_{AttitudeARappelContrat}$), at "moderately agree" and "agree" levels; very rare threats from the client ($X_{AttitudeAMenace}$)—i.e. the supplier "total[ly] disagree[s]" or "disagree[s]" that the client "threatens to disrupt our business relationship if we fail to do our part"—as well as having roles and responsibilities well or very well defined on the innovation project ($X_{RoleDefini}$).

We can also observe that the frequency of contact from the purchasing function of the client firm ($X_{ContactAchatA}$) has to be "often" to be a necessary condition for success—but it can also be a necessary condition for failure when it is "always." Finally, there are three necessary conditions for failing relationships: rare or absent contact from the purchasing function of the supplier firm ($X_{ContactAchatM}$), a mean or low frequency of contact from the general management of the client firm ($X_{ContactDGA}$), and rare or absent contact from the sales function of the client firm ($X_{ContactVenteA}$).

Table 8 Positive results of NCA on discrete coordination practices

Necessary conditions for Y_1							
$X_{ContactDGM}$	Never	Rarely	Sometimes	Often	Always		
Y_1	2	6	8	8	10		
Y_2	7	9	27	26	9		
Y_3	8	10	14	8	2		
$X_{ContactOperatM}$	Never	Rarely	Sometimes	Often	Always		
Y_1	2	4	5	11	9		
Y_2	12	16	17	12	13		
Y_3	10	6	12	8	5		
$X_{ContactQlteM}$	Never	Rarely	Sometimes	Often	Always		
Y_1	1	4	4	14	3		
Y_2	10	9	15	23	8		
Y_3	8	8	9	11	3		
$X_{ContactMkgM}$	Never	Rarely	Sometimes	Often	Always		
Y1	3	5	11	6	2		
Y2	11	13	12	16	9		
Y3	15	4	9	8	2		
$X_{ContactFinA}$	Never	Rarely	Sometimes	Often	Always		
Y_1	3	11	11	3	2		
Y_2	24	17	16	7	2		
Y_3	10	9	7	4	3		
$X_{AttitudeAEncourag}$	Totally disagree	Disagree	Moderately disagree	Neither agree nor disagree	Moderately agree	Agree	Totally agree
Y_1	0	0	0	5	2	20	7
Y_2	1	5	3	10	23	39	2
Y_3	1	3	3	11	11	11	4

(continued)

Table 8 (continued)

$X_{AttitudeARappelContrat}$	Totally disagree	Disagree	Moderately disagree	Neither agree nor disagree	Moderately agree	Agree	Totally agree
Y_1	2	3	3	4	8	15	1
Y_2	1	11	7	16	19	23	4
Y_3	4	6	3	7	4	16	3
$X_{AttitudeAMenace}$	Totally disagree	Disagree	Moderately disagree	Neither agree nor disagree	Moderately agree	Agree	Totally agree
Y_1	9	9	4	5	3	3	1
Y_2	18	23	5	12	10	11	3
Y_3	10	6	0	9	8	6	3
$X_{RoleDefini}$	Totally disagree	Disagree	Moderately disagree	Neither agree nor disagree	Moderately agree	Agree	Totally agree
Y_1	0	2	1	0	5	18	8
Y_2	3	3	5	13	16	33	9
Y_3	3	1	4	5	12	13	6
Necessary condition for Y_1 and for Y_3							
$X_{ContactAchatA}$	Never	Rarely	Sometimes	Often	Always		
Y_1	0	0	7	17	6		
Y_2	11	2	8	29	25		
Y_3	4	2	5	8	20		
Necessary conditions for Y_3							
$X_{ContactAchatM}$	Never	Rarely	Sometimes	Often	Always		
Y_1	7	7	5	5	2		
Y_2	22	20	9	6	4		
Y_3	21	9	4	3	1		
$X_{ContactDGA}$	Never	Rarely	Sometimes	Often	Always		
Y_1	6	8	4	8	7		
Y_2	17	16	23	13	2		
Y_3	19	9	8	2	0		

$X_{ContactVenteA}$	Never	Rarely	Sometimes	Often	Always		
Y_1	9	7	1	7	2		
Y_2	38	7	4	6	4		
Y_3	24	7	2	1	2		

6 Discussion and Implications

Our study takes up a recent observation from a panel of supply chain management and Open Innovation scholars on the need to identify the best managerial practices—and the worst, or least productive ones—to use in innovation collaborations (Le Dain et al. 2011; Petersen et al. 2005; van der Valk et al. 2016; Vanhaverbeke et al. 2014; West and Bogers 2014). Our aim was to provide a new vision of the best and worst discrete practices used in innovation cooperation between client and supplier firms, without considering sets of governance practices or governance modes. To that end, we mobilized a new methodology, Necessary Condition Analysis (Braumoeller and Goertz 2000; Dul 2015), to systematically review a large panel of governance practices identified by the literature and mobilized in French vertical innovation cooperations. However, our theoretical aim was to provide, under the paradigm of necessary condition for outcomes, quantitative validation of qualitative findings, and confirmation of quantitative findings that were obtained under traditional paradigms of multi-causality of outcomes.

The results offer a list of critical practices: governance practices that by their presence, absence, or level necessarily lead to highest or lowest relationship performance levels (Table 9). Our study makes seven notable academic contributions. First, the finding that considering the other firm as a strategic client/supplier is a necessary condition for success underlines the relevance of firms taking a strategic approach for managing their supplier/client portfolios, panels and providing dedicated treatment to their strategic supplier/client. This result is in line with ESI literature on the importance of being the preferred customer or supplier in achieving innovative relationships (Nollet et al. 2012; Schiele 2012). It is also congruent with the result that indicates that the involvement of the supplier's general management is a necessary condition for success, and that the absence of involvement of the client's general management is a necessary condition for failure: these practices are manifestations of considering, or not, the other firm as a key account.

Second, the result showing that having a single coordinator only in the supplier firm is a necessary condition for successful relationship can be considered coherent with the finding that (1) regular involvement of the purchasing function in the client firm is a necessary condition for success, while (2) a *permanent* involvement of the same function is a necessary condition for failure. The explanation could be that it is important for the supplier to provide a single coherent vision of innovation and its exploitation to their client. At the same time, for client firms, it has been demonstrated that a partial integration of the purchasing and R&D functions, particularly over longer periods, is positive for the relationship (Melander and Lakemond 2014): In case of problems, they can play a "good cop, bad cop" game that facilitates problem-solving (Brattström and Richtnér 2013). Our results reinforce these findings.

Third, the result that indicates that clearly defining roles and responsibilities is a necessary condition for successful relationship is eye-catching. In the turbulent environment of innovation, we could expect that too rigid an organization would

Table 9 Critical governance practices for client-supplier innovation cooperation

Governance mechanisms		Identified necessary conditions	Promote	Avoid	At what level
Contractual provisions	Safeguard provisions	Absence of contract		×	No contract
		Safeguard provision that allows the client to audit the supplier		×	Absence of provision
	Sharing provisions	Provision for sharing of responsibilities	×		My firm significantly participate to conception of a part of the innovation and produces it
Coordination	Information sharing mechanisms	Having a central coordinator in the supplier firm	×		Presence of a central coordinator in the supplier firm
		Co-location of client's people in supplier's place		×	Absence of co-location of client's people in supplier's place
		Co-location of supplier's people in client's place		×	Absence of co-location of supplier's people in client's place
		Access to a part of the other firm IS		×	Absence of an access to a part of the other firm IS
		Shared Product Lifecycle Management system		×	Absence of shared PLM system
	Functions frequency of involvement	General Management of supplier	×		At least sometimes
		Operation function of supplier	×		At least often
		Quality function of supplier	×		Often
		Marketing function of supplier	×		Sometimes
		Purchasing function of supplier		×	Never or rarely
		Finance function of client	×		Rarely or sometimes
		Purchasing function of client	×	×	Often for highest / Always for lowest
		General Management of client		×	From never to sometimes
		Sales function of client		×	Never or rarely

(continued)

Table 9 (continued)

Governance mechanisms		Identified necessary conditions	Promote	Avoid	At what level
	Control mechanisms	Client is considered strategic for the supplier		×	Client is not considered strategic for the supplier
		Supplier is considered strategic for the client		×	Supplier is not considered strategic for the client
		Encouraging attitude from the client	×		From regular
		Recall to the contract	×		Regular but not always
		Threats from the client	×		Never
		Definition of roles and responsibilities on the project	×		Well to fully defined

rather limit agility and relationship success. This result shows that even though the object of the cooperation—the targeted innovation—may evolve, it is still best to define the *manner* of cooperating at the beginning of the partnership.

Fourth, our results concerning the forms of influence exerted by supplier firms confirm and sharpen the findings from (Hausman and Johnston 2010): Frequent encouraging attitudes and the absence of threats are necessary conditions for success, but so are occasional references back to legal agreements. These results are in line with works that underline the necessity for an existing but limited constraint in order to build a trusting relationship (Reuer and Ariño 2007; Reuer et al. 2002; Ring and van de Ven 1992); these studies extend the recommendation from contractual to relational practices.

Fifth, another interesting result is that a necessary condition for an unsuccessful relationship is the absence of co-location of members of both firms—even though co-location itself is not a necessary condition for success. This can be explained by the fact that in innovation projects between client and supplier, co-location of innovation teams from both companies is recommended only during certain phases of the project. At other times, physical separation appears to be the best practice (Lakemond and Berggren 2006; Le Dain and Merminod 2014).

Sixth, our results indicate the necessary involvement of operation and quality functions from the supplier firm for success. This underlines the importance of considering, in an innovation project within a vertical dyad, the future exploitation of the innovation—and this is best done through the early involvement of these two functions. These results are coherent with the other positive results indicating that the task allocation necessary for success is one where the supplier is involved in both the conception and production of the innovation. It means that, for a successful relationship, the client firm must anticipate the exploitation of the innovation project by sharing exploration responsibilities with the supplier, as well as by involving the supplier's exploitation-dedicated functions early on.

Finally, this study contributes to NCA studies proposing a first application of systematic screening of items, and not aggregates, to determine critical independent variables. NCA is a new method that can be applied to dichotomous variables, as well as to those that have been measured using ordinal, interval, or continuous scales (Dul 2015; Dul et al. 2010). It has not yet been applied in such a way to such a data set. The proposed process for crunching and analyzing discrete and dichotomous variables could facilitate new versions of the R extension developed by Dul (2015). It also suggests another way of considering the size effect in the upstream rather than downstream phases of data analysis to evaluate the validity level of the results.

The managerial implication of this research is to give a straight answer to firms' dilemma over the best way to govern vertical innovation cooperations with their clients/suppliers. Indeed, this study offers managers a directly actionable list of best and worst practices (Table 9)—keeping in mind that using the best practices (and avoiding the worst) is necessary, but not sufficient, to get the expected performance.

The findings and limitations of this study and its method suggest several directions for further research into the governance of interfirm relationships—in particular, those limitations relating to model specification, econometric identification, and variable measurement. For example, the moderating roles of relational capital and relational governance should be studied to check whether or not they impact, complement, or substitute the identified necessary conditions (Clauss and Spieth 2016; Kohtamäki et al. 2012; Melander et al. 2014; Poppo and Zenger 2002). This should be specifically tested in an industrial context other than France, which is considered a low-trust country (Fukuyama 1995). Further, some governance practices might have not been included in the survey, and the aggregation of the discrete practices should be tested to check whether there are some associations that can also be considered as necessary for successful or unsuccessful innovation relationships. And last but not least, for practitioners, the limit of positive results based on NCA is that it is very rare to find sufficient conditions—that is, those that can guarantee positive results from a governance approach.

Acknowledgements The support of the firm innov'& is gratefully acknowledged for the realization of this study. We also appreciate the support of Jan Dul for helping us apply the NCA method to our data, and the support of Thibaud Guedon and Georges-Edouard Sarkis, students of Ecole polytechnique, in this process. First versions of this work were presented at the EURAM and IPSERA annual conferences in 2016 and 2017 respectively.

References

Alreck PL, Settle RB (1985) The survey research handbook, vol 2. Irwin, Homewood IL

Bogers M (2012) Knowledge sharing in open innovation: an overview of theoretical perspectives on collaborative innovation. *SSRN eLibrary*

Brattström A, Richtnér A (2013) Good cop—bad cop: trust, control, and the lure of integration. J Prod Innov Manag 31(3):584–598

Braumoeller BF, Goertz G (2000) The methodology of necessary conditions. Am J Polit Sci 44 (4):844–858

Bstieler L, Hemmert M (2015) The effectiveness of relational and contractual governance in new product development collaborations: evidence from Korea. Technovation 45–46:29–39

Chesbrough H, Vanhaverbeke W, West J (2014) New frontiers in open innovation. Oxford University Press, Oxford

Churchill GA Jr (1979) A paradigm for developing better measures of marketing constructs. J Mark Res 16(1):64–73

Clauss T, Spieth P (2016) Treat your suppliers right! Aligning strategic innovation orientation in captive supplier relationships with relational and transactional governance mechanisms. R&D Manag 46(S3):1044–1061

Das TK, Teng B-S (2000) A resource-based theory of strategic alliances. J Manag 26(1):31–61

Dekker HC (2004) Control of inter-organizational relationships: evidence on appropriation concerns and coordination requirements. Acc Organ Soc 29(1):27–49

Dillman DA (2007) Mail and internet surveys: the tailored design method. Wiley, Hoboken

Doz YL (1987) Technology partnerships between larger and smaller firms: some critical issues. Int Stud Manag Organ 17(4):31–57

Doz YL (1996) The evolution of cooperation in strategic alliances: initial conditions or learning processes? Strateg Manag J 17:55–83

Dul J (2015) Necessary condition analysis (NCA): logic and methodology of "necessary but not sufficient" causality. Organ Res Methods 19(1):10–52

Dul J, Hak T, Goertz G, Voss C (2010) Necessary condition hypotheses in operations management. Int J Oper Prod Manag 30(11):1170–1190

Dussauge P, Garrette B, Mitchell W (2000) How to get the best results from alliances. Eur Bus Forum 3:41–46

Dyer JH, Singh H (1998) The relational view: cooperative strategy and sources of interorganizational competitive advantage. Acad Manag Rev 23(4):660–679

Fukuyama F (1995) Trust: the social virtues and the creation of prosperity. Free Press, New York

Goertz G, Hak T, Dul J (2013) Ceilings and floors where are there no observations? Sociol Methods Res 42(1):3–40

Hair JF, Black WC, Babin BJ, Anderson RE, Tatham RL (2006) Multivariate data analysis, 6th edn. Prentice Hall, New Jersey

Håkansson H, IMP Project Group (1982) International marketing and purchasing of industrial goods: an interaction approach. Wiley, Chichester

Hausman A, Johnston WJ (2010) The impact of coercive and non-coercive forms of influence on trust, commitment, and compliance in supply chains. Ind Mark Manag 39(3):519–526

Jap SD (2001) "Pie Sharing" in complex collaboration contexts. J Mark Res 38(1):86–99

Johnsen TE (2009) Supplier involvement in new product development and innovation: taking stock and looking to the future. J Purch Supply Manag 15(3):187–197

Johnsen TE, Johnsen RE, Lamming RC (2008) Supply relationship evaluation: the relationship assessment process (RAP) and beyond. Eur Manag J 26(4):274–287

Kauppila O-P (2014) Alliance management capability and firm performance: using resource-based theory to look inside the process black box. Long Range Plan 48(3):151–167

Kohtamäki M, Vesalainen J, Henneberg S, Naudé P, Ventresca MJ (2012) Enabling relationship structures and relationship performance improvement: the moderating role of relational capital. Ind Mark Manag 41(8):1298–1309

Lakemond N, Berggren C (2006) Co-locating NPD? The need for combining project focus and organizational integration. Technovation 26(7):807–819
Lakemond N, Berggren C, van Weele A (2006) Coordinating supplier involvement in product development projects: a differentiated coordination typology. R&D Manag 36(1):55–66
Laursen LN, Andersen PH (2016) Supplier involvement in NPD: a quasi-experiment at Unilever. Ind Mark Manag 58:162–171
Le Dain M, Merminod V (2014) A knowledge sharing framework for black, grey and white box supplier configurations in new product development. Technovation 34(11):688–701
Le Dain M, Calvi R, Cheriti S (2011) Measuring supplier performance in collaborative design: proposition of a framework. R&D Manag 41(1):61–79
Littler D, Leverick F, Bruce M (1995) Factors affecting the process of collaborative product development: a study of UK manufacturers of information and communications technology products. J Prod Innov Manag 12(1):16–32
Maniak R, Midler C (2008) Shifting from co-development to co-innovation. Int J Automot Technol Manag 8(4):449–468
MacNeil IR (1980) The new social contract: an inquiry into modern contractual relations. Yale University Press, New Haven
Melander L, Lakemond N (2014) Variation of purchasing's involvement: case studies of supplier collaborations in new product development. Int J Procurement Manag 7:103–118
Melander L, Rosell D, Lakemond N (2014) In pursuit of control: involving suppliers of critical technologies in new product development. Supply Chain Manag 19(5/6):722–732
Nollet J, Rebolledo C, Popel V (2012) Becoming a preferred customer one step at a time. Ind Mark Manag 41(8):1186–1193
Nonaka I, Takeuchi H (1995) The knowledge creating company. Oxford University Press, New York
OECD and Statistical Office of the European Communities (2005) Oslo manual: guidelines for collecting and interpreting innovation data, The measurement of scientific and technological activities, 3rd edn. OECD Publishing, Luxembourg
Petersen KJ, Handfield RB, Ragatz GL (2005) Supplier integration into new product development: coordinating product, process and supply chain design. J Oper Manag 23(3–4):371–388
Poppo L, Zenger T (2002) Do formal contracts and relational governance function as substitutes or complements? Strateg Manag J 23(8):707–725
Provan KG, Skinner SJ (1989) Interorganizational dependence and control as predictors of opportunism in dealer-supplier relations. Acad Manag J 32(1):202–212
Ragatz GL, Handfield RB, Scannell TV (1997) Success factors for integrating suppliers into new product development. J Prod Innov Manag 14(3):190–202
Reuer JJ, Ariño A (2007) Strategic alliance contracts: dimensions and determinants of contractual complexity. Strateg Manag J 28(3):313–330
Reuer JJ, Zollo M, Singh H (2002) Post-formation dynamics in strategic alliances. Strateg Manag J 23(2):135–151
Ring PS, van de Ven AH (1992) Structuring cooperative relationships between organizations. Strateg Manag J 13(7):483–498
Ring PS, van de Ven AH (1994) Developmental processes of cooperative interorganizational relationships. Acad Manag Rev 19(1):90–118
Säfsten K, Johansson G, Lakemond N, Magnusson T (2014) Interface challenges and managerial issues in the industrial innovation process. J Manuf Technol Manag 25(2):218–239
Schiele H (2012) Accessing supplier innovation by being their preferred customer. Res Technol Manag 55(1):44–50
Sluyts K, Matthyssens P, Martens R, Streukens S (2011) Building capabilities to manage strategic alliances. Ind Mark Manag 40(6):875–886
Smith A (2002) Screening for drug discovery: the leading question. Nature 418(6896):453–459
Sobrero M, Roberts EB (2002) Strategic management of supplier–manufacturer relations in new product development. Res Policy 31(1):159–182

Stanko MA, Calantone RJ (2011) Controversy in innovation outsourcing research: review, synthesis and future directions. R&D Manag 41(1):8–20

Thomas E (2013) Supplier integration in new product development: computer mediated communication, knowledge exchange and buyer performance. Ind Mark Manag 42(6):890–899

van der Valk W, Sumo R, Dul J, Schroeder RG (2016) When are contracts and trust necessary for innovation in buyer-supplier relationships? A necessary condition analysis. J Purch Supply Manag 22(4):266–277

Vanhaverbeke W, Du J, Leten B, Aalders F (2014) Exploring open innovation at the level of R&D projects. In: New frontiers in open innovation. Oxford University Press, Oxford

West J, Bogers M (2014) Leveraging external sources of innovation: a review of research on open innovation. J Prod Innov Manag 31(4):814–831

Wheelwright SC, Clark KB (1992) Organizing and leading "heavyweight" development teams. Calif Manag Rev 34(3):9–28

Williamson OE (1975) The limits of organization. J Bus 48(3):452–453

Romaric Servajean-Hilst PhD, is an associate researcher at the Management Research Center of Ecole Polytechnique (CRG-i3) and an entrepreneur. As a specialist in innovation breakthrough management and open innovation management, he teaches, researches, consults, and practices around these subjects. The core of his research works, conducted together with entrepreneurs and with large firms, centers on two questions: What is the best way to build sustainable innovation cooperation? What is the role of purchasing in open innovation? The specificity of his approach is to combine ethnographic-inspired research with quantitative inquiries—using the same models to analyze data and develop management methods. He was awarded best French doctoral thesis for purchasing and best French doctoral thesis for innovation management.

Collaborative New Product Development in SMEs and Large Industrial Firms: Relationships Upstream and Downstream in the Supply Chain

Filipe Silva and António Carrizo Moreira

Abstract The aim of this chapter is to compare collaborative new product development (CNPD) established by industrial companies with their suppliers and customers, according to their size and the type of innovation generated. To do so, eight in-depth case studies were analyzed, based on semi-structured interviews. The findings show that CNPD with suppliers in more active than with clients. The results also show that firm size is important in CNPD activities namely when product differentiation and large scale production activities are at stake. From another perspective, the results show that the development of processes and management methodologies in upstream activities are not extensively used. The chapter contributes to knowledge about CNPD by comparing how upstream and downstream are affected based on firm size and the type of innovation generated.

1 Introduction

The competitiveness of the global market, the complexity of products and the need for specialization promote collaborative new product development (CNPD) in upstream as well in downstream activities in the supply chain (Hoegl and Wagner 2005; Nieto and Santamaria 2010; Moreira and Karachun 2014; Tuli and Shankar 2015; Un and Azakawa 2015). Therefore, firms' involvement in value creation in CNPD represents a competitive advantage (Soosay et al. 2008; Nieto and Santamaria 2010; Büyüközkan and Arsenyan 2012; Un and Azakawa 2015; Lager 2016).

New product development activities have been the primary objective of many firms. After reviewing 461 articles Moreira and Karachun (2014) found that cooperative strategies are heavily dependent on knowledge transfer and management in order to integrate innovation, supplier integration, user involvement, new product and new process development processes, the management of internal and external teams and the launch of new products. In this context, the literature highlights the

F. Silva (✉) · A. C. Moreira
University of Aveiro, Aveiro, Portugal
e-mail: filipereissilva@ua.pt

A. C. Moreira et al. (eds.), *Innovation and Supply Chain Management*, Contributions to Management Science, https://doi.org/10.1007/978-3-319-74304-2_5

importance of CNPD in firms' positioning in the market leading to the debate about collaboration with both suppliers and customers (Faems et al. 2005; Soosay et al. 2008; Van de Vrande et al. 2009; Brettel and Cleven 2011; Theyel 2012). For example, Powell et al. (1996), Soosay et al. (2008), Wynarczyk et al. (2013) and Tuli and Shankar (2015) mention that collaboration refers to activities carried out between firms to conceive new products, manufacturing processes and management methodologies. Other authors address the subject of collaboration between firms to analyze its effect on CNPD (Nieto and Santamaria 2007; Parida et al. 2012; Tuli and Shankar 2015; Hossain and Karaunen 2016). Finally, most of studies on CNPD do not compare the involvement between industries of different technological intensity, namely involving small and medium-sized firms (SMEs) operating in low-tech industries (Silva and Moreira 2017).

The dynamics of CNPD emphasizes the characteristics of firms in their collaborative activities upstream and downstream in the supply chain, which enables the understanding and the determination of what their differences are and how they are operationalized. Clearly, most of the studies analyze CNPD based on supplier-client relationships; however, there is a lack of empirical evidence in what pertains to the comparison of upstream and downstream activities involving large firms vis-à-vis SMEs.

The aim of this research is to compare firms' involvement in CNPD in upstream and downstream activities of the supply chain, based on firm size and the innovation generated, and determine the differences according to those characteristics. For this purpose, a framework was developed serving to study firms' involvement in CNPD in various circumstances. The study aims to answer the following questions: How is CNPD operationalized between industrial firms and their suppliers and customers? How do firms intervene in CNPD according to their size and the innovation created?

The chapter is divided in five sections. After this introduction, that comprises the first section, the theory on CNPD is depicted in the second section, when collaboration in upstream and downstream activities are explored. Section 3 addresses the method and data used throughout the article. Section 4 discusses the results. Finally, Sect. 5 presents the main conclusions of the chapter.

2 Theory on Collaborative New Product Development

2.1 Involvement Between Firms in CNPD

CNPD is based on sharing information, risks and rewards (Takeishi 2001; Emden et al. 2006; Wagner and Hoegl 2006; Inauen and Schenker-Wicki 2012; Un and Azakawa 2015), being common practice in high-tech industries, as is the case of the pharmaceutical, electronics and auto industries (Powell et al. 1996; Takeishi 2001; Inauen and Schenker-Wicki 2012), and subsequently spreading to other industries. Chesbrough (2003) studied this collaborative practice in large firms in high-tech industries, calling it open innovation. Various authors conclude that CNPD is operationalized by large firms, because they are highly specialized and able to make major investments (Lecocq and Demil 2006; Lichtenthaler 2008; Ferrary 2011; Gay 2014). However, SMEs also participate in CNPD, despite their limitations regarding R&D investment and production capacity (Laursen and Salter 2006; Van de Vrande et al. 2009; Lee et al. 2010; Theyel 2012; Silva and Moreira 2017). Other studies claim that collaboration between companies promotes the development of new products, manufacturing processes or management methodologies according to firms' size and the innovation generated in CNPD (Laursen and Salter 2006; Nieto and Santamaria 2007; Parida et al. 2012; Theyel 2012; Hossain 2015). Nevertheless, there is no literature concerning the comparison between CNPD carried out by SMEs and large firms in upstream and downstream activities.

2.1.1 Objectives in Resorting to External Technology

CNPD has been analyzed to distinguish objectives when resorting to external technology (Freel 2003; Lecocq and Demil 2006; Van de Vrande et al. 2009; Lee et al. 2010; Hossain 2015). Large firms operationalize CNPD aiming to develop differentiated products and processes, contributing to the externalization of more fruitful results from their activity (Chesbrough and Crowther 2006; Gassmann 2006; Ferrary 2011). In the case of SMEs, the aim is to diversify their product portfolio (Lecocq and Demil 2006; Van de Vrande et al. 2009; Wynarczyk et al. 2013; Hossain 2015) and extend their network of partners to access diversified markets (Lichtenthaler 2008; Lee et al. 2010). In the view of Freel (2003), Madrid-Guijarro et al. (2009) and Parida et al. (2012), the CNPD undertaken by SMEs aims to increase the efficiency of their resources, mainly in relation to manufacturing processes. The above studies reveal it is large firms that involve suppliers in CNPD with the aim of creating differentiated products. However, there are differences in the interaction between firms in CNPD, according to the characteristics of those involved and the innovation created (Nieto and Santamaria 2007; Theyel 2012; Spithoven et al. 2013; Hossain and Karaunen 2016). Consequently, firm size and the innovation created in CNPD can be expected to lead to different results regarding their influence on the objectives of resorting to external technology.

2.1.2 CNPD Operationalized According to Its Innovation Nature

Various studies classify innovation according to its radical nature in order to understand its influence on CNPD (Van de Vrande et al. 2009; Parida et al. 2012; Inauen and Schenker-Wicki 2012; Hossain 2015). Radical innovation, understood as the development of new products and processes, is created through collaboration between firms operating in high-tech industries to promote differentiation (Garcia and Calantone 2002; Koberg et al. 2003; Diedericks and Hoonhout 2007; Inauen and Schenker-Wicki 2012), whereas incremental innovation, understood as the development of improved products and processes, emerges from collaboration between firms to reposition their products or increase the efficiency of their activity (Inauen and Schenker-Wicki 2012; Parida et al. 2012). Other studies claim that radical innovation is created in CNPD between large firms, because they have major technological resources, whereas incremental innovation is created by SMEs (Koberg et al. 2003; Van de Vrande et al. 2009; Inauen and Schenker-Wicki 2012). However, the literature is controversial, because SMEs also have the capacity to develop new products, despite their limited resources (Lee et al. 2010; Hossain 2015). From another perspective, radical innovation emerges from upstream CNPD activities among large firms and between those large firms and consultancy firms and academic institutions, due to their capacity to develop emerging technologies (Gassmann 2006; Ferrary 2011; Un and Azakawa 2015), while incremental innovation emerges from the downstream collaboration undertaken by SMEs (Bonner and Walker 2004; Faems et al. 2005; Roy and Sivakumar 2010). However, these approaches do not compare the circumstances in which CNPD occurs in relation to the industries in which firms operate.

2.2 *Vertical Collaboration*

2.2.1 CNPD Operationalized Upstream

CNPD undertaken upstream corresponds to suppliers' involvement with firms (Powell et al. 1996; Hoegl and Wagner 2005; Laursen and Salter 2006; Brunswicker and Vanhaverbeke 2014; Tuli and Shankar 2015), through participation in B2B networks, outsourcing contracts and acquiring patents (Gomes-Casseres 1997; Laursen and Salter 2006; Van de Vrande et al. 2009). The involvement of universities and consultants in CNPD activities is another form of collaboration upstream, when firms need to incorporate emerging technologies (Tether 2002; Freel 2003; Faems et al. 2005; Fontana et al. 2006; Bruneel et al. 2010; Brunswicker and Vanhaverbeke 2014). According to Cohen and Levinthal (1990), Tether (2002) and Inauen and Schenker-Wicki (2012), suppliers' specialization is crucial for CNPD dynamics upstream.

Frequent Versus Occasional Collaboration Frequent collaboration occurs with regular industrial suppliers operating in high-tech industries (Peng et al. 2014). However, the great complexity of CNPD favors collaboration between firms and new suppliers (Croom 2001; Bueno and Balestrin 2012), principally in the case of large firms (Boehe 2007). New suppliers' collaboration in CNPD, which can occur occasionally, is due to firms' need for specialization in differentiated products (Phillips et al. 2006; Bueno and Balestrin 2012; Raluca 2013). In general, firms collaborate frequently with regular suppliers, involving competences which are specific and complementary to the diversity of requirements for new products (Tidd et al. 2001; Bueno and Balestrin 2012).

Objectives of the Innovation Generated Collaboration between suppliers and industrial firms contributes to the development of products and processes (Fritsch and Lukas 2001; Lager 2016) when firms need to incorporate, simultaneously, specialized material, manufactured technologies or management methodologies (Clark and Fujimoto 1991; Knudsen 2007; Faems et al. 2005; Soosay et al. 2008; Brettel and Cleven 2011). In addition, universities collaborate with large firms to develop new processes (Faems et al. 2005; Laursen and Salter 2006; Un and Azakawa 2015) for more agile development of new-to-the-market differentiated products (Tether 2002). However, CNPD undertaken upstream, mainly by SMEs, creates improved products and processes (Verhees and Meulenberg 2004; Faems et al. 2005; Soosay et al. 2008; Lager and Frishammar 2012; Theyel 2012) to increase the efficiency of their activity (Soosay et al. 2008). The facts described reveal that radical innovation promotes collaboration between service suppliers and large firms for the development of new processes. From another perspective, the innovation created in CNPD is more significant in firms operating in high-tech industries (Faems et al. 2005; Tether 2002). However, Schiele (2010) defends that suppliers' specialization is what matters the most, which underpins radical innovation. Most studies claim that the radical innovation is created mainly in the CNPD undertaken by large firms, when compared to the less radical nature of products developed by SMEs (Clark and Fujimoto 1991; Laursen and Salter 2006; Un and Azakawa 2015). However, these views are limited because they do not compare the innovation created in CNPD according to the industries in which firms operate.

Typology of Firms Involved The literature states that large firms are more active in CNPD than SMEs (Chesbrough 2003; Spithoven et al. 2013; Hossain and Karaunen 2016). Upstream CNPD is mostly carried out by large industrial firms operating in high-tech industries (Wynstra et al. 2010; Gay 2014). Another perspective claims that firm size does not affect upstream collaboration in CNPD (Lee et al. 2010; Parida et al. 2012; Wynarczyk et al. 2013) and that SMEs intervene in CNPD with firms of varying sizes (Parida et al. 2012; Theyel 2012) due to their flexibility and capacity to adapt to the market (Lecocq and Demil 2006; Lee et al. 2010), despite their limitations regarding major R&D investment and production capacity (Van de Vrande et al. 2009; Lee et al. 2010).

Although SMEs develop new products in collaboration with industrial suppliers (Nieto and Santamaria 2010; Hossain 2015; Hossain and Karaunen 2016), the need for SME specialization encourages collaboration with small and medium-sized service suppliers, mainly when they operate in specific niche markets (Verhees and Meulenberg 2004; Tether and Tajar 2008; Hossain 2015). In addition, growing technological development has led to firms acquiring academic knowledge about new materials and manufacturing processes (Faems et al. 2005; Nieto and Santamaria 2010; Un and Azakawa 2015). In this context, university and consultant involvement upstream in CNPD with large firms promotes the development of new processes and management methodologies (Cohen and Levinthal 1990; Tether 2002; Faems et al. 2005; Un and Azakawa 2015; Brettel and Cleven 2011). These facts show the importance of the contribution of service suppliers to CNPD when companies need to develop new processes (Faems et al. 2005; Laursen and Salter 2006; Un et al. 2010; Nieto and Santamaria 2010).

2.2.2 CNPD Operationalized Downstream

Downstream CNPD (Powell et al. 1996; Brockhoff 2003; Lagrosen 2005; Theyel 2012) has been undertaken through agreements (Gomes-Casseres 1997; Laursen and Salter 2006; Lager and Frishammar 2012), granting rights to use technology (Powell et al. 1996, Wynarczyk et al. 2013) and collaboration between companies (Van de Vrande et al. 2009; Lee et al. 2010; Gay 2014). Downstream collaboration has been addressed in various studies due to customers' importance in spreading the innovation created upstream (Freel 2003; Knudsen 2007; Lee et al. 2010; Un and Azakawa 2015). However, the literature on CNPD involving customers is controversial and needs a deeper analysis.

Frequent Versus Occasional Collaboration When the main objective is the creation of customized products, frequent relations involving SMEs and customers in CNPD are more common than those involving large firms and customers (Lee et al. 2010). For Tether (2002) and Knudsen (2007), frequent relations in CNPD occur between companies and customers operating in high-tech industries when developing new products. Stephanie and Shulman (2011) and Peng et al. (2014) add that it is clients' need for specialization in manufacturing processes that stimulates firms' frequent involvement in CNPD, due to product differentiation. Therefore, frequent relations between SMEs and clients operating in high-tech industries are due to the radical nature of CNPD.

Objectives of the Innovation Generated The aim of collaboration is to respond with differentiation to market needs (Tether 2002; Van de Vrande et al. 2009; Inauen and Schenker-Wicki 2012; Lager 2016). In this context, CNPD involving SMEs generates disruptive outcomes (Parida et al. 2012), although Brockhoff (2003), Bonner and Walker (2004), Faems et al. (2005) and Roy and Sivakumar (2010) conclude that collaboration between companies and clients creates new products based on incremental innovation when the collaboration occurs with SMEs. These approaches reveal that downstream collaboration is affected by the technological intensity of the industries in which firms operate and by the innovation created in CNPD.

Typology of Firms Involved In upstream interaction, clients normally involve large industrial firms in CNPD, due to their technological specialization and production capacity (Tether 2002; Lichtenthaler 2008). In turn, large firms promote more active collaboration downstream, because they have technology and experience, resulting from the frequent interaction with clients (Lagrosen 2005; Lee et al. 2010; Moghaddam and Tarokh 2012). In addition, large firms operating in high-tech industries are more likely to be involved by industrial clients because they introduce greater specialization in their activity (Tether 2002). Therefore, downstream CNPD is mostly undertaken between large firms operating in high-tech industries

(Lichtenthaler 2008). On the contrary, other studies conclude that SMEs are also involved with clients in CNPD because their technical know-how and market knowledge upstream stimulate innovation (Verhees and Meulenberg 2004; Van de Vrande et al. 2009; Lee et al. 2010; Nieto and Santamaria 2010; Silva and Moreira 2017). Consequently, firm size is not necessarily a limitation for collaboration with clients (Johnsen and Ford 2006; Van de Vrande et al. 2009; Lee et al. 2010), when CNPD does not require a large production scale.

3 Methodology

The particularities of CNPD allow the exploration of several research areas, due to its diversity of analysis. This fact leads to a comparative study of the collaboration carried out by firms in upstream and downstream activities, by SMEs and large firms, to identify the asymmetries found in interaction with suppliers and customers and the circumstances in which they occur.

This research is an exploratory study, based on qualitative research to describe the situation studied, from an inductive perspective (Yin 2003; Baxter and Jack 2008; Heath and Tynan 2010). The methodology used here is similar to that used in other studies on similar subjects (Lagrosen 2005; Emden et al. 2006; Eslami and Lakemond 2016; Silva and Moreira 2017).

The field research was carried out in three stages. The first relates to collecting data from industrial firms undertaking CNPD. The next stage corresponds to the analysis, treatment and description of the information gathered. The third involves joining the information in cases, according to a set of variables analyzed during the literature review. The case study is used to describe the situation observed, in accordance with the methodology of this research (Miles and Huberman 1994; Yin 2003; Baxter and Jack 2008) and with the study's objective. Cases are analyzed individually to describe the situation of each firm. Then the cases are compared to determine the differences in the study variables portraying each firm (Yin 2003; Baxter and Jack 2008). The information about the cases is subsequently summarized in tables to analyze the results, provide the "big picture" of the situation studied and answer the research questions (Miles and Huberman 1994; Yin 2003).

Based on the characterization of CNPD carried out by firms, this research used purposive sampling whose elements were chosen from a number of previously selected industrial firms to ensure maximum information about the topics of analysis in the study (Malhotra 2007; Black 2010). Purposive sampling was constructed based on the information gathered about firms' innovation activities. The sample is composed of four SMEs and four large industrial companies that carry out CNPD with suppliers and clients, the European Commission (2002) criterion having been used to classify firms according to number of employees. This classification lets us describe and compare the situation of SMEs and large firms. Table 1 presents the firms selected for this study, which will be designated as focal firms.

Table 1 Profile of the companies (*focal firms*)

Description	C1	C2	C3	C4	C5	C6	C7	C8
Type of activity	Electronic products	Automotive fastening systems	Metal components and accessories	Precision metallurgy	Exposition and hotel trading equipments	High-end appliances	Bathrooms furniture	Lighting products
Technological intensity	High	High	Moderate	Moderate	Moderate	Moderate	Low	Moderate
Main products	Electronic boards and circuits; Miscellaneous electronic equipment	Fastening systems; shafts; by-products	Locks; handles; hinges; keys; bath accessories	Metal boxes, racks, metallic computer accessories, other precision accessories	Stainless steel furniture, shelves, check-outs	Microwave ovens; Steam ovens; Cooker hoods; Hobs; Stoves; Dishwashers	Furniture; Mirrors; Washbasins; Accessories	Lamps; Luminaires; Lighting rails; Wall lights
Sales volume (10^6 €)	8	22	11.5	0.75	5.7	80	0.3	10.9
Export intensity (% exports/ sales)	70%	80%	80%	2,5%	8%	85%	5%	60%
Number of workers (SME/Large firm)	200 SME	253 LF	252 LF	18 SME	85 SME	347 LF	27 SME	256 LF
Main markets/ industries	Electronics; Telecommunications; Automotive; Transportation; miscellaneous industry	Automotive	Automotive, Hardware stores and Stockists	Metal mechanics; Telecommunications; Electronics; miscellaneous industry	Retailers; Stores; Stockists	Distributors; Appliance industry	Hotels, Building materials retailers	Retailers; distributors; Importers
Interviewee	CEO	Industrial Manager	Industrial Manager; Commercial Manager	Manager; Technical Director	Industrial Manager; New Product development Manager	R&D Manager	Manager	Manager, Technical Manager

The diagnosis of firms' situation was made through holding semi-structured interviews with the leaders of the focal firms, according to their structure and lines of command, as shown in Table 1. Interviews allow us to determine how focal firms undertake CNPD regarding the objectives of resorting to external technology upstream, the innovation created and the typology of firms involved in CNPD both upstream and downstream. The semi-structured interviews were carried out using a script, aiming to guide the central themes of the study (Malhotra 2007). They lasted between a minimum of 1 h 5 min and a maximum of 2 h 15 min. They were recorded and then transcribed to paper to be treated individually as case studies. Subsequently, the cases are compared (Yin 2003; Baxter and Jack 2008), in order to distinguish the difference in CNPD carried out by the focal firms upstream and downstream according to firm size. The confidentiality of the information gathered was assured by the identity of the firms interviewed remaining anonymous (being designated as C1–C8). The interviews took place on the firms' premises, allowing the situation and innovative practices implemented to be checked. Additional information about the firms' activity was also supplied, such as catalogues of their product portfolio, company profile, information on the technology used, markets, sources of supply, and manufacturing bases and process, among others. During the interviews, themes related to the innovative projects developed by the focal firms in the last 2 years were dealt with, to understand the innovation created in CNPD with their suppliers and clients. Additional contacts were also made with the focal firms by e-mail and telephone to complement and confirm the information collected.

The unit of analysis is the dyadic relationship in focal firms' CNPD with suppliers and customers.

A framework was drawn up to analyze the collaboration carried out between firms and their suppliers and clients in CNPD, as shown in Fig. 1.

The results of the cases studied regarding firms' objectives when resorting to external technology upstream are presented in Table 2. The results shown in Tables 3 and 4 describe the CNPD undertaken with upstream and downstream partners, according to firm size and the innovation generated.

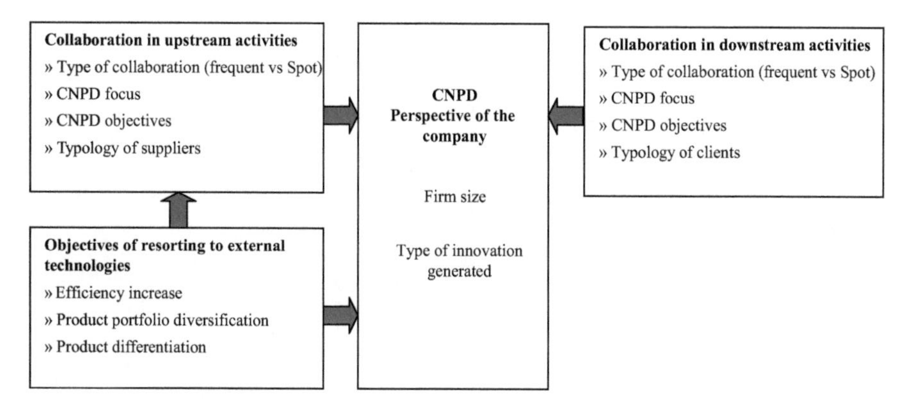

Fig. 1 Framework analysis

Table 2 Search of external technologies and supplier intervention by focal firms

Description	C1	C2	C3	C4	C5	C6	C7	C8
Reasons for searching external technologies								
Product differentiation/ Product portfolio diversification/ Efficiency increase	Efficiency increase	Product differentiation	Product portfolio diversification	Product portfolio diversification	Efficiency increase	Product differentiation	Product portfolio diversification	Product portfolio diversification
Supplier intervention in CNPD								
Product development	*Project* (idea), *Components* (specifications, functionality, design, conception): electronic parts, metal parts and boxes, plastic parts and boxes	*Project* (idea), *Raw materials* (specifications, functionality, applicability design, conception, finishing): metal alloys	*Components* (specifications, applicability, conception, finishing): metal alloys, plastics and plastic parts	*Raw materials* (specifications, conception, finishing): metal alloys, paints, *Components* (specifications, conception, finishing): cables, plastics, hinges and locks	*Raw materials* (specifications, applicability, conception): metal plates *Components* (specifications, applicability, conception): metal parts, plastics, carbon, coating films	*Project* (idea), *Components* (specifications, applicability, design, conception): modules and electronic parts, modules and plastic parts	*Raw materials* (specifications, conception, finishing): paints and varnishes, *Components* (specifications, applicability, conception): hinges, handles, coating skins, lighting systems	*Components* (specifications, applicability, conception): lighting parts, plastics, electronic parts, aluminum modules, metal plate modules
Supplier intervention in CNPD								
Process development	*Manufacturing process* (software programming, laboratory tests)	*Manufacturing process* (machines, tools, software programming)	*Manufacturing process* (tools, tests)	*Manufacturing process* (tools, software programming)	*Manufacturing process* (tools)	*Manufacturing process* (tools, software programming) *Management methodologies* (business models)	*Manufacturing process* (tools)	*Manufacturing process* (tools)

Table 3 Collaboration in upstream activities

Description	C1	C2	C3	C4	C5	C6	C7	C8
Frequent versus spot cooperation (regular and new suppliers)								
Frequent collaboration	Regular suppliers	Regular suppliers	Regular suppliers	Regular suppliers	×	Regular suppliers	Regular suppliers	Regular suppliers
Spot cooperation	×	×	×	×	Regular suppliers	Regular suppliers	New suppliers	Regular suppliers
CNPD focus								
Product development	Yes	Yes	Yes	Yes	Yes	Yes	Yes	Yes
Process development	Yes	Yes	Yes	Yes	Yes	Yes	Yes	Yes
CNPD objective								
New product	Yes	Yes	Yes	Yes	Yes	Yes	Yes	Yes
Improved product	Yes	Yes	Yes	Yes	Yes	Yes	Yes	Yes
New production process	Yes	Yes	Yes	×	Yes	Yes	×	Yes
Improved production process	Yes	×	Yes	Yes	×	Yes	Yes	×
Management methodology	×	×	×	×	×	Yes	×	×
Size of suppliers								
SME	SME	×	SME	SME	SME	SME	SME	×
Large firm (LF)	LF	LF	LF	LF	LF	×	LF	LF
Typology of suppliers								
Industrial firms (Ind)	Ind	Ind	Ind	Ind	Ind	Ind	Ind	Ind
Service firms (Serv)	Serv	Serv	Serv	×	Serv	Serv	Serv	×
Institutions (Univ)	×	×	×	×	×	Univ	×	×

Table 4 Collaboration in downstream activities

Description	C1	C2	C3	C4	C5	C6	C7	C8
Type of collaboration								
Frequent collaboration	Yes	Yes	Yes	Yes	×	Yes	×	×
Spot cooperation	×	×	×	Yes	Yes	×	Yes	Yes
CNPD focus								
Product development	Yes	Yes	Yes	Yes	Yes	Yes	Yes	Yes
Process development	Yes	×	×	Yes	×	×	×	×
CNPD objective								
New product	Yes	Yes	Yes	Yes	Yes	Yes	Yes	Yes
Improved product	Yes	Yes	Yes	Yes	Yes	Yes	Yes	Yes
New production process	Yes	×	×	×	×	×	×	×
Improved production process	×	×	×	Yes	×	×	×	×
Size of suppliers								
SME	SME	×	SME	SME	SME	SME	SME	SME
Large firm (LF)	LF	LF	LF	LF	LF	LF	×	LF
Typology of suppliers								
Industrial firms (Ind)	Ind	Ind	Ind	Ind	×	Ind	×	×
Commercial firms (Comm)	×	×	Comm	×	Comm	Comm	Comm	Comm

4 Discussion

4.1 Collaboration Between Suppliers and Firms in CNPD

Firms' Objective in Resorting to External Technology The diversification of firms' activity is the main reason for collaborating with suppliers, as shown by cases C3 and C8 in relation to large firms, and cases C4 and C7 in the case of SMEs. In addition, SMEs' need to increase the efficiency of their activity promotes involvement with suppliers, as shown by cases C1 and C5. In the case of large firms operating in high-tech industries, it is differentiation that stimulates collaboration with suppliers in NPD, as revealed by case C2. The cases studied show that more radical product innovation is created in the CNPD undertaken upstream by large firms.

Frequent Versus Occasional Collaboration Upstream In most cases, regular suppliers collaborate frequently in the CNPD carried out by firms. However, case C5 shows that SMEs could involve regular suppliers in CNPD. These results complement other studies (Hartley et al. 1997; Wasti and Liker 1999; Peng et al. 2014), showing that CNPD may result from the occasional involvement of suppliers. This difference in suppliers' frequent/occasional collaboration in NPD is due to the different technological evolution of the industries and the slower pace of CNPD. Another perspective reveals that large firms frequently collaborate with regular suppliers, as shown by cases C2, C3, C6 and C8, and also occasionally with new suppliers to develop differentiated products or implement diversification, as shown by cases C6 and C8. These results complement prior research (Phillips et al. 2006; Bueno and Balestrin 2012; Raluca 2013) as beyond differentiated products there are cases of collaboration with suppliers to diversify product portfolio. In particular, cases C3 and C6 also show that frequent collaboration occurs between focal firms and small and medium-sized suppliers.

Focus of CNPD Upstream Collaboration between suppliers and firms in CNPD creates new products (as cases C3, C5 and C6 exemplify) and also new processes (C2, C2, C6 and C8). In this respect, the cases studied reveal that the upstream CNPD undertaken by both SMEs and large firms is not limited to the physical development of products. These results complement several studies (Clark and Fujimoto 1991; Tether 2002; Lager 2016) showing that upstream collaboration is not affected by firm size. In particular, cases C1, C4, C5 and C7 show that SMEs have competences to intervene in CNPD upstream, and complement some studies about SMEs' tendency to engage in CNPD downstream (Theyel 2012; Hossain 2015).

Then again, large firms' need to conceive new product concepts and management methodologies in the form of new business models, as exemplified by case C6, favors upstream collaboration with consultants and universities, as referred in prior studies (Faems et al. 2005; Laursen and Salter 2006; Un and Azakawa 2015). This collaboration results from large firms' need for specialization in the conception of differentiated products, as referred by Tether (2002). Therefore, the differentiation of activity in large firms favors collaboration with service suppliers to develop new manufacturing processes and management methodologies that go towards conceiving radical new products, as exemplified by cases C2 and C6. On the other hand, the less radical nature of manufacturing processes created in CNPD results from the diversification of the activity of SMEs operating in low-tech industries, of which cases C4 and C7 are examples. However, case C1 shows that SMEs also develop new production processes in collaboration with suppliers to increase the efficiency of their activity. Therefore, comparing cases C1 and C2 with the other cases shows that development of new manufacturing processes is due to the radical nature of CNPD and the high-tech industries of activity in which firms operate. These results differ from previous research (Clark and Fujimoto 1991; Laursen and Salter 2006; Un and Azakawa 2015) and show that firm size does not condition CNPD.

Typology of Suppliers Involved The involvement of SMEs and large firms in upstream activities is quite widespread. From this point of view, firms' need of specialization promotes upstream collaboration with medium-sized service suppliers. This interaction generated upstream is due to large firms' differentiation activity (case C6), increased efficiency of SME activity (cases C1 and C5) and also the high-tech industries in which companies operate (cases C1 and C6). In this context, case C6 differs from the others concerning the type of supplier and the specialization required at the initial stage of CNPD. In this respect, C6 involves consultants and research centers in CNPD to idealize differentiated products and develop the business model adapted to its target. These facts show that, in general,

CNPD between industrial companies and their suppliers is not affected by their size, but rather by their specialization, as referred in other studies (Laursen and Salter 2006; Lee et al. 2010). These results complement prior research (Chesbrough and Crowther 2006; Bianchi et al. 2011; Ferrary 2011; Gay 2014) about the influence of firm size in upstream CNPD. Nevertheless, some large firms collaborate with suppliers of a similar size because of the type of industry they operate and their manufacturing process requires a high scale of production, and only large firms have such processes, as referred in prior studies that analyzed large firms (Chesbrough 2003; Christensen et al. 2005; Wynstra et al. 2010), and quoted from cases C2 and C8: "*we involve large firms because they are more able to respond with the specialization and capacity for our demands*" (case C2), "*the suppliers operating in this industry and involved in the CNPD are large firms operating in high-tech industries*" (case C8). So the difference between cases C2 and C8 and the others shows that supplier size affects their collaboration when a high specialization and scale of production is required by the CNPD.

4.2 Collaboration Between Firms and Clients in CNPD

Frequent Versus Occasional Collaboration Downstream Frequent downstream collaboration in CNPD occurs between industrial companies, as shown by cases C1, C2, C3, C4 and C6. In this respect, cases C1 and C4 show that SMEs also collaborate with industrial customers because of their capacity to develop disruptive products, as previous studies has mentioned (Tether 2002; Knudsen 2007). Thus, these results complement the studies of Brockhoff (2003), Bonner and Walker (2004), Faems et al. (2005) and Roy and Sivakumar (2010) about the innovation generated downstream in the CNPD involving SMEs. On the other hand, cases C5, C7 and C8 show that occasional downstream collaboration in NPD occurs between industrial and commercial companies. Comparison between cases shows that frequent downstream collaboration results from more active involvement between industrial firms, as opposed to involvement between industrial and commercial firms, and that firm size does not affect CNPD interaction.

Focus of Downstream CNPD In collaboration between large firms (cases C2, C3, C6 and C8) firms' need for specialization downstream is limited to the physical development of products. However, some SMEs also collaborate with clients to develop new processes, case C1 being an example. In this case, collaboration arises from the fact that the firms operate in high-tech industries, as referred by Parida et al. (2012), and use similar technology, and not due to their size. Furthermore, SME specialization favors collaboration with medium-sized clients operating in the same industry to develop new processes, as in case C4, as mentioned in prior research (Brockhoff 2003; Bonner and Walker 2004; Faems et al. 2005; Roy and Sivakumar 2010). Comparing cases C1 and C4 with the others shows, in the first place, that collaborative development of new processes downstream is carried out between firms and clients operating in the same industry and that firm size does not affect that collaboration, and secondly, that SMEs also have competences to develop products and processes in collaboration with clients, complementing previous studies (Lagrosen 2005; Un and Azakawa 2015; Eslami and Lakemond 2016).

Objective of Downstream CNPD Collaboration with clients is not affected by size or by the industry type, as shown by cases C1, C4 and C5. These results confirm previous studies (Tether 2002; Lee et al. 2010; Parida et al. 2012) and complement the literature (Bjerke and Johansson 2015; Eslami and Lakemond 2016) about how SMEs intervene in downstream CNPD. Moreover, the collaborative atmosphere generated between SMEs and clients operating in the industry favors the development of manufacturing processes, as shown by cases C1 and C4. Here, the more radical nature of manufacturing processes conceived by focal firms C1 and C4 is due to the collaborative development of new products with large clients operating in the same industry, as in the quote from C1: "*because the complexity of the new products idealized by our clients in the electronics industry requires the development of new programming with specific functions*". From another perspective, the development of improved manufacturing processes between SMEs and medium-sized clients operating in the same industry is due to the less radical nature of the products developed, as is quoted from case C4: "*we improved that product's finishing process to reposition it*". Finally, the cases show that firms involved in downstream CNPD do not develop any management methodology, because their objective concerns the physical conception of products.

5 Conclusions

This study concluded that CNPD between industrial firms and suppliers is more frequent and active than that with clients, giving rise to asymmetry in upstream and downstream collaboration, and that fact is due to the innovation created in CNPD and those taking part in it. This phenomenon has various implications for theory. Firstly, it shows that asymmetry is the result of greater upstream collaboration between industrial companies, and between these and service suppliers, compared to downstream collaboration. From another perspective, it reveals that collaboration between SMEs and suppliers is more active than that with clients. Consequently, more active collaboration in upstream CNPD is due above all to the frequent interaction between industrial companies and between these and service suppliers to develop new products, manufacturing processes and management methodologies, when compared to the less active downstream collaboration between industrial companies and the occasional collaboration between these and commercial firms for physical development of products. These facts complement the literature on CNPD by showing that more active collaboration between firms upstream in the supply chain is the result of extending the range of suppliers involved in CNPD to service firms, due to the need for specialization at the early stages of CNPD.

Generally speaking, industrial companies involve suppliers of different types in CNPD, mainly when they aim to diversify their product portfolio. However, in certain cases, differentiation is generated by collaboration between large firms operating in the same high-tech industry, to create new products, as exemplified by case C2. These facts corroborate the view of large industrial firms' tendency to collaborate, motivated by great specialization, available technology and high production capacity, and show that the more radical nature of CNPD stimulates collaboration between suppliers and large clients operating in the same industry. In addition, this study shows that differentiation in the activity of large industrial firms also promotes upstream collaboration with SMEs. This argument contradicts the perspective of collaboration restricted to large firms, allowing the conclusion that SMEs collaborate with large firms, when the latter turn to specialization in other industries and a high scale of production is not required. Another view infers that SMEs involve suppliers of varying sizes in CNPD when they aim to increase the efficiency of their activity. So the various interactions generated between industrial companies and suppliers in CNPD let us conclude that firs size does not limit their intervention in CNPD when the aim is to diversify their product portfolio or increase the efficiency of their activity, while sustained differentiation in great specialization and production capacity is generated between large firms.

Collaboration between firms and suppliers to develop new processes is more active than that carried out with clients. This phenomenon is due to the limited collaboration between industrial firms and clients to develop new manufacturing processes, which is non-existent in developing new management methodologies, and allows the conclusion that downstream collaboration is focused mostly on the physical development of new products. So the asymmetry found in upstream and downstream collaboration in NPD is due to more active collaboration between large firms and suppliers to develop new manufacturing processes and management methodologies. Nevertheless, the results show that SMEs take part in CNPD, but more actively in upstream activities to develop new manufacturing processes.

The asymmetry found in collaboration between companies according to their industry is due to the innovation generated in CNPD. In general, differentiation is created through collaboration between firms operating in the same industry, whereas CNPD carried out between firms operating in different industries creates diversified products or promotes increased efficiency in firms' activity. However, the interaction between firms is not formed linearly, because collaboration between firms and service suppliers to develop new processes is more active upstream than downstream. The situation studied revealed that downstream collaboration between industrial firms and service suppliers is more limited, because in most cases it is restricted to physical development of products. On the other hand, industrial firms collaborate more actively with service suppliers in upstream CNPD. In this context, upstream collaboration between large firms and research centers is due to the more radical nature of the CNPD. This goes against the view of some authors regarding service suppliers' collaboration only in post-production activities and shows the importance of their collaboration in the early stages of the NPD undertaken by industrial companies.

From another perspective, this study concludes that CNPD is carried out between large firms operating in high-tech industries and generating large production scales (as in case C2). However, this study also reveals that SMEs operating in high-tech industries (exemplified in case C1) involve other large firms in CNPD. These facts show, in the first place, that upstream and downstream CNPD is influenced by the technological intensity of firms' operating industries, and secondly, that firm size affects their intervention in CNPD only when a high scale of production is required.

Then again, the asymmetry of upstream and downstream collaboration between firms is due to the diversity of specialization required by CNPD, particularly service suppliers' upstream collaboration to create manufacturing processes and management methodology adapted to new product concepts. In turn, the differentiation in large firms' activity and the specialization of the industry in which they operate promotes collaboration between specialist firms (case C2 being an example). Therefore, greater symmetry in upstream and downstream collaboration between firms occurs when those taking part in the CNPD are specialist firms operating with similar technology in the same industry.

The framework developed in this study explains the possible interactions between firms and their suppliers and clients, according to their size and the innovation created. Generally speaking, the CNPD undertaken by industrial companies aims to diversify their activity or increase their efficiency. However, large industrial firms operating in specific niche markets undertake CNPD to differentiate their portfolio. In this context, firm size affects their intervention in CNPD, when differentiation requires great specialization and large-scale production. These facts have implications for defining firms' innovation strategy and the strategic options adapted to their business environment. In this context, firms' strategy should consider their competences and the objectives of innovation to occupy a favorable position in CNPD. Study of the collaboration between firms and their suppliers and clients reveals that collaborative development of manufacturing processes and management methodologies are little explored areas of business, particularly in downstream activities of the supply chain. The framework developed explains the interactions firms can establish in CNPD to expand their activity.

References

Baxter P, Jack S (2008) Qualitative case study methodology: study design and implementation for novice researchers. Qual Rep 13(4):544–559

Bianchi M, Cavaliere A, Chiaroni D, Frattini F, Chiesa V (2011) Organizational modes for open innovation in bio-pharmaceutical industry: an exploratory analysis. Technovation 31(1):22–33

Bjerke L, Johansson S (2015) Patterns of innovation and collaboration in small and large firms. Ann Reg Sci 55(1):221–247

Black K (2010) Business statistics: contemporary decision making. Wiley, Chichester

Boehe DM (2007) Os papéis de subsidiárias brasileiras na estratégia de inovação de empresa multinacionais estrangeiras. RAUSP Rev Admin 42(1):5–18

Bonner JM, Walker OC (2004) Selecting influential business-to-business customers in new product development: relational embeddedness and knowledge heterogeneity considerations. J Prod Innov Manag 21(3):155–169
Brettel M, Cleven NJ (2011) Innovation culture, collaboration with external partners and NPD performance. Creat Innov Manag 20(4):253–272
Brockhoff K (2003) Customers' perspectives of involvement in new product development. Int J Technol Manage 26(5–6):464–481
Bruneel J, D'Este P, Salter A (2010) Investigating the facts that diminish the barriers to university-industry collaboration. Res Policy 39(7):858–868
Brunswicker S, Vanhaverbeke W (2014) Open innovation in small and medium-sized enterprises (SMEs): external knowledge sourcing strategies and internal organizational facilitators. J Small Bus Manag 53(4):1241–1263
Bueno B, Balestrin A (2012) Inovação colaborativa: uma abordagem aberta no DNP. ERA Rev Admin Empres 52(5):517–530
Büyüközkan G, Arsenyan J (2012) Collaborative product development: a literature overview. Prod Plan Control 23(1):47–66
Chesbrough H (2003) The era of open innovation. Sloan Manage Rev 44(3):35–41
Chesbrough H, Crowther AK (2006) Beyond high-tech: early adopters of open innovation in other industries. R&D Manag 36(3):229–236
Christensen JF, Olesen MH, Kjaer JS (2005) The industrial dynamics of open innovation: evidence from the transformation of consumer electronics. Res Policy 34(10):1533–1549
Clark KB, Fujimoto T (1991) Product development strategy, organization, and management in the world auto industry. Harvard Business School Press, Boston
Cohen WM, Levinthal DA (1990) Absorptive capacity: a new perspective on learning and innovation. Adm Sci Q 35(1):128–152
Croom SR (2001) The dyadic capabilities concept: examining the process of key supplier involvement in collaborative product development. Eur J Purch Supply Manag 7(1):29–37
Diedericks E, Hoonhout H (2007) Radical innovation and end-user involvement: the ambilight case. Knowl Technol Policy 20(1):31–38
Emden Z, Calantone RJ, Droge C (2006) Collaborating for new product development: selecting the partner with the maximum potential to create value. J Prod Innov Manag 23(4):330–341
Eslami MH, Lakemond N (2016) Knowledge integration with customers in collaborative product development projects. J Bus Ind Mark 31(7):889–900
European Commission (2002) Observatory of European SMEs: European Committee
Faems D, Van Looy B, Debackere K (2005) Interorganizational collaboration and innovation: toward a portfolio approach. J Prod Innov Manag 22(3):238–250
Ferrary M (2011) Specialized organizations and ambidextrous clusters in the open innovation paradigm. Eur Manag J 29(3):181–192
Fontana R, Geuna A, Matt M (2006) Factors affecting university-industry R&D projects: the importance of searching and signalling. Res Policy 35(2):309–323
Freel MS (2003) Sectoral patterns of small firm innovation, networking and proximity. Res Policy 32(5):751–770
Fritsch M, Lukas R (2001) Who cooperates in R&D? Res Policy 30(2):297–312
Garcia R, Calantone R (2002) A critical look at technology innovation typology and innovativeness terminology: a literature review. J Prod Innov Manag 19(2):110–132
Gassmann O (2006) Opening up the innovation process: towards an agenda. R&D Manag 36(3):223–228
Gay B (2014) Open innovation, networking, and business model dynamics: the two sides. J Innov Entrep 3:2. https://doi.org/10.1186/2192-5372-3-2
Gomes-Casseres B (1997) Alliance strategies of small firms. Small Bus Econ 9(1):33–44
Hartley JL, Zinger BJ, Kamath RR (1997) Managing the buyer-supplier interface for on-time performance in product development. J Oper Manag 15(1):57–70
Heath T, Tynan C (2010) Crafting a research proposal. Mark Rev 10(2):147–168
Hoegl M, Wagner SM (2005) Buyer-supplier collaboration in product development projects. J Manag 31(4):530–548

Hossain M (2015) A review of literature on open innovation in small and medium-size enterprises. J Glob Entrep Res 5(6). https://doi.org/10.1186/s40497-015-0022-y
Hossain M, Karaunen I (2016) Open innovation in SMEs: a systematic literature review. J Strateg Manag 9(1):58–73
Inauen M, Schenker-Wicki A (2012) Fostering radical innovations with open innovation. Eur J Innov Manag 15(2):212–231
Johnsen RE, Ford D (2006) Interaction capability development of small suppliers in relationships with larger customers. Ind Mark Manag 35(8):1002–1015
Knudsen (2007) The relative importance of interfirm relationships and knowledge transfer for new product development success. J Prod Innov Manag 24(2):117–138
Koberg CS, Detienne DR, Heppard KA (2003) An empirical test of environmental, organizational and process factors affecting radical innovation. J High Technol Manag Res 14(1):21–45
Lager T (2016) Managing innovation and technology in the process industries: current practices and future perspectives. Proc Eng 138:459–471
Lager T, Frishammar J (2012) Collaborative development of new process technology/equipment in the process industries: in search of enhanced innovation performance. J Bus Chem 9(2):67–84
Lagrosen S (2005) Customer involvement in new product development: a relationship marketing perspective. Eur J Innov Manag 8(4):424–436
Laursen K, Salter A (2006) Open for innovation: the role of openness in explaining innovation performance among U.K. manufacturing firms. Strateg Manag J 27(2):131–150
Lecocq X, Demil B (2006) Strategizing industry structure: the case of open systems in a low-tech industry. Strateg Manag J 27(9):891–899
Lee S, Park G, Yoon B, Park J (2010) Open innovation in SMEs: an intermediated network model. Res Policy 39(2):290–300
Lichtenthaler U (2008) Open innovation in practice: an analysis of strategic approaches to technology transactions. IEEE Trans Eng Manag 55(1):148–157
Madrid-Guijarro A, Garcia D, Van Auken H (2009) Barriers to innovation among Spanish manufacturing SMEs. J Small Bus Manag 47(4):465–488
Malhotra N (2007) Marketing research – an applied orientation. Person Prentice Hall, New Jersey
Miles MB, Huberman AM (1994) Qualitative data analysis: an expanded source book. Sage, Thousand Oaks, CA
Moghaddam P, Tarokh MJ (2012) Customer involvement in innovation process based on open innovation concepts. Int J Res Ind Eng 1(2):1–9
Moreira AC, Karachun HL (2014) Uma revisão interpretativa sobre o desenvolvimento de novos produtos. Cuad Admin 27(49):155–182
Nieto MJ, Santamaria L (2007) The importance of diverse collaborative networks for the novelty of product innovation. Technovation 27(6–7):367–377
Nieto MJ, Santamaria L (2010) Technological collaboration: bridging the innovation gap between small and large firms. J Small Bus Manag 48(1):44–69
Parida V, Westerberg M, Frishammar J (2012) Inbound open innovation activities in high-tech SMES: the impact on innovation performance. J Small Bus Manag 50(2):283–309
Peng DX, Heim GR, Mallick DN (2014) Collaborative product development: project complexity on the use of information technology tools and new product development practices. Prod Oper Manag 23(8):1421–1438
Phillips W, Lamming R, Bessant J, Noke H (2006) Discontinuous innovation and supply relationships: strategic alliances. R&D Manag 36(4):451–461
Powell WW, Koput KW, Smith-Doerr L (1996) Interorganizational collaboration and the locus of innovation: networks of learning in biotechnology. Adm Sci Q 41(1):116–145
Raluca B (2013) Trust, partner selection and innovation outcome in collaborative new product development. Prod Plan Control 24(2–3):145–157
Roy S, Sivakumar K (2010) Innovation generation in upstream and downstream business relationships. J Bus Res 63(12):1356–1363

Schiele H (2010) How to distinguish innovative suppliers? Identifying innovative suppliers as new task for purchasing. Ind Mark Manag 35(8):925–935

Silva LF, Moreira AC (2017) Collaborative new product development and the supplier/client relationship: cases form the furniture industry. In: Garcia Alcaraz JL, Alor-Hernández G, Maldonado Macias AA, Sanchez-Ramírez C (eds) New perspectives on applied industrial tools and techniques. Springer International Publishing, Heidelberg, pp 175–195

Soosay CA, Hyland PW, Ferrer M (2008) Supply chain collaboration: capabilities for continuous innovation. Supply Chain Manag 13(2):160–169

Spithoven A, Vanhaverbeke W, Royjakkers N (2013) Open innovation practices in SMES and large enterprises. Small Bus Econ 41(3):537–562

Stephanie CS, Shulman AD (2011) A comparison of new services versus new product development: configurations of collaborative intensity as predictors of performance. J Prod Innov Manag 28(4):521–535

Takeishi A (2001) Bridging inter-and intra-firm boundaries: management of supplier involvement in automobile product development. Strateg Manag J 22(5):403–433

Tether BS (2002) Who Co-operates for innovation, and why. An empirical analysis. Res Policy 31(6):847–967

Tether BS, Tajar A (2008) Beyond industry-university links: sourcing knowledge for innovation from consultants, private research organizations and public science-base. Res Policy 37(6/7):1079–1095

Theyel N (2012) Extending open innovation throughout the value chain by small and medium-sized manufacturers. Int Small Bus J 31(3):256–274

Tidd J, Bessant J, Pavitt K (2001) Managing innovation: integrating technological, market and organizational change. Wiley, Chichester

Tuli P, Shankar R (2015) Collaborative and lean new product development approach: a case study in the automotive product design. Int J Prod Res 53(8):2457–2471

Un CA, Azakawa K (2015) Types of R&D collaborations and process innovation: the benefit of collaborating upstream in the knowledge chain. J Prod Innov Manag 32(1):138–153

Un CA, Cuervo-Cazurra A, Azakawa K (2010) R&D collaborations and product innovation. J Prod Innov Manag 27(5):673–689

Van de Vrande V, de Jong JPJ, Vanhaverbeke W, de Rochement M (2009) Open innovation in smes: trends, motives and management challenges. Technovation 29(6/7):423–437

Verhees F, Meulenberg M (2004) Market orientation, innovativeness, product innovation, and performance in small firms. J Small Bus Manag 42(2):134–154

Wagner SM, Hoegl M (2006) Involving suppliers in product development: insights from R&D directors and project managers. Ind Mark Manag 35(8):936–943

Wasti NS, Liker JK (1999) Collaborating with suppliers in product development: a U.S. and Japan comparative study. IEEE Trans Eng Manag 46(4):444–461

Wynarczyk P, Piperopoulos P, McAdam M (2013) Open innovation in small and medium-size enterprises. An overview. Int Small Bus J 31(3):240–255

Wynstra F, von Corswant F, Wetzels M (2010) In Chains? An empirical study of antecedents of supplier product development activity in the automotive industry. J Prod Innov Manag 27(5):625–639

Yin RK (2003) Case study research: design and methods. Sage, Thousand Oaks, CA

Luis Filipe Silva is a PhD candidate in marketing and strategy at Aveiro University, Portugal. He obtained his master's degree in management also at Aveiro University. He has a bachelor's degree in management by the University of Beira Interior, Portugal. He is an active trainer and management consultant in management where he has been working with several firms and institutions in strategy, accountancy, and finance. He has been involved in lecturing activities at Coimbra Polytechnic Institute.

António Carrizo Moreira obtained a bachelor's degree in electrical engineering and a master's degree in management, both from the University of Porto, Portugal. He received his PhD in management from UMIST—University of Manchester Institute of Science and Technology, England. He has a solid international background in industry leveraged by working for a multinational company in Germany as well as in Portugal. He has also been involved in consultancy projects and in research activities. He is assistant professor at the Department of Economics, Management, Industrial Engineering, and Tourism, University of Aveiro, Portugal, where he headed the bachelor and master degrees in management for 5 years. He is member of GOVCOPP research unit.

It's Time to Include Suppliers in the Product Innovation Charter (PIC)

Subroto Roy

Abstract The role of supply chain relationships in innovation is being recognized and researched increasingly in recent times. However, the focal buying firm that is trying to innovate for New Product Development (NPD) does not seem to have specific guidelines on how and when to involve suppliers in innovation. The Product Innovation Charter (PIC) is the mission statement of innovation that can offer guidelines to the managers in the buying organization about how and when to involve suppliers in innovation. The chapter explains the PIC and builds the argument that suppliers need to be explicitly mentioned. Such mention should consider the role and capability of new and existing suppliers for innovation that is radical or incremental, early stage versus later stage NPD while defending the intellectual property of the innovating focal organization. Guidelines for mentioning the supplier in the PIC are offered.

1 Introduction

The role of supply chain relationships in innovation is being recognized and researched increasingly in recent times as evidenced by a comprehensive recent review by Zimmermann et al. (2016). Despite the rising scholarly and managerial interest in the role of supply chain relationships in innovation there is a lack of strategic direction in buying organizations with respect to involving suppliers in innovation. Specifically, the mission statement for organizational innovation or Product Innovation Charter (PIC) fails to mention the role of suppliers in innovation.

The Product Innovation Charter (PIC) is a mission statement for innovation and new product/service development originally introduced in the 1980s by C. Merle Crawford (1980). It answers the "who, what, where, when, and why" of the new product development (NPD) project (PDMA glossary). Surprisingly, since the 1980s although the world has changed drastically in terms of globalization and technology and there is

S. Roy (✉)
University of New Haven, West Haven, CT, USA
e-mail: sroy@newhaven.edu

A. C. Moreira et al. (eds.), *Innovation and Supply Chain Management*, Contributions to Management Science, https://doi.org/10.1007/978-3-319-74304-2_6

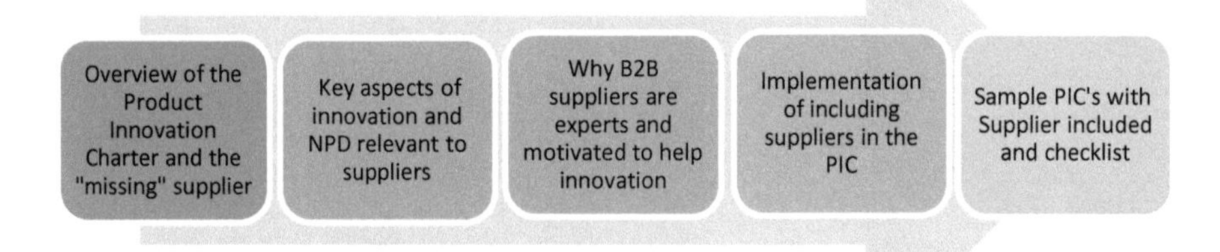

Fig. 1 It's time to include suppliers in the PIC

burgeoning academic research on supply chain and innovation (see Bart 2002 for a review), there is no mention of the supplier in the PIC in the extant literature.

The gap in the literature explicitly mentioning the supplier in the PIC is surprising for at least three reasons. First, compared to 1980s today suppliers are involved in all stages and tasks of NPD. Second the process of NPD via buyer-seller interactions globally has become easy and inexpensive via the Internet. Finally, in the absence of strategic direction on involving suppliers via the PIC, organizations improvise interactions for NPD with suppliers as best as they can with mixed results.

This book chapter advocates the need for PICs to have an explicit mention of involving suppliers in NPD. It specifically lays out aspects on innovation and NPD as they relate supplier involvement.

The rest of the chapter proceeds as follows as depicted in Fig. 1. First, the chapter provides an overview of Product Innovation Charter (PIC) and the "missing" supplier. Second, the chapter identifies key aspects on innovation and NPD that inform supplier involvement viz. new versus existing suppliers, the "S Curves of Innovation" and suppliers, intellectual property and suppliers and startups as suppliers. Third, the chapter discusses why Business-to-Business (B2B) suppliers are experts and are highly motivated for innovation. Fourth, the chapter covers implementation of supplier involvement via PIC. Fifth, the chapter presents sample PICs with supplier included and concludes with a summary checklist of considerations for including suppliers in the PIC.

The following sections expand upon the flow-chart in Fig. 1.

2 Overview of the Product Innovation Charter (PIC)

Introduced by Crawford (1980), "Product Innovation Charter" (PIC) is the mission and vision statement for product innovation for an organization. Just as department or functional mission-vision statements, it must help achieve the company mission and vision, similarly the PIC points to the overarching mission of the company. However, unlike a company mission and vision, the PIC is confidential, and not put on the website, as there are both competitor and intellectual property issues that may arise.

The Product Development Management Association (PDMA) glossary has this definition:

> **Product Innovation Charter (PIC)**: A critical strategic document, the Product Innovation Charter (PIC) is the heart of any organized effort to commercialize a new product. It contains the reasons the project has been started, the goals, objectives, guidelines, and boundaries of the project. It is the "who, what, where, when, and why" of the product development project. In the Discovery phase, the charter may contain assumptions about market preferences, customer needs, and sales and profit potential. As the project enters the Development phase, these assumptions are challenged through prototype development and in-market testing. While business needs and market conditions can and will change as the project progresses, one must resist the strong tendency for projects to wander off as the development work takes place. The PIC must be constantly referenced during the Development phase to make sure it is still valid, that the project is still within the defined arena, and that the opportunity envisioned in the Discovery phase still exists. (from the PDMA Glossary)

In the above definition of PIC is that it is more relaxed in the earlier creative and low cost parts of the innovation process viz. idea generation, concept development and concept testing. It is only at the big money development, prototype, manufacturing marketing and launch that the PIC is used to stay on track.

Here is an example:

> Let us assume that an organization retails its range of food products between $4–6/unit and has the marketing and distribution costs pretty much figured out for line extensions. They have a PIC drawn up for new flavors that specifies a target cost of manufacturing that should not exceed $2/unit. Now if a new flavor costs $2.50 or 25% more to manufacture the PIC should be sending out a red flag and sales projections, marketing messages and alternative supply sources should come under intense scrutiny instead of allowing the project to just float along and disappoint eventually. In other words, the team working on the "new flavor" project should know upfront that if they have a bunch of flavor ideas they need to keep the manufacturing cost under $2. Let us say the full focus of the supply chain folks helps to bring manufacturing costs down to $2.20 and the Market Insight folks are able to re-confirm the sales volume projections—guess what—it's OK to proceed! You at least know where you are going and post launch sales efforts at the retail end might just up those sales numbers, making the manufacturing cost affordable. In other words the PIC helps you to know where you want to go and serves as a road map to deploy the organizations efforts effectively. (Source: www.StratoServe.com)[1]

[1]www.StratoServe.com is the author's blog and website since 2006 and Supply Chain and Innovation are important topics covered in the blog that enjoy popularity with a global audience. Forbes and Harvard Business Review, among others cite www.StratoServe.com. It is inaugural winner of "Most Valuable Blogger Award" by CBS Television, Connecticut, USA. Several references are made in this chapter to content on www.StratoServe.com

2.1 Contents of the Product Innovation Charter (PIC)

The innovation literature realized early on that given the inter-disciplinary nature of innovation some kind of strategy guidelines were necessary. While organizations did tend to have a NPD strategy, this tended to be in-formal and ad-hoc and "back of the envelope". Accordingly, Crawford (1980) coined the term "Product Innovation Charter" (PIC) that allowed NPD teams from a variety of company functions like Production, Marketing, Finance, and R&D to stay on track with NPD, consistent with company goals. The use of the word "Charter" denotes an emphasis on both direction and activity of the company innovation process (Page 4, Crawford 1980).

Since the introduction of PIC as a strategy guideline for all innovation projects within an organization a variety of scholars have referenced the term and explored questions such as content and impact of PIC's (Bart 2002) and content, specificity and impact (Bart and Pujari 2007). The PIC contents can be schematically seen in Fig. 2. The *background* includes a situation analysis of the company including its strengths and weaknesses and the opportunities and threats, managerial mandates or dicta including expectations of the shareholders and stock market and reasons for preparing the PIC. The *focus* of the PIC includes at least one technology dimension and one market dimension. The *goals/objectives* of the PIC outlines what the project hopes to achieve in the short and long run. Finally, it has a *guidelines* section, that provides rules of the road that has cost/quality guidelines, innovativeness etc. that has senior management intent in operationalizing the Product Innovation Charter (PIC).

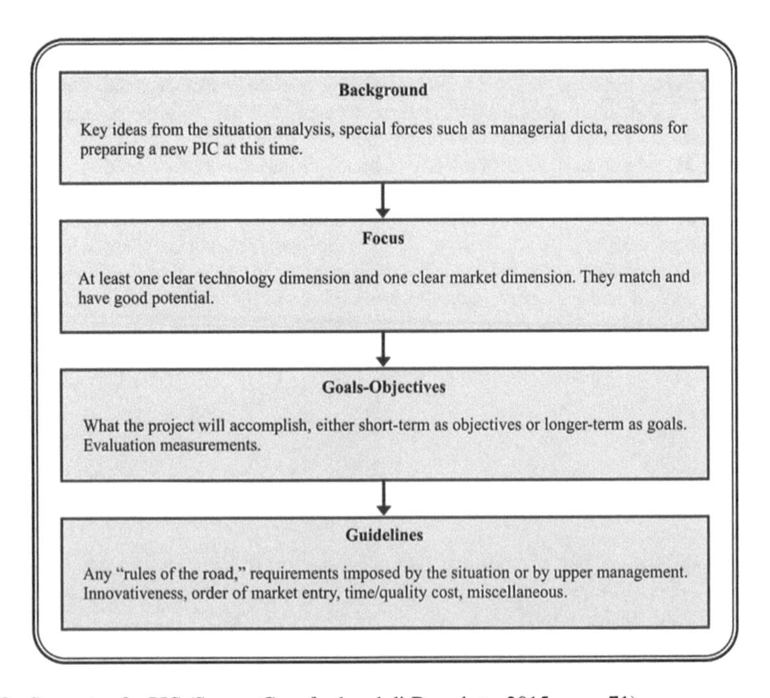

Fig. 2 Contents of a PIC (Source Crawford and di Benedetto 2015, page 71)

It is in the *guidelines* section that suppliers need to be mentioned explicitly, a case for which is developed in this chapter. These include considerations for new versus existing suppliers, the "S Curves of Innovation" and suppliers, intellectual property and suppliers and startups as suppliers,

2.2 The "Missing" Supplier in the Product Innovation Charter

In research relating to supplier in the PIC, I could find only two "footnotes" mentioning the supplier. These supplier references include 14.1% respondents of a PIC content survey respondents who mention "concern for suppliers" among other variables like "concern for society" (Bart 2002). This "concern" is categorized with concern for employees, shareholders and public image as a Corporate Social Responsibility (CSR) type component (Bart and Pujari 2007). The supplier is completely missing in Durmuşoğlu et al. (2008). *In other words, the PIC being the organization's strategy document for innovation is silent on the role of the supplier.*

The expertise and potential knowledge contribution of suppliers is completely ignored in the literature on the PIC.

Bart (2002) illustrates a sample PIC (desirable) and the actual PIC's and these can be seen in Fig. 3a, b.

3 Key Aspects of Innovation and NPD Relevant to Suppliers

3.1 Existing Versus New Suppliers

Existing suppliers or the so-called "supply base" are vendors who are currently supplying or have supplied in the past to the focal company. There is some history of performance and capabilities and there are existing relationships between personnel between the buying and selling company. The existing supplier is at the core of the vast buyer-seller relationship literature involving trust and commitment (Morgan and Hunt 1994). Both buyer and seller are committed to the relationship in the positive sense. However, such a relationship was formed in the first place due to specific knowledge, resources and skill sets of the supplier following the resource based theory (Verwaal 2017; Wernerfelt 1984). The best suppliers can help with incremental innovation and will be more reliable with protecting intellectual property.

On the other hand, "new suppliers" have not done business with the focal company and lack the advantage of being a "known" entity. They are however very easy to find today because of Google and the Internet. There is no internal history of performance and quality. The advantages of new suppliers are that they can bring in new knowledge, skills and resources to the relationship. Thus they will

a

Corning Glass Works develops pyroceramin in the 1950s and up front homework looks into market opportunities. After winnowing these down, a kitchenware team is given the following charter.

Product innovation charter

Background

Women are entering the workforce in greater numbers, and are very stressed for time at home. The advent of frozen foods and other conveniences are changing the way people prepare and serve food. There is an opportunity for an attractive vessel that can go from the freezer to the stove top to the dinner table.

Focus

(a) Technology: Utilize the unique thermal properties of pyroceram. Use current glass product manufacturing technologies.
(b) Markets: Home makers in specified income bracket. Must also appeal to large retailers that will be used as channels. Benefit segment is characterized by those who value practical convenience and affordability, without compromising table appearance.

Goals and objectives

Cookware must be attractive and affordable. We intend to build a long-term market, so the sales objectives (specify) in early years will outweigh near term ROI on this launch. We should seek to launch, or be ready to follow up with, an entire line of cookware.

Guidelies

Use current distribution channels. The cookware should fit seamlessly into the kitchen environment (freezer space, stovetop limitation, cleaning, service at the table). Expect to incur large advertising expenditures (specify) to build awareness.

b

Black & Decker US power tools

Raise customer expectations and industry standards with respect to speed (reduced cycle time), improved quality and reduced costs ($/day). Integrate plans and resources. Improve the Product Development process. Maximize B&D preference time. Understand end user requirements and expectations. Personnel development.

3M

30% of products must be new within the last 4 years.

Bausch & Lomb global eyewear

To assure the timely and successful introduction of new products meeting and exceeding marketplace VOC (voice of the consumer) requirements.

Partial PIC for NewProd Corporation (disguised at the company's request)

The NewProd Corporation is dedicated to a program of new product development in metal fabricated sports equipment for the high-performance skiing, tennis and golfing markets. Our goals are to become the world market leader in all of our product categories (as measured by units sold); to maintain and build our reputation for outstanding quality and uniqueness; and to earn at least a 50% ROI from all new product activities. We will develop new products with the aim of being first to market and with superior offerings. R&D will be given resources commensurate with the projects approved and every effort will be expended to provide an environment in which talented scientists and engineers can feel appreciated, respected and rewarded for their contribution to company's new product performance success. With our new products, we seek to be the envy of our competitors, the delight of our customers and the pride of our employees.

Fig. 3 Samples of Product Innovation Charter. Source: Bart (2002, p. 24)

be more likely as a source for radical innovation, more risky for intellectual property unless counter balanced by the desire to build a long term relationship.

The following subsections explain in more detail the specifics of involving new and existing supplies in the NPD process in the digital age.

Early supplier involvement (ESI) for NPD success is a stream of literature (e.g. Sjoerdsma and van Weele 2015) that was motivated by the success of Japanese Auto Industry innovations (Clark and Fujimoto 1991) and has grown since then. Most research has focused on the buyers' perspective with exceptions that take the seller's perspective (e.g. Yeniyurt et al. 2014).

The perspective of the buyer in supplier NPD involvement literature is natural as suppliers are generally considered highly motivated to supply and provide value in business markets (Chesbrough and Rosenbloom 2002) while caution and risk averse behavior is the hallmark of buyers within the procurement function (Kraljic 1983; Quinn and Hilmer 1994). For example, every website attempts to market its products and services while only some large corporations have "supplier portals" to welcome suppliers. Similarly the marketing literature is all about getting orders with concepts such as buying center while the supply chain literature cautions the buyer when dealing with suppliers and particularly new suppliers.

Given that successful NPD is a risky endeavor, approaches to NPD such as the stage gate process (Cooper 2008) are extensively taught and practiced. A large literature on NPD teams attempts to help manage teams that have different functional reporting within the organization and are globally dispersed (e.g. McDonough et al. 2001). Without clear strategic direction via the PIC to involve suppliers in the NPD process, managers have no guidance as to whether and how to involve suppliers.

Further considering that supplier motivation is generally high and information on supplier capabilities have become far more available online, an explicit direction to involve suppliers will be beneficial for NPD.

3.2 "S" Curves of Incremental and Radical Innovation

The "S Curve" innovation thinking is attributed to Richard Foster (1986) and made famous by Clayton Christensen in the book on the Innovator's dilemma (Christensen 2013) where he discusses how each successive computer hard drive industry got wiped out.

Think of each "S" (see Fig. 4) curve as a technology platform. Movement up an "S" curve is incremental innovation while stepping down on a lower new "S" curve now, can lead to radical innovation, as the new "S" curve surpasses your existing "S" curve. The music industry, following some of the timeline of audio formats, is a great example.

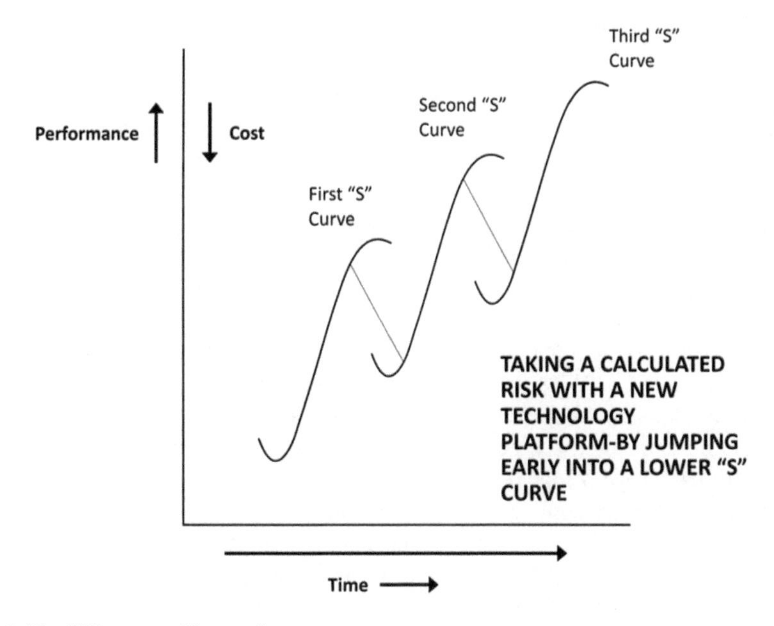

Fig. 4 The "S" curves of innovation

You had workers who specialized on manufacturing cassette tapes, there were specialized suppliers and of course the Sony Walkman that made music cassettes so special. Cassettes came in 60 min and then 90 min formats. Avid listeners (the final consumers) tried getting the 90-min cassette that must have involved a lot of incremental innovation by suppliers and personnel in the plastic music cassette industry. You can visualize six-sigma and total quality programs at cassette factories that reduced waste and defects in the product.

Suddenly you had music on CDs that improved quality a whole lot and "Sony Discman" became popular as the cassette industry started dying, just as vinyl records had died before that. The CD industry had its own players and supply chain.

Next off course you have the MP3 player, iPod and literally thousands of songs on your device and then the iTunes store on the cloud. The MP3 players and cloud also require a new set of employee skills and a differently skilled supply base.

If you think about each industry, it ignored the march of technology and refused to get started on the next technology "S" curve from the current technology "S" curve. This reluctance was because at the early stages, each new "S" curve looked unattractive from the existing "S" curve.

You see that the dominant players in each technology type became extinct just because they thought that the upcoming technology was too much behind—*and will never catch up*. By the time the new technology (second and third curves) became really comparable in performance and cost—the incumbents of older "S" curves were too far behind.

3.3 Suppliers and "S" Curves

The tragedy in Christensen's account or the numerous cases of industries in decline (e.g. brick retail) is that e-commerce related suppliers did reach out to try and move the focal firm to the next phase of the market. Thus we can expect that various vendors went to pitch to brick retailers like Circuit City and Radio Shack that have closed and Sears and Walmart who are in trouble, soon after Amazon was becoming a success. However, without a clear management directive to stay ahead of market trends, executives at these companies did not put e-commerce on the top of their priorities. In other words, the next "S" curve looked too difficult and too inefficient at the time. As a result, companies that appeared unshakable have disappeared while legends like Sears and Walmart are considered in trouble.

The technology space itself is not immune to the "S" curves of innovation (see Sood 2017). Large enterprise solutions like SAP and Oracle are facing competition from start-ups. The old model (traditional "S" curve) involved large expenses in setting up an enterprise application like SAP with enormous investments in customization and training. Today, startups specializing in narrow niches are able to integrate their solutions to the backbone of large enterprise software. This makes weak solutions of large enterprise software less relevant and these large companies are struggling to become more agile. A major factor in entrepreneurs being able to come up with specialized, more user friendly applications is the availability of "cloud" computing that is both affordable and reliable from services like Amazon web services (AWS).

3.4 *Startups and Acquisitions as Resources for the Next "S" Curve of Radical Innovation*

Since radical innovation in large organizations is so difficult (McDermott and O'Connor 2002) organizations like IDEO (Thomke and Nimgade 2000) offer consulting for design of new products to large organizations.

Given the growth in startups (Say 2016) enormous opportunities exist for larger organizations to collaborate with startups that have the agility to innovate much faster. However, such collaboration remains sporadic depending on the sales efforts of the startups and the receptivity of larger organizations given the current ambiguity in the Product Innovation Charter. Typically, integrating a start-ups offering as a purchased item can involve disruption in the routine and a temporary lower "S" curve for the buying firm leading to the "innovator's dilemma" in a world of rapid growth of startups and disruptive innovation.

Without the ability to try out an innovation from start-ups at relatively lower levels of investment and risk, companies tend to go into acquisition straight away. Acquisitions can be far more difficult to manage (Zollo and Singh 2004) than contractual buyer seller relationships (Dyer and Singh 1998). I argue that the inability of organizations to tap into the creativity of startups without the "wait and see" approach of acquisitions is a result of no clear strategy guidelines with respect to involvement of suppliers in innovation in a Product Innovation Charter.

3.5 *Early Stage Versus Later Stage of NPD and Finding Suppliers Online*

The stage-gate model posits that the NPD process has several interrelated stages (e.g., Cooper 1979; Cooper and Kleinschmidt 1987). To achieve parsimony, I use the terminology of early stage of NPD that includes idea generation, concept development and concept testing and later stage NPD that includes prototype development, production, product testing, market testing and launch. Intuitively early stage NPD can be visualized in the hiring of a design firm such as IDEO (Thomke and Nimgade 2000) or the contracting of an advertising agency for developing a new marketing campaign. Periodically, advertising agencies are fired (Davies and Prince 2010; Kulkarni et al. 2003) because the outsourcing firm and the old advertising agency are no longer able to come up with new ideas critical to the initiation stage of NPD. Similarly, at the implementation stage of NPD, organizations in the global software development business have realized the importance of having onshore teams that interface with the client and offshore teams to ensure that implementation is exactly as the client requires (Rai et al. 2009).

The early stages of NPD involve idea generation and conceptualization including drawing up of designs and drawings. With the long tail of the Internet (Anderson 2004; Brynjolfsson et al. 2011) searching for potential suppliers for any skills the internal NPD team needs is possible. Thus for design services for a particular machine part idea can be searched online and several resources would appear including companies that are fairly well developed, university researchers who already work in the particular domain, market research companies that specialize in concept testing (e.g. ACNeilsen Bases) and freelancers who would be willing to join the NPD team. Beyond locating potential *new* suppliers globally, the early stage NPD skill suppliers would also help locate online feedback based on reviews of the supplier. Thus, some amount of assurance of quality is frequently available through reviews online. Most suppliers would be willing to work on a pilot basis till results are seen.

Similarly, it is easy today to find *new* suppliers online for the later stage of NPD. In fact, aggregators of manufacturers are available at Alibaba.com while for example if you provided drawing and specifications a supplier can be found easily on Alibaba.com including performance guarantee by Alibaba.com. Similarly, for a variety of digital tasks including software development, crowdsourced resources are available on portals such as Amazon Mechanical Turk, Fiverr, Upwork etc. Surprisingly, it is not only small businesses with lower resources that use these services but also some big brands are listed among the clients of such services. For example, Upwork lists Airbnb and Dropbox among its users (Accessed UpWork.com August 31, 2017).

3.6 Intellectual Property Concerns in the Early Stage Versus Late Stage NPD

Protection of intellectual property is a concern in NPD (Roy and Sivakumar 2011). This could happen as the same supplier would be supplying to other competitors. Thus a software supplier for managing clinical trials at one pharmaceutical company would gather knowledge about a new drug being tested and intentionally or unintentionally share information with a competing pharmaceutical firm.

However, intellectual property is more of a concern in the early stages of NPD that is more amenable to patent protection. For this reason, in pharmaceutical research, compounds are sourced without the supplier being made aware of the pharmaceutical product being developed. In fact, pharmaceutical sourcing departments have a protocol of not allowing the supplier's scientists to meet the concerned scientists at the pharmaceutical firm at any time so that inadvertent intellectual property leaks may not occur (Roy and Sivakumar 2011).

In the later stages of NPD, counterfeiting is possible and occurs (Minagawa et al. 2007) primarily as suppliers either leak the manufactured product through the gray market or allow other manufacturers access to designs and tools so that the branded product is now sold at lower cost than the genuine brand. Surprisingly the sellers of counterfeit luxury goods are self-declared as "counterfeit" that is acceptable in foreign markets (Ahuvia et al. 2013).

3.6.1 Intellectual Property and Early Stage NPD

Intellectual property concerns are important in outsourcing innovation (Roy and Sivakumar 2011). Generally, early stage NPD i.e. idea generation, concept development and concept testing are most sensitive to intellectual property theft. It is critical that the outsourcing firm and supplier have an NDA i.e. Non-Disclosure Agreement. Particular attention must be paid upfront to the legal environment of the country of the supplier (Pai and Basu 2007). "Trust but verify" should be the watchword at this stage of NPD.

3.6.2 Intellectual Property and Later Stage NPD

The later stages of NPD generally involves organizations that area specialized and do work for multiple upstream businesses. Here intellectual property is less of a concern compared to leakage of plans and progress to competitors. Appropriate safeguards for "self -seeking with guile" (Morgan and Hunt 1994) must be discussed and implemented with close monitoring by the focal firm.

4 Why B2B Suppliers Are Experts and Highly Motivated

In Business-to-Business markets, suppliers are experts (Melander et al. 2014) and respond to Request for Proposals (RFP's) based on their expertise. Once suppliers develop expertise in a particular domain, they seek to expand growth by finding new applications for their product or expertise. Examples include Baking Soda applications for refrigerator de-odorizing and in laundry as whitener and 3M's post application in painter's masking tape. While these examples are visible to consumers, there are numerous industrial products in the upstream supply chain like zinc oxide that can be used with appropriate refining in rubber products, pharmaceuticals etc.

It may seem obvious that suppliers are naturally motivated because they want to sell their expertise. Thus, if either the buyer wanted to make changes in an upstream input from a supplier, the supplier would be willing to make changes for the purposes of developing a new market. An indirect measure of the motivation of sellers in the number of jobs in B2B sales in the world and compare it to the number of jobs in purchasing or innovation. In addition, the head of marketing and sales is directly responsible to the CEO if the CEO does not herself/himself directly manage marketing. The reason might seem obvious, i.e. sales and marketing bring in the money that keeps the organization running. A famous quote from Peter Drucker (as quoted by Jack Trout in Forbes 2006) illustrates the above.

> Because the purpose of business is to create a customer, the business enterprise has two–and only two–basic functions: marketing and innovation. Marketing and innovation produce results; all the rest are costs. Marketing is the distinguishing, unique function of the business.
> Peter Drucker

Apart from the motivation of bringing in money to the organization, there are frequently incentives and commissions for sales people that have spawned a large literature on sales force motivation and compensation (e.g. Franke and Park 2006).

Paradoxically, organizations that give high priority to marketing and sales appear highly closed to new suppliers. Purchasing or supply managers actively avoid sales calls even when there might be something innovative that the supplier might offer. The purpose of this chapter is to thus enshrine the supplier in the PIC so as to encourage active supplier involvement.

4.1 *Suppliers at the Input Boundary of the Organization*

If we think about the organization in "Input-Process-Output" model terms then the supplier is at one end of the value chain as in Fig. 5.

The input-process-output is a way of looking at the firm's value chain. Supply managers handle the input coming in and the marketing folks handle the output coming out.

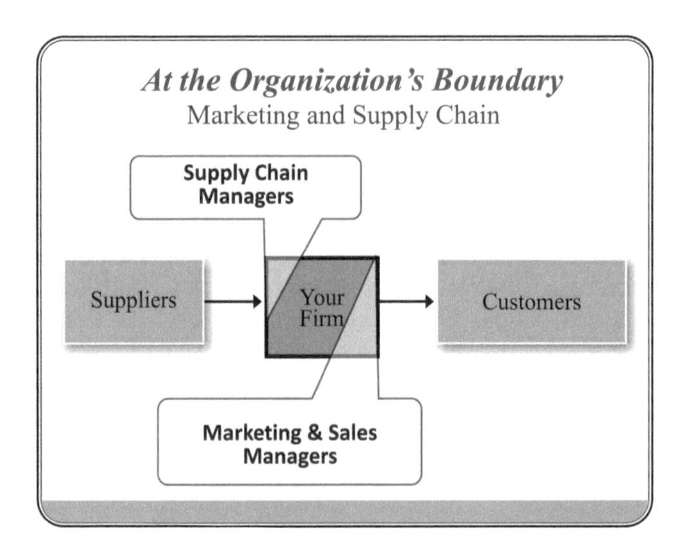

Fig. 5 Boundaries of the organization

Like:

global suppliers -> supply chain -> [firm-] -> marketing -> global customers

Supply and marketing folks are people who sit at either end of the firm and look at the world outside in the firm's value chain. The marketing manager reaches out to customers and supply chain managers reaches out to suppliers and also need to reach "in" to internal firm users. Both off course do not formally talk to one another—organizations are not set up to encourage the talking. Unless there are operations review meetings, that can become mindless and boring. ERP systems can help but ERP speaks to only data and not the gut feel of these important folks.

If you look at the firm as just a processor—focused on creating superior value for its customers—you start realizing how much firms miss out in tapping on to the combined knowledge and expertise of their input and output side teams.

Based on my involvement with professional marketers (American Marketing Association Professional Chapter) and supply/purchasing managers (Institute for Supply Management) over several years here are five observed differences between professionals in each group (From StratoServe blog)

1. Marketing managers being outward facing are constantly looking for opportunities sometimes without regard to what their organization can really do. Supply chain managers first look inside before looking outside at suppliers—to ensure a good fit.
2. Marketing and sales managers are more social compared to supply chain managers who are more conservative. The former does the chasing of prospects while the latter need to stave off marketing people who are the firm's upstream suppliers.
3. Marketing and sales people are measured by sales (volume and price) while supply managers are assessed first on availability of goods and services required by the firm, then on cost.
4. Marketing managers work on an open canvas of the market and prospective customers and use techniques like segmentation, targeting, market research and the 4 Ps. Supply Chain managers also have an open canvas of suppliers but they need to make supplies work in their firms value add process- any snags and they are in the direct line of fire! Supply managers are therefore much more risk averse and deliberative.
5. On a more fun and more social note you will find many more marketing groups on the web and off it than supply chain managers. And when it comes to professional meetings marketing people may have extremes of no food to open bar at a high price ticket. Supply chain meetings will stay steady with modest burgers, pizza, strict cash bars, and a predictable member fee.

Since supply managers and marketing and sales managers are so different in their orientation, it is critical for the focal organization to give clear guidance to its supply managers. This guidance is in the Product Innovation Charter as proposed in this chapter.

4.2 Internet Search and Changes in the Input-Process-Output Model

Before the Internet, it was difficult to locate a supplier globally. Traditional approaches included contacts at international trade shows and conferences.

Today due to the massive and instant search capability on the Internet (e.g. Google) the searching ability of all members of the organization has enhanced tremendously. If you have a problem or want to research something, you can simply "Google" it and are likely to find results (including videos) that speak to your problem. And this is early days for the Internet, and things are likely to get much better.

Highly specialized functions can do their own searches online. Thus a highly specialized pharmaceutical research scientist can identify leading thinkers based on specialized conferences and Google Scholar. They are also able to identify suppliers and collaborators that might be able to provide inputs and support for a new product being developed by the research scientists' team. Here, pharmaceutical companies working on drug discovery encourage scientists to work with the supply department to communicate with the supplier. This way direct communication between a scientist and supplier are reduced, thus reducing the inadvertent loss of Intellectual Property.

The search functions for supply/procurement managers tend to be primarily in indirect spend items (Cox et al. 2005) like stationery, travel etc.

Thus, when innovation teams within organizations need ideas or resources, they tend to find their own sources and involve the supply function to place formal orders.

However, due to a lack of clear direction to all employees to seek ideas and inputs that can help them innovate, there is a great deal of variability among employees who actively try to integrate external resources (i.e. suppliers) to enhance the speed and impact of innovative efforts. This chapter advocates such a clear mandate by including suppliers in the Product Innovation Charter.

5 Discussion and Tips on Execution Issues of Supplier in PIC

Once enshrined in the PIC, suppliers need to be managed for innovation just as innovation teams within the organization. For example, in pharmaceutical NPD the expert (e.g. scientist in pharmaceutical research) working closely with an officer specialized in purchasing in the supply department. This way the intellectual property risks of using suppliers can be mitigated. Supply professionals have the skills and resources to put in contractual and behavioral safeguards in working with suppliers, particularly global suppliers.

A supplier portal on the website can be a useful mechanism to enlist suppliers with various capabilities and resources that are useful to the focal firm's innovation efforts.

Finally, a formal process of review of progress and supplier performance needs to be put in place so that the contribution of the supplier can be tracked and improved upon based on the feedback from internal personnel.

It is also critical (Yan and Kull 2015) that interaction and communication is kept up on a regular basis. Such interactions help in keeping channels of communication with suppliers open.

Figure 6 presents in red how the supplier might be included in the PIC depicted in Fig. 3. Following the above sections, if suppliers were to be included in the Bart (2002, p. 24) samples of PIC's they would have content as mentioned in red.

a

Corning Glass Works develops pyroceram in the 1950s and up front homework looks into market opportunities. After winnowing these down, a kitchenware team is given the following charter.

Product innovation charter

Background

Women are entering the workforce in greater numbers, and are very stressed for time at home. The advent of frozen foods and other conveniences are changing the way people prepare and serve food. There is an opportunity for an attractive vessel that can go from the freezer to the stove top to the dinner table.

Focus

(a) Technology: Utilize the unique thermal properties of pyroceram. Use current glass product manufacturing technologies. We will look for external suppliers who can help with extending the capabilities of pyroceram and newer materials that help us with our market goals. Our R&D will work closely with our supply department and external suppliers to bring in materials that can reach our market goals outlined below. Appropriate intellectual property and patent safeguards for Corning must be in place in co-ordination with legal and supply management.

(b) Market: Home makers in specified income bracket. Must also appeal to large retailer; that will be used as channels. Benefit segment is characterized by those who value practical convenience and affordability, without compromising table appearance.

Goals and objectives

Cookware must be attractive and affordable. We intend to build a long-term market, so the sales objectives (specify) in early years will outweigh near term ROI on the launch. We should seek launch, or be ready to follow up with, an entire line of cookware.

Guidelines

Use current distribution channel. The cookware should fit seamlessly into the kitchen environment (freezer space, stovetop limitation, cleaning, service at the table). Expect to incur large advertising expenditures (specify) to build awareness.

b

Black & Decker US power tools

Raise customer expectations and industry standards with respect to speed (reduced cycle time), improved quality and reduced costs ($/day). Integrate plans and resources. Improve the Product Development process. Maximize B&D preference time. Understand end user requirements and expectations. Personnel development.

3M

30% of products must be new within the last 4 years.

Bausch &Lomb global eyewear

To assure the timely and successful introduction of new products meeting and exceeding marketplace VOC (voice of the consumer) requirements.

Partial 1MC for NewProd Corporation (disguised at the company's request)

The NewProd Corporation is dedicated to a program of new product development in metal fabricated sports equipment for the high-performance skiing, tennis and golfing markets.

We will look for suppliers in performance plastics, other composite materials in industries far and wide including aerospace. Looking for knowledge and resources outside our firm for the goals below will be a priority. Our innovation team and Supply team will work closely together to develop these new and innovative suppliers that can start with pilot projects. Appropriate intellectual property and patent safeguards for Corning must be in place in co-ordination with legal and supply management.

Our goals are to become the world market leader in all of our product categories (as measured by units sold): to maintain and build our reputation for outstanding quality and uniqueness: and to earn at least a 50% ROI from all new product activities. We will develop new products with the aim of being first to market and with superior offerings.. R&D will be given resources commensurate with the projects approved and every effort will be expended to provide an environment in which talented scientists and engineers can feel appreciated, respected and rewarded for their contributions to the company's new product performance success. With our new products, we seek to be the envy of our competitors, the delight of our customers and the pride of our employees.

Fig. 6 Supplier included PIC sample

5.1 *Guidelines for Interaction with Suppliers Detailed in the PIC*

Today global interactions for innovation with suppliers (Roy et al. 2004) have become easier with free social media (e.g. WhattsApp) and mobile based project management tools (e.g. Trello) instantly. The focal organization must take the initiative to keep interacting with the supplier on a regular basis. Some concluding tips for framing supplier relationships in innovation are as follows.

1. Suppliers are motivated globally to be involved in new product development and innovation.
2. Existing suppliers come up with new ideas (Christensen) but because all systems are working well, organizations are reluctant to try something new because of lower “S” curve cost/efficiency considerations. The product innovation charter should encourage the buying firm managers to be welcoming of new ideas from existing suppliers. These suppliers are often hesitant to offer new idea because of the reluctance of buying managers to work off a lower “S” curve.
3. For radical innovation new suppliers like startups tend to have high motivation and energy to make a success of innovation at the buying firm.
4. Acquisitions of companies (e.g. Biotech firms) at the radical early stage of NPD can be a pathway for growth of established firms (e.g. pharmaceuticals). Here a supply relationship is transformed to a part of the company in the ownership sense.
5. Particular care is needed for protecting intellectual property at the early stage of innovation from potential leakage via suppliers to potential competitors.

In summary, by explicitly including the supplier in the PIC managements can leverage the huge global resources, skills and motivation of suppliers for innovation that has become possible today.

References

Ahuvia A et al (2013) What is the harm in fake luxury brands? Moving beyond the conventional wisdom. Luxury Mark. Gabler Verlag, pp 279–293

Anderson C (2004) The long tail. Wired Magazine, October 2004

Bart CK (2002) Product innovation charters: mission statements for new products. R&D Manag 32(1):23–34

Bart C, Pujari A (2007) The performance impact of content and process in product innovation charters. J Prod Innov Manag 24(1):3–19

Brynjolfsson E, Hu Y, Simester D (2011) Goodbye pareto principle, hello long tail: the effect of search costs on the concentration of product sales. Manag Sci 57(8):1373–1386

Chesbrough H, Rosenbloom RS (2002) The role of the business model in capturing value from innovation: evidence from Xerox Corporation’s technology spin-off companies. Ind Corp Chang 11(3):529–555

Christensen CM (2013) The innovator’s dilemma: when new technologies cause great firms to fail. Harvard Business Review Press, Boston, MA

Clark KB, Fujimoto T (1991) Product development performance: strategy, organization, and management in the world auto industry. Harvard Business Press, Boston, MA
Cooper RG (1979) The dimensions of industrial new product success and failure. J Mark 43(3):93–103
Cooper RG (2008) Perspective: the stage-gate® idea-to-launch process—update, what's new, and nexgen systems. J Prod Innov Manag 25(3):213–232
Cooper RG, Kleinschmidt EJ (1987) New products: what separates winners from losers? J Prod Innov Manag 4(3):169–184
Cox A, Chicksand D, Ireland P, Davies T (2005) Sourcing indirect spend: a survey of current internal and external strategies for non-revenue-generating goods and services. J Supply Chain Manag 41(2):39–51
Crawford CM (1980) Defining the charter for product innovation. Sloan Manag Rev 22(1):3–12
Crawford CM, di Benedetto A (2015) New products management. McGaw-Hill Education, New York
Davies M, Prince M (2010) Advertising agency compensation, client evaluation and switching costs: an extension of agency theory. J Curr Issues Res Advert 32(1):13–31
Durmuşoğlu SS, McNally RC, Calantone RJ, Harmancioglu N (2008) How elephants learn the new dance when headquarters changes the music: three case studies on innovation strategy change. J Prod Innov Manag 25(4):386–403
Dyer JH, Singh H (1998) The relational view: cooperative strategy and sources of interorganizational competitive advantage. Acad Manage Rev 23(4):660–679
Foster RN (1986) Working the S-curve: assessing technological threats. Res Manag 29(4):17–20
Franke GR, Park J-E (2006) Salesperson adaptive selling behavior and customer orientation: a meta-analysis. J Mark Res 43(4):693–702
Kraljic P (1983) Purchasing must become supply management. Harv Bus Rev 61(5):109–117
Kulkarni MS, Vora PP, Brown TA (2003) Firing advertising agencies-possible reasons and managerial implications. J Advert 32(3):77–86
McDermott CM, O'Connor GC (2002) Managing radical innovation: an overview of emergent strategy issues. J Prod Innov Manag 19(6):424–438
McDonough EF, Kahn KB, Barczak G (2001) An investigation of the use of global, virtual, and colocated new product development teams. J Prod Innov Manag 18(2):110–120
Melander L, Rosell D, Lakemond N (2014) In pursuit of control: involving suppliers of critical technologies in new product development. Supply Chain Manag Int J 19(5/6):722–732
Minagawa T, Trott P, Hoecht A (2007) Counterfeit, imitation, reverse engineering and learning: reflections from Chinese manufacturing firms. R&D Manag 37(5):455–467
Morgan RM, Hunt SD (1994) The commitment-trust theory of relationship marketing. J Mark 58(3):20–38
Pai AK, Basu S (2007) Offshore technology outsourcing: overview of management and legal issues. Bus Process Manag J 13(1):21–46
Quinn JB, Hilmer FG (1994) Strategic outsourcing. Sloan Manag Rev 35(4):43–55
Rai A, Maruping LM, Venkatesh V (2009) Offshore information systems project success: the role of social embeddedness and cultural characteristics. MIS Q 33(3):617–641
Roy S, Sivakumar K (2011) Managing intellectual property in global outsourcing for innovation generation. J Prod Innov Manag 28(1):48–62
Roy S, Sivakumar K, Wilkinson IF (2004) Innovation generation in supply chain relationships: a conceptual model and research propositions. Acad Mark Sci J 32(1):61–79
Say M (2016) The state of the Startup Accelerator Industry. Forbes, June 29. https://www.forbes.com/sites/groupthink/2016/06/29/the-state-of-the-startup-accelerator-industry/#3aacd38c7b44
Sjoerdsma M, van Weele AJ (2015) Managing supplier relationships in a new product development context. J Purch Supply Manag 21(3):192–203
Sood R (2017) Will consumer tech companies rule the enterprise world? YourStory.com, August 2017 https://yourstory.com/2017/08/will-consumer-tech-rule-the-enterprise/
Thomke S, Nimgade A (2000) IDEO product development. Harvard Business School, Cambridge, MA

Trout J (2006) Peter Drucker on Marketing. Forbes, July 3, 2006. https://www.forbes.com/2006/06/30/jack-trout-on-marketing-cx_jt_0703drucker.html
Verwaal E (2017) Global outsourcing, explorative innovation and firm financial performance: a knowledge-exchange based perspective. J World Bus 52(1):17–27
Wernerfelt B (1984) A resource-based view of the firm. Strateg Manag J 5(2):171–180
Yan T, Kull TJ (2015) Supplier opportunism in buyer–supplier new product development: a China-US study of antecedents, consequences, and cultural/institutional contexts. Decis Sci 46(2):403–445
Yeniyurt S, Henke JW, Yalcinkaya G (2014) A longitudinal analysis of supplier involvement in buyers' new product development: working relations, inter-dependence, co-innovation, and performance outcomes. J Acad Mark Sci 42(3):291–308
Zimmermann R, Ferreira LMDF, Moreira AC (2016) The influence of supply chain on the innovation process: a systematic literature review. Supply Chain Manag Int J 21(3):289–304
Zollo M, Singh H (2004) Deliberate learning in corporate acquisitions: post-acquisition strategies and integration capability in US bank mergers. Strateg Manag J 25(13):1233–1256

Subroto Roy is a Professor of Marketing at the University of New Haven, Connecticut, USA. He was a visiting scholar at the Yale University School of Management where he continues his research. Prior to academe and PhD from Australia, Dr. Roy was the head of Marketing and Sales at the Indian joint venture of Tetra Pak, Sweden, where he was lead/member for numerous innovation/NPD projects. He has published in journals such as *Journal of Academy of Marketing Science*, *Journal of Business Research*, *Marketing Science*, *European Journal of Innovation Management*, and *Journal of Product Innovation Management.* Dr. Roy is on the editorial board of four journals, and his web and blog is www.StratoServe.com, and he can be followed on Twitter @StratoServe.

Mission Impossible: How to Make Early Supplier Involvement Work in New Product Development?

Arjan J. van Weele

Abstract Innovation is paramount to survive in today's rapidly changing world. Developing and marketing new technologies, solutions and products successfully requires the concerted actions of many stakeholders in global value chains. Large companies have embraced the idea of open innovation. They realize that in order to speed up development and reduce risk, they need to collaborate with supply and knowledge partners. However, mobilizing partner specialist knowledge seems problematic. Academic research demonstrates contrasting results. In some cases, supplier collaboration in new product development i.e. early supplier involvement may create large benefits. In other cases, it may lead to detrimental and even devastating results. This chapter discusses why these contrasting results are found. It draws on over 30 years of academic research, that was conducted and/or supervised by the author. The chapter concludes that, as the drivers and enablers of early supplier involvement today are clear, fostering effective human interaction aimed at sensitive knowledge and information exchange on behalf of organizations with conflicting interests is crucial in early supplier involvement. As the human factor in technology driven organizations is often undervalued, more research is needed to understand how to mobilize interorganizational knowledge sharing in such exchanges.

1 Introduction

In order to survive in today's rapidly changing global economies, companies need to innovate. Products, processes and business models need to be adapted continuously to meet the ever-changing business requirements and consumer needs. As most products and services today have a large supplier content, companies need to rely on knowledge and expertise of their supply partners. Mobilizing their supply partners to share and integrate their knowledge and expertise, allows global

A. J. van Weele (✉)
School of Industrial Engineering, Eindhoven University of Technology, Eindhoven, The Netherlands
e-mail: a.j.v.weele@tue.nl

A. C. Moreira et al. (eds.), *Innovation and Supply Chain Management*, Contributions to Management Science, https://doi.org/10.1007/978-3-319-74304-2_7

manufacturers to speed up new product development (NPD) time and reduce risk. However, early supplier involvement is not a guarantee for NPD success. Hartley et al. (1997) reported in her research that was conducted among 79 companies in the electromechanical industry that engaging suppliers early in NPD did not result in lower product cost, better products or reduce cycle times. On the contrary. Her findings were identical to those of Birou (1994), who reported even higher product and development cost as a result of early supplier involvement, combined with a lower product quality and longer time-to-market. Eisenhardt and Tabrizi (1995) completed this picture with similar results, contrasting earlier research that presented successful cases on early supplier involvement.

Clark (1989) reported in his study on Japanese car manufacturers, who were able to reduce engineering hours significantly as a result of earlier and extensive supplier involvement. A finding that was equal to the landmark study on the American and Japanese car industry by Womack et al. (1990). Also, Ragatz et al. (1997) reported positive effects of engaging suppliers early in new product development. These studies reported on significant improvements in terms of product quality and cycle time. These results were substantiated later by Primo and Amundson (2002), who found positive effects when studying 38 projects in the electronics industry.

Engaging suppliers early in new product development seems not without trouble and may easily lead to disputes and even court cases. A recent example is Apple which ran into problems in its relationship with Qualcomm,[1] that sued Apple because of infringements of intellectual property (on Force Touch and energy management). Moreover, being aware of Apple's impressive profit margins, Qualcomm wanted, as a compensation for its development work, to change its revenue model from a fixed price per chip to a percentage of Apple's XPhone sales price. Which was unacceptable to Apple. Qualcomm therefore sued Apple in China to stop the sales of its new XPhone immediately.

This example shows problems that may occur when working with suppliers in new product development: conflict of interest, knowledge misappropriation, a unfair return on development cost, and gain and pain sharing of new product development outcomes. However, many other problems may exacerbate successful collaboration in new product development.

Based on these observations during the mid-nineties, the questions emerged: 'why are results and outcomes of early supplier involvement so controversial? What explains the different outcomes of these studies? What truths and threats are lying behind this often advocated practice of early supplier involvement? How to optimize supplier engagement in new product development or is this a mission impossible?' These questions have been leading many research projects that we conducted and supervised over the past decades. As it will become clear, there was not a single study that was able to cover all of these questions. On the contrary: previous studies were necessary to create a fair understanding of early supplier

[1] See: https://www.digitaltrends.com/business/apple-vs-qualcomm-news/

involvement as a phenomenon. Research findings were fed into new research designs, leading to additional insight.

In the remainder of this chapter we will discuss the outcomes of the main (PhD)-research projects which we initiated and supervised during the past decades. First, we draw on previous work conducted by Wynstra (1998), who revealed the main areas and processes underlying early supplier involvement. Next, we will discuss research that was conducted in assessing effects of supplier involvement in new product development. Here, we will base our discussion on work conducted by Van Echtelt (2004). And finally, we will discuss the effects of both contractual and relational governance on innovation outcomes. In doing so, we will draw upon recent studies conducted by De Vries (2017). These studies show that early supplier involvement is about creating both careful contractual and relational governance aimed at fostering intercompany human interaction to foster knowledge sharing behavior among technology experts. After our discussion of these studies, we will put these into perspective, and discuss several managerial implications.

2 Obstacles Preventing Early Supplier Involvement in New Product Development

Many obstacles prevent effective early supplier involvement in new product development. These are partly due to limitations of the actual theories in use (Argyris 1990)[2] within companies. For another part these are due to ill-defined processes on how to engage suppliers effectively. These problems may relate to the manufacturer organization, the supplier organization and to the manufacturer supplier relationship. We will discuss these topics shortly.

Supplier involvement is defined here as: 'the contributions (capabilities, resources, information, knowledge and ideas) that suppliers provide, the tasks that they carry out and the responsibilities that they assume regarding the development of a part, process or service for the benefit of a current and/or future buyer's product development projects' (Van Echtelt 2004, p. 27).

When studying early supplier involvement in new product development, different theories can be used. A popular theory is transaction cost economics (see e.g. Williamson 1979), which holds that buyers will predominantly seek for transactions resulting in lowest total transaction cost. Transaction cost include development and manufacturing costs, logistics and transportation costs, administrative cost etc. Companies that operate from a transaction cost perspective will predominantly

[2]Theories in use: 'Those theories that are implicit in what we do as practitioners and managers. They govern actual behavior and tend to be tacit structures' (Argyris and Schön 1974, p. 30). Argyris et al. argue that people have mental maps with regard to how to act in situations. This involves the way they plan, implement and review their actions. It is these maps that guide people's actions rather than the theories they explicitly espouse.

consider cost and financial aspects as the prime consideration in supplier selection and decision-making. This usually results in a short-term orientation: investments made in suppliers should preferably generate a short-term return. This transaction cost orientation seems in conflict with new product development, as supplier investments will only materialize on a longer-term.

Another important theoretical perspective which we feel relevant here is the Principal-Agent theory (Eisenhardt 1989). This theory assumes that, in commercial relationships, business partners will suffer from four basic problems. First, conflict of interest will arise as the buyer wants to spend as little as possible and the supplier intends to generate as much income from the relationship as possible. Secondly, the relationship may suffer from information asymmetry. Usually, the buyer is not aware of the problems that a supplier may incur in developing, testing and actually manufacturing and delivering a component. Whereas the supplier may not have a complete picture of the environment in which his component is embedded in the final product. The supplier may also not be informed about how the final product is being used by the buyer's end-user. Thirdly, parties may suffer from risk. In commercial relationships, the buyer attempts to shift most of the risk to the supplier, whereas the supplier attempts to do the same in the relationship with the buyer. Usually this problem of risk allocation is solved by negotiating complex contracts, where duties, risks, liabilities, indemnities and guarantees are described in a high level of detail. It is assumed in such situations that all risks can be identified and arranged for beforehand. However, in practice risks may occur that were not foreseen. This is general practice in innovation and new product development projects, which by definition are surrounded by risks and uncertainty. Finally, agency theory holds that parties may suffer from moral hazard. This relates to a lack of trust and respect that the other party will have for the interests of the other party.

We may conclude that holding a transaction cost theory perspective will lead to an overly short-term and financial orientation toward engaging suppliers in new product development projects. Whereas the agency perspective (the supplier needs to act in the interest of the buyer) may lead to a situation where a buyer will try to mitigate its risk and liabilities by shifting these to the supplier. Both the transaction cost perspective and agency perspective may not be optimal to guide buyer supplier collaboration in new product development.

The transaction cost and agency perspectives are reflected in a few main problems and challenges that have been reported in previous research to relate to early supplier involvement (Van Echtelt 2004, pp. 34–35):

- Loss of knowledge and skills: intensive collaboration with suppliers in product development poses potential risks for loss of proprietary knowledge and the loss of skills crucial for future product development.
- Supplier technology lock-in: in fast-changing high-tech environments, companies risk becoming locked into a supplier's technology (as is the case with Apple in its relationship with Qualcomm).
- High relationship costs: companies that involve a supplier earlier in the product development process or that collaborate in technology development need to

spend more time and bring together different management styles and budgeting processes. This implies time and effort being spent on coordinating the work between the two collaborative parties.
- Reduced product development speed: involving suppliers can even slow down the overall development process, since several design iterations and technology alignments may be necessary before arriving at the final design and product.
- Diverging objectives, interests and levels of commitment: already at the beginning, parties may have different objectives and interests. This may be because of views on how to recapture past investments may differ among stakeholders. Moreover, expected results may change over time and unforeseen circumstances may arise which could give rise to relationship conflicts. Another challenge is related to the free rider problem: how to prevent that suppliers, that take part in a product development project, take it easy and wait for others to take the initiative?

Apart from these problems other problems are that both parties may be unwilling to take risks in establishing relationships, may have limited experience in new product development, may embark on a project without clear agreements, may have misunderstandings about how each organization functions and may have different cultures. Furthermore, disagreement may occur about sharing the pains and gains of the collaboration. Some problems may be related to the supplier organization, where the supplier conveys a need to capture and secure business on the short-term, be overly price sensitive and may work with incapable engineering staff and sub suppliers. Other problems may relate to the manufacturer's organization, where the different business functions (research and development, purchasing, production) are insufficiently aligned, where the culture is characterized by a not-invented-here syndrome, and supplier knowledge is seen as a threat to jobs in the research and development organization.

We conclude here, that effective early supplier involvement suffers from many challenges and problems that are not easy to overcome. These problems may be due to the manufacturer organization, the supplier organization and the relationship between parties involved. Next, these may originate due to an ineffective theory in use. We would argue here that other theoretical perspectives, other than the Transaction Cost Theory and Agency Theory, such as the Resource Based View of the firm, Resource Dependence Theory and Stewardship Theory (Davis et al. 1999) may be more useful lenses to understand complex interorganizational collaborations.

3 From Early Purchasing Involvement to Timely Supplier Involvement

Wynstra (1998) observes that increasing specialization in European industry explained the manufacturers' decreasing share in the added-value of their own products (p. 1). Since part of the production activities, that were previously carried out by the manufacturer, were outsourced to suppliers, manufacturers became more dependent on the resources of their suppliers. As a result, the impact of the purchasing function of a company on its production value increased. This would lead to a different role of purchasing within those organizations. Following Axelsson and Hakansson (1984), Wynstra distinguishes three different roles for purchasing, i.e. a rationalization role, a network or structure role and a development role. The first relates to purchasings' task to contribute to the firm's competitive strength by minimizing total cost of production, logistics, prices of inputs, etc. The second role relates to handling the firm's supplier network and managing the degree of dependency of the firm on specific suppliers. The third role concerns systematically matching the firm's technological development with the capabilities of suppliers and the supplier network. Based on this, the author defines purchasing involvement in product development as: 'contributing knowledge, taking part in managerial processes and participating in decisions with regard to product development, from a perspective of purchasing, i.e. striving towards lowest possible total product cost, well-balanced dependencies on suppliers, and an optimal technological match with suppliers' (p. 65).

As Wynstra intends to explore purchasing's contribution to new product development, he elaborates on this development role. In his view this role, essentially, consists of four key processes: prioritizing, mobilizing, coordinating and timing (p. 67). *Prioritizing* concerns the choices manufacturers have to make on how and where to invest available resources. Following Hakansson (1989), *prioritizing* not only concerns the choice of actual collaboration partners, but also the choice for a specific form and intensity of supplier involvement. *Mobilizing* involves motivating suppliers to start working on a particular development. Whilst *coordinating* involves the adjustment and adaptation of development activities and resources between suppliers and manufacturer. Without coordination, joint development will result in poor integration of components, double work, incompatible technical solutions, etc. Of course, this need for coordination grows as a result of increasing specialization and fragmentation of development activities across different supply chain partners. Finally, timing requires the meticulous coordination and adaptation of development activities and resources across time. Without timing, product development will suffer from unexpected bottlenecks, unnecessary delays and missed deadlines. Having defined these four key processes, Wynstra argues that these are to be applied in three areas i.e. suppliers, technologies and projects. The challenge for companies is how to manage these processes across these three areas. The author concludes that therefore early purchasing involvement essentially is a cross functional activity, which should not be exacerbated by functional boundaries within organizations. Following Dowlatshahi (see Table 1), he argues that silo thinking in organizations, especially between purchasing, and research and development, is a major risk when collaborating with suppliers in new product development. The author then sets out to explore the mechanisms underlying these processes and areas in nine comprehensive, longitudinal case studies.

Based upon these nine in-depth case studies, the author identifies four management areas that should be covered when engaging suppliers in new product development processes (see Fig. 1):

Table 1 Purchasing and development orientations lead to contrasting interests (source: Dowlathahi 1992)

Purchasing orientation	Development orientation
• Minimum acceptable margins of quality, safety and performance • Use of adequate materials • Lowest ultimate cost • High regard for availability • Practical and economical contributors, specifications, features and tolerances • General view of product quality • Cost estimation of materials • Concern for just-in-time deliveries and supplier relationships	• Wider margins of quality, safety and performance • Use of ideal materials • Limited concern for cost • Limited regard for availability • Close or near perfect para meters, specifications, features and tolerances • Conceptual abstraction of product quality • Selection of materials • Concern for overall product design

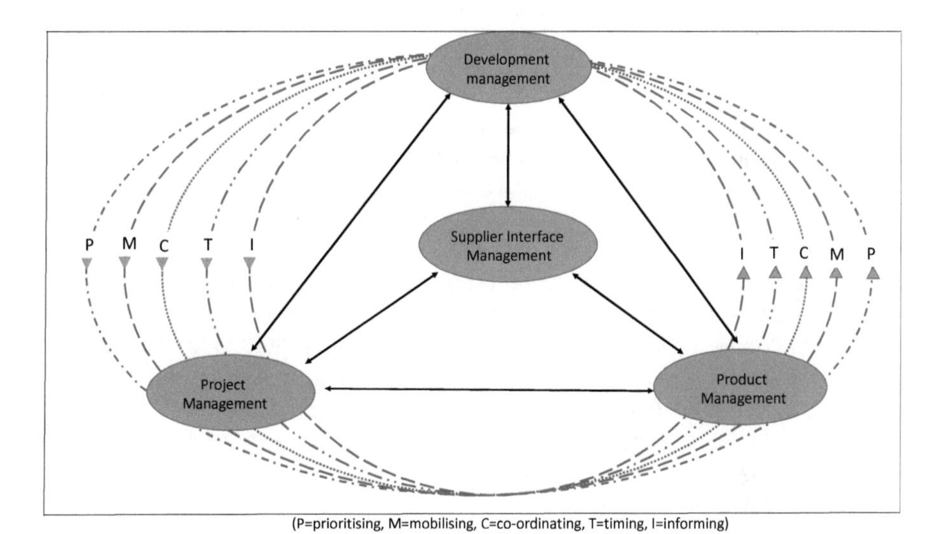

Fig. 1 Interrelations between the four management areas and underlying key processes (Wynstra 1998, p. 199). *P* prioritising, *M* mobilising, *C* co-ordinating, *T* timing, *I* informing

- Development management. Development management includes a clear strategy on what technologies to keep or develop in-house and which ones to outsource to suppliers. It also relates to policies that are in place on how to effectively engage suppliers in new product development in the relationship with internal business domains within the firm. Without development management, NPD projects will suffer from a lack of guidance and suffer from politics.
- Supplier interface management. This includes monitoring supplier markets for new technological developments that may be relevant for the company. It also includes pre-selecting suppliers per key technology area and exploiting their

technical capabilities. Finally, it includes also the monitoring and evaluation of a supplier's development capabilities and performance.
- Project management. This activity can be distinguished into two sub activities i.e. planning and execution. Project planning includes the actual decisions on a project level on what to develop inside or outside the company. In case of the former, the actual decision is made on what suppliers to engage in the project and when to do that. Project execution includes introducing and onboarding the supplier with the firm's business strategy, domains, projects and engineers. It also includes the actual orchestration of the activities of first tier suppliers in their relationship with second tier suppliers. Finally, it includes the ordering and chasing of prototypes and managing technical changes and variations.
- Product management. This activity includes evaluating product designs in terms of part availability, manufactureability, leadtime, quality and cost and promoting standardization and simplification of designs and parts across products and suppliers.

To be able to engage suppliers in new product development effectively, each of these four activities need to be managed by the firm to some extent. Preferably through a concerted action by all internal stakeholders involved. Enabling factors that foster a successful execution of each of these four management areas are: the internal organization of the purchasing department just as the development team, the access to and availability of human resources information including quality performance. The last finding leads the author to conclude to a fifth key process: *informing*.

When summarizing, the author argues that the involvement of purchasing in a product development project should aim to realize or contribute to five (instead of four) key processes: prioritizing, mobilizing, coordinating, timing and informing. These five key processes should focus on four management areas including new product development, supplier interface management, project management and product management. Valuable suggestions are: not to talk about early purchasing involvement but, rather, stress the role of suppliers in fostering and improving new product development success. Next, we recommend to talk about *timely* supplier involvement rather than *early* supplier involvement as it is important to engage suppliers, based upon their capabilities, at the right time and the right level of responsibility in a new product development process (see Box 1). The role of the human factor should not be underestimated.

Box 1 The Supplier Involvement Portfolio (Wynstra and Ten Pierick 2000)

The objective of the supplier involvement portfolio is to provide guidance for setting priorities with regard to the involvement of suppliers in new product development. It will help companies to mobilize supplier expertise in the best possible way. As not all suppliers are equally important, only very few need to

(continued)

Box 1 (continued)

be engaged early. Other suppliers may be involved later, whereas most of the suppliers will be involved when the product design has been fully tested and is frozen. The portfolio distinguishes four types of supplier involvement based upon two variables: (1) the degree of responsibility for product development that is contracted out to the supplier, and (2) the development risk involved (see Fig. 2). Suppliers may assume responsibility for component design in four ways:

- Functional specifications. Based upon functional specifications for a component or module, the supplier is responsible for conceptual design, detailed design, prototype, testing and setting up its production and assembly process.
- Global design. Here, the buyer communicates a rough design to the supplier, who needs to work out a detailed design and submit this for approval to the buyer. When approved, the supplier is responsible for prototyping, testing and manufacturing.
- Detailed design. The supplier is responsible for submitting a prototype or sample to the buyer for approval, which is tested. Next, the supplier is responsible for setting up production and assembly.
- Standard design. Here the buyer decides to integrate a standard component in their product design. After the product design has been tested and is frozen, the supplier is requested to submit a price proposal and production planning.

Development risk is related to a number of factors. Examples are: the component is new to the buyer, the buyer is unfamiliar with the functionality of the component, criticality of the component for the buyer's product functioning, the component is on the critical path of planning and, the number of technologies represented in the supplier's component. Based upon these criteria, an assessment on part-level can be made per project to assesses whether the buyer falls short in terms of knowledge and expertise. Those are the parts where specialized suppliers will be engaged early in the new product development project. Standard parts come with low risk in general and low technical complexity. Suppliers of standard parts therefore can be engaged late in the process. This is how the supplier development portfolio may guide buyer decision-making on ESI, which is better referred to as timely supplier involvement (see Fig. 2).

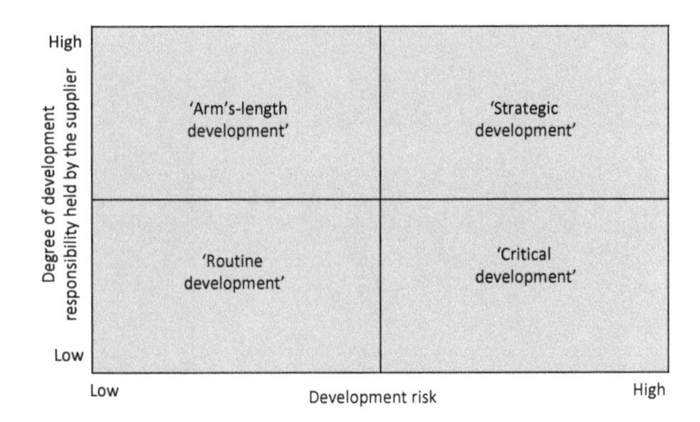

Fig. 2 The purchasing development portfolio (Wynstra and Ten Pierick 2000)

The value of Wynstra's work is that he provides a detailed insight into what it takes to engage suppliers within the product development processes of the firm, distinguishing the degree and timing of supplier involvement. A lot of processes need to be in place, in order to align the supplier contributions and match technology needs. Furthermore, enabling factors need to be organized in order to put these processes in place. As companies do probably not have all these processes in place equally, it may partly explain the different outcomes of early supplier involvement among companies and sectors.

4 Shifting Suppliers into Gear: Effective Supplier Collaboration

Building on Wynstra's groundwork, Van Echtelt (2004) investigates the way in which intercompany collaboration, and specifically vertical collaboration between the manufacturer and its supplier, can strengthen a company's capability to develop new products. Again, the 'aim of this study was to identify what the critical processes are for managing the involvement of suppliers to lead to improved performance in product development' (p. 1). Van Echtelt observes that companies are being forced to develop and implement new strategies just as ways to organize their product development function. In general, companies may pursue three different strategic and organizational responses: (1) outsourcing new product development, (2) concurrent development and cross functional collaboration, and (3) intercompany collaboration. The author observes that companies increasingly engage in collaborative arrangements with other companies in the area of technology and product development (p. 7). He further argues that this was in contrast to the more traditional arm's length supplier relationships in combination with a complete reliance on suppliers' development capabilities. Collaboration with other companies became a mechanism for tapping into external resources of knowledge to speed up development. Next, through supplier collaboration financial risks could be shared in developing new products. Supplier collaboration can however assume different, hybrid forms as well, such as mergers, acquisitions, joint ventures, strategic alliances, license agreements and collaborative arrangements with suppliers. With regard to the latter, the author observes that there is a lack of sufficient empirical understanding of critical processes and conditions underlying effective supplier involvement that allow companies to attain short-term product development targets and long-term business goals. Next, he developed a framework that identifies the objectives, critical activities and conditions for effectively leveraging suppliers' in product development. Again, this researcher builds his research on eight in-depth case studies, which are conducted at a global high-tech manufacturer. These case studies were followed by four additional case studies taken from companies in other business sectors.

Based upon his extensive work, a number of adaptations to Wynstra's framework for analyzing management of supplier involvement are suggested. First, three instead of four relevant managerial arena's are suggested, i.e. the strategic management arena, the project management arena and the collaboration management arena (p. 264). A major reason for this is that companies may need to focus on managing individual collaborations with suppliers. As Van Echtelt (2004) argues (p. 264): 'adopting a relationship view actually is like black boxing a phenomenon that itself is driven by events in different coloration episodes that together drive an evolving relationship'. As development projects aim at realizing both short-term and long-term objectives, the author maintains Wynstra's original idea of the strategic management arena and project management arena. Here, only few ideas are added. However, the collaboration management arena includes activities aimed at designing the appropriate collaboration form, executing development activities in an individual collaboration and learning from each collaboration episode. For each of the three arenas, critical management processes are identified, which need to be managed collectively (see Fig. 3). More specifically, the three management arenas seem to follow basic iterative cycles, rather than being sequential in nature. Based upon his extensive work, the author concludes that important enablers for making this integrated new product development framework work, are: cross functional orientation purchasing—research and development, human resource quality, recording and availability of information. Here, we conclude that most of the enabling factors that were identified by Wynstra, are confirmed.

The implications of this study are clear: if companies have made many efforts and have spent significant time in defining and describing critical management processes, these will not be successful if projects suffer from silo thinking, political plays between purchasing and research and development, lack of information support and management reporting, and lack of human resources. The value of Van Echtelt's work is that he translates his research results into a coherent and pragmatic audit tool, which can be used to assess the maturity of both the manufacturing and supplier organization for collaboration. The outcomes enable companies to either improve their processes and/or enablers for successful future collaborative innovations. In doing so they might focus on putting things right first before embarking in joint collaboration in new product development.

Both Wynstra's (1998) and Van Echtelt's (2004) research reveal the key areas and critical processes that need to be in place to allow for effective supplier

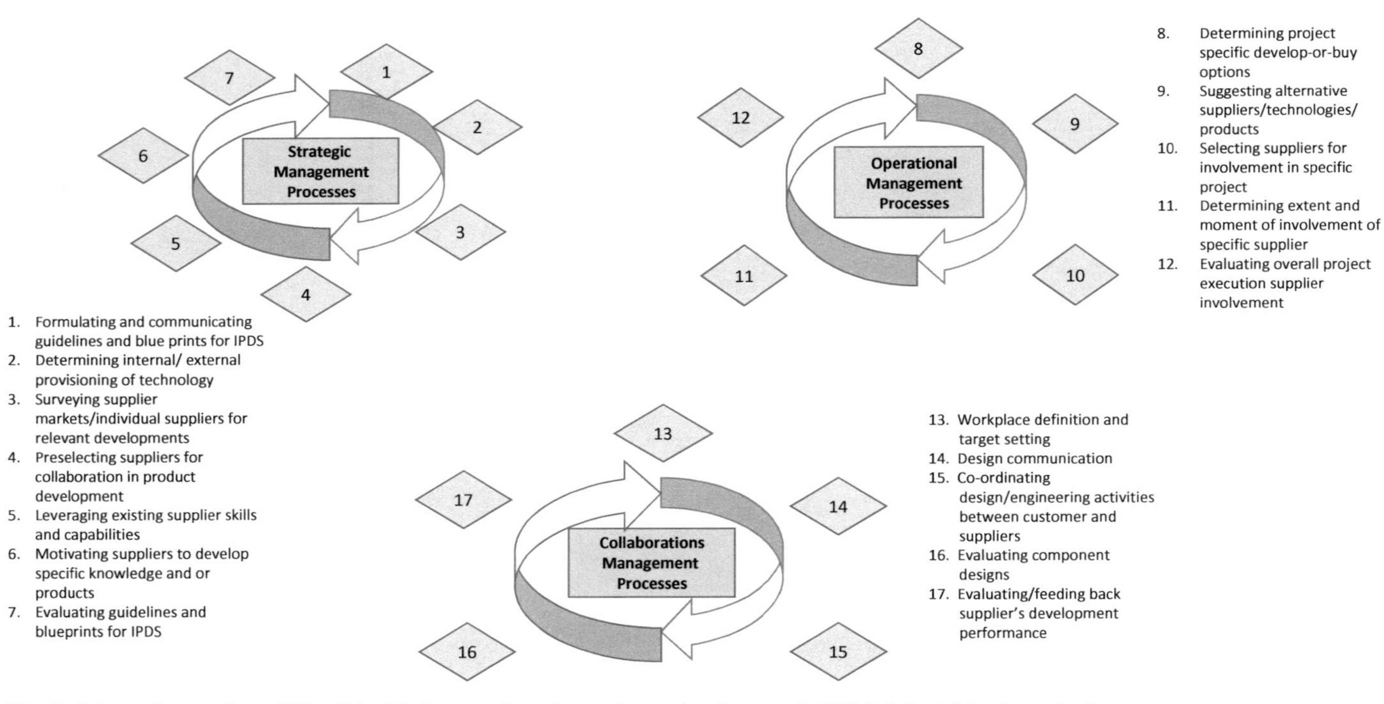

Fig. 3 Schematic overview of Van Echtelt's Integrated product and sourcing framework (2004) (adapted by the author)

collaboration in new product development projects. Both studies emphasize the importance of cross-functional teamwork among research and development engineers and purchasing professionals: an open atmosphere to exchange information between parties involved and the role of human resources. Enabling cross-functional collaboration in essence occurs at a person-to-person level, and thus may explain why supplier involvement is successful in certain cases and in other cases not. This insight was the reason to further deepen our initial research questions to include: what exactly motivates supplier specialists to contribute to the innovation goals and objectives of a manufacturer during collaborative innovations projects? Here, the inter-human dynamics within innovation teams needed to be more clearly understood such as the psychological factors and processes that affect inter-organizational and intra-organizational knowledge exchange. As a result, we became particularly interested in the social and human factors behind supplier involvement, which was the trigger to the next series of studies.

5 How to Release External Expertise: When Do Suppliers Care to Share?

Following upon the previous research, De Vries (2017) conducted three studies aimed at understanding the psychological factors that influence actual knowledge sharing in interorganizational collaboration in new product innovation. These studies were different from the previous ones. Through Wynstra's and Van Echtelt's work we intended to obtain in-depth insight in the mechanisms, processes and enablers underlying early supplier involvement. Hence, these studies were explorative, qualitative and case based. Through De Vries' research we intended to explore and assess the effects of human interaction in conflicting interorganizational settings. More specifically, we were interested in discovering the actual drivers underlying effective inter-human information and knowledge exchange, as information management and exchange emerged as a key enabler from the previous studies. Therefore, De Vries's research is of a different nature, i.e. quantitative and more specific in terms of independent and dependent variables. Here, we report on two of the three studies.

The first study, which was conducted among 70 experienced relationship managers at a large, global electronics manufacturer, was aimed at investigating how contractual and relationship characteristics enhance exploitative and exploratory knowledge sharing by service partners to whom manufacturers have outsourced customer facing services (De Vries 2017, p. 18). As many manufacturing firms today outsource after sales services to third-party service providers, these service providers have become crucial for the knowledge exchange about quality and usage behavior by end-users. Manufacturers can greatly benefit by integrating knowledge on post sales experiences by end-users into their new product designs. However, how could manufacturers capture such knowledge from these service providers? In order

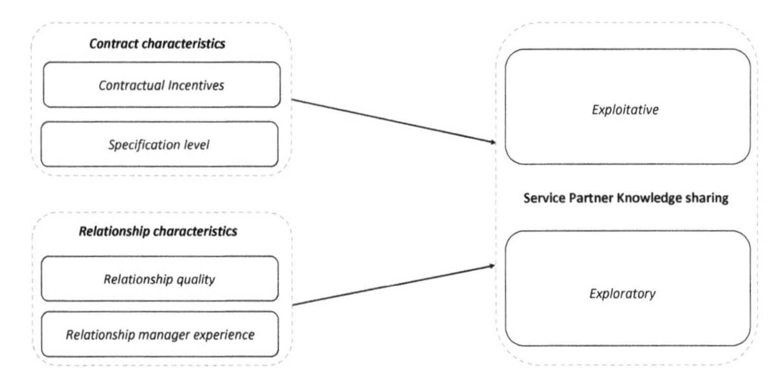

Fig. 4 Initial research framework on effects of contract characteristics and relationship characteristics on service partner knowledge sharing (De Vries 2017, p. 28)

to address this question, the effects of both contractual governance and relational governance on knowledge sharing behavior were investigated. A distinction was made between exploitative knowledge sharing (aimed at obtaining knowledge to improve existing products and processes) and explorative knowledge sharing (aimed at obtaining knowledge to create entirely new products and processes). Effective knowledge sharing was, in line with organizational learning theory, (e.g. Bell et al. 2002) deemed necessary for delivering successful innovation projects. Figure 4 provides a schematic overview of the research model. Contractual governance was explained by using two sets of variables: one was related to contractual incentives. The other set was related to contract specifications. It was hypothesized that both variables would positively affect service partner knowledge sharing behavior. With regard to relationship characteristics the differentiation was made between relationship quality and relationship manager experience. Relationship quality was measured by assessing the level of cooperation, responsiveness, empathy, assurance and trust among parties. Also for these variables a positive effect on knowledge sharing behavior was assumed. A limitation of the study was that data gathering was gained only from the manufacturer's relationship managers, and was based upon self-administered questionnaires.

Using multivariate analysis, the findings were the following. First, clearly defined contracts seem to be characterized by higher levels of knowledge sharing. Positive relationships were found between the level of contract specification and knowledge sharing. Which confirmed earlier research that unclear contract specifications hinder knowledge transfer. Unarticulated expectations leave service partners guessing for desired performance levels, resulting in disappointing service performance. We conclude that clear contractual specifications provide a frame of reference that makes a service partner share those insights that provide value to a manufacturer. Secondly, a strong negative relationship between contractual incentives and exploratory knowledge sharing was found. This negative relationship was not found in the relationship with exploitative knowledge sharing. Clearly, contractual incentives focus suppliers on realizing short-term objectives and gains and reducing risks that could jeopardize the realization of agreed incentives. Contractual incentives avoid suppliers to think out-of-the-box. However, they may be useful when pursuing exploitative knowledge sharing aimed at continuous improvement or improving existing product and process designs. Thirdly, a positive relationship was found between relationship quality and both types of knowledge sharing. Relationship quality builds a long-term commitment among parties and as a result, both are willing to make idiosyncratic investments to the relationship. Experienced relationship managers may also trigger explorative knowledge sharing as they seem to better manage and guide the interactions with the service partner. This is important to note, as changing relationship managers too frequently and within a too short period of time, will be detrimental to exploratory knowledge exchange and, hence, collaborative innovation outcomes.

Concluding, the value of this research is that it shows that both contractual governance and relational governance seem to affect knowledge sharing between partners in collaborative innovation projects. Of course, with regard to contractual governance, this research only tested the effects of contract specification and contractual incentives. Clear contractual specifications are required to guide the development activities among innovation partners. Contractual incentives, in general, foster incremental innovation. However, they do not seem to foster radical innovation. On the contrary. Relationship quality is positively related to knowledge sharing and therefore a key variable for driving collaborative innovation. Therefore, it doesn't come as a surprise that tenured relationship managers, who build on their past experiences, seem important in building trustful relationships.

6 Exploring the Drivers of Knowledge Sharing: Aligning Interests and Rewards

The previous study indicated that relational governance seems important in collaborative innovation projects. Especially cooperation, responsiveness, empathy and trust seem to correlate positively with knowledge sharing. This all may be true, but what makes supplier engineers actually share their knowledge? This was the major research question underlying a second, quantitative follow-up study among 187 supplier technical engineers, who were engaged in collaborative innovation projects of seven global high-tech manufacturers (De Vries 2017, p. 42). This research was deemed relevant since evidence from practice showed that engaging suppliers and integrating supplier knowledge into new product designs was not without problems. As an example may serve here Goodyear, the global tire manufacturer. It's tire engineers imitated a technological innovation of the supplier, whose employees had been involved in a R&D project. Next, this supplier's technology was pushed out of the new product and the supplier did not get a fair yield of its contribution (De Vries 2017, p. 41). Similar experiences have been reported in the relationship between Compuware and IBM (Cowley and Larson 2005) and Lexar Media versus Toshiba

(Thomas 2003). These cases suggest that suppliers do not always get a fair return for their development work in collaborative innovation projects. This makes technical engineers cautious, as they have to consider both the responsibility to advance the manufacturer's business as well as obtaining a fair reward for the supplier's development efforts.

Clearly, knowledge misappropriation and unfair distribution of rewards in collaborative innovation projects, may lead to misalignment of interests, and hence may demotivate supplier engineers to share their knowledge with their clients. Therefore, this follow-up study aimed at investigating the importance of alignment of interests and economic rewards in collaborative innovation projects. Here, fairness theory (Fehr and Schmidt 1999) was used to explain how individuals balance their invested efforts against expected outcomes. Fairness theory holds that when the balance is assessed as fair, self-regulation motivates individuals to contribute to collaborative innovation projects. Next, stewardship theory (Davis et al. 1999) was used which defines stewardship as an individual's (here: supplier engineer) felt ownership and responsibility for the manufacturer's overall welfare. Based upon this, it was hypothesized that customer stewardship would positively influence knowledge sharing behavior, and, hence, affect collaborative innovation project outcomes positively. In line with Golden and Raghuram (2010), knowledge sharing behavior was defined as 'knowhow relayed to others on an impromptu basis, whereby individuals feel comfortable to spontaneously disclose personal experiences'. This definition reflects that knowledge sharing is rather an interpersonal, spontaneous activity, than a planned or programmed activity. Knowledge sharing behavior was defined as the actual disclosure of information, whereas knowledge sharing intentions were defined as the willingness to engage in knowledge sharing behavior in the near future. Based upon these ideas, a preliminary research framework was built and tested (see Fig. 5).

The framework assumes that actual knowledge sharing behavior in conflicting, interorganizational settings is determined by knowledge sharing intention. Whereas knowledge sharing intention is affected by customer stewardship, and perceived

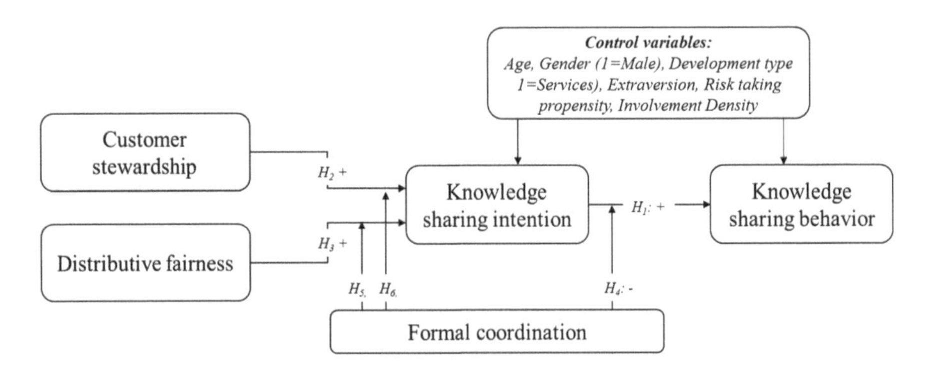

Fig. 5 Research model for investigating effects of fairness and rewards on knowledge sharing behavior (De Vries 2017, pp. 45–46)

distributive fairness. Through this framework we wanted to assess to what extent a supplier expert's care for the manufacturer's interest and his perception of a fair return of his development efforts would affect his willingness to share sensitive knowledge.

The author found that the hypothesized research model explained 33.5% of the variance in knowledge sharing behavior. The results indicate a positive effect between knowledge sharing intention and knowledge sharing behavior. Next, customer stewardship related positively to knowledge sharing intention. Furthermore, a positive interaction effect was found of formal coordination on the relationship between knowledge sharing intention and knowledge sharing behavior. An interesting finding was that a supplier engineers' risk-taking propensity and extraversion related positively to knowledge sharing behavior and intention. These findings were in line with earlier works, which reported that risk averse individuals tend not to engage in risky behavior such as knowledge sharing. Strange enough, no evidence was found for the effect of distributive fairness on knowledge sharing intention. However, the moderating effects of formal coordination on the relationship between fairness and intention did turn out to be significant. Therefore, it was investigated at what levels of formal coordination (i.e. high, medium or low) could affect the effect of distributive fairness on knowledge sharing behavior. It appeared that under conditions of low formal coordination, both the alignment of interests (stewardship) and alignment of rewards (fairness) drive knowledge sharing behavior through knowledge sharing intention. However, under circumstances of high formal coordination only alignment of rewards drives knowledge sharing behavior (De Vries 2017, p. 63).

Hence, the degree of formal coordination in collaborative R&D project seems an important influencer of supplier engineers' motivations to share knowledge. Based upon this research, R&D managers need to make a conscious choice with regard to the level of formal coordination in collaborative R&D projects. They should instruct their employees either to heavily rely on informal coordination mechanisms such as trust and mutual understanding, or to provide strict guidance to supplier engineers by frequently referring to what has been agreed.

When formal coordination is high, managers are encouraged to emphasize the benefits for the supplier of being involved in this project. They should convince the supplier engineer that his/her knowledge contribution allows both parties to attain their business interests. Furthermore, they should secure that future revenues will be fairly distributed over the parties involved. When formal coordination in R&D projects is low, managers may have to do much more. In such a situation they have to make sure that supplier engineers experience a sense of stewardship for the manufacturers well-being. Supplier engineers should be allowed a fair degree of autonomy to work and share know-how in the R&D team. In order to realize this, the manufacturer should explicitly deploy onboarding practices to make the supplier engineers feel at home. Next, they should be appreciative of the unique competences that supplier engineers bring to the table.

Concluding, the value of this study is that it reveals the intricate and sensitive mechanisms underlying knowledge sharing behavior between individuals in collaborative innovation projects. It shows that, in essence, collaborative innovation is a process of human interaction. If this human interaction is not guided through both formal and informal mechanisms, knowledge exchange will suffer and, hence, will affect collaborative innovation project outcomes negatively.

7 Early Supplier Involvement: A Mission Impossible?

After many years of research, we finally have found the answer to the contrasting results of early supplier involvement. The large differences found in terms of early supplier involvement outcomes indeed can be explained by the fact that some strategic management processes, operational management processes and collaboration management processes are not or not sufficiently in place. However, more importantly, these differences may be due to the manufacturer's inability to engage and mobilize supplier expertise effectively in the relationship with the manufacturer's engineers. It is the human interface that seems to make the difference. The main reason for this is that in collaborative innovation, it is all about sensitive knowledge sharing and information exchange between engineers that usually need to operate in a setting with conflicting interests.

Knowledge and information exchange essentially occurs between people, is by definition person to person exchange. In order to make people share sensitive information both contractual and relational governance are important. Contractual governance is necessary in order to formally align business interests and expectations and to provide for rewards and incentives for the work that has been delivered by suppliers. Relational governance relates to cooperation, responsiveness, communication, consistency, empathy and trust. It is needed to make supplier engineers feel respected and rewarded for their inputs. Depending on the type of innovation which the company wants to pursue (incremental versus radical innovation) the manufacturer should adapt its rewards and incentives in the relationship with suppliers. In both situations, careful selection of suppliers and their representatives, and deployment of effective on boarding practices are important. When working with suppliers in new product development projects, the manufacturer should secure a fair return on the supplier's inputs and efforts in order to generate sufficient stewardship and alignment from suppliers. As collaborative innovation projects are unique, so are the teams and the individuals that need to work on these. Different innovation projects represent different technical and commercial challenges that need to be overcome by different teams and people.

What can we learn from the previous studies? What should managers do to benefit from early supplier involvement in collaborative innovation projects? What should be avoided to prevent failure? The learnings are many, as are the challenges ahead. Based our research on collaborative innovation projects, we conclude the following:

- Informal and informal governance mechanisms. Engaging suppliers early in new product development successfully requires a fair mix of both formal and informal governance. It seems that such interorganizational collaborations cannot do without formal contracts. As contracts guide and provide the context of the future collaboration. Development contracts, which stipulate how to deal with intellectual property and how investments will be recouped by both parties, are necessary to manage the expectations of the parties involved. Next, the manufacturer can opt to manage the project formally or informally. The manufacturer should avoid an unclear mix of both, as this will be confusing to the supplier partner. Formal governance mechanisms are important; however, informal governance mechanisms may make the difference. As cooperation, responsiveness, empathy, assurance and trust seem to determine the motivation of supplier engineers to actually share knowledge and contribute to the manufacturer's new product development goals. Here, seniority of the manufacturer's relationship managers and their tenure is important. Contractual incentives may be used to stimulate supply partners to share their knowledge. Incentives are useful when pursuing incremental innovation. They should better not be used when pursuing radical innovation.

- Arenas and key processes. Following Van Echtelt (2004), apart from contracts, formal governance mechanisms should be created around three important processes

  - Strategic management processes
  - Operational management processes
  - Collaboration management processes.

  Each of these processes is to be worked out in several sub processes (17 in total) which need to guide internal and external stakeholders, taking part in the innovation project. Regular audits should secure that most of these processes are defined and followed in practice. In many product development projects, we have observed that most of these processes were not in place. In such cases collaborative development teams need to make up their own decisions.
- Cross functional teamwork, information and human resources. As many studies have shown, these are important enablers to foster interorganizational innovation projects. Cross functional teamwork should be encouraged and should be in place in order to avoid suppliers to be confronted with political plays and differences of opinion among the manufacturer's representatives. Information management is necessary to create common IT-platforms for design,- planning-, and data sharing. It is of utmost importance that information systems among stakeholders are compatible and connectable. Human resources seem to be a key asset in collaborative innovation projects. Not only in terms of the expertise that is required from both parties, which makes it mandatory that team members are highly qualified and experienced. However, also in terms of the ability to collaborate and operate in teams. Which makes it necessary in any innovation project to invest significant time and money in project startup and onboarding, to allow team members to get acquainted with the project, what is expected from them, with their own roles and with their colleagues. Having witnessed many collaborative innovation projects, we observe that the necessary investments in these three important enablers are often insufficient.

- Knowledge sharing and innovation. As De Vries's research shows, collaborative innovation in essence seems to be a human process i.e. a process of human interaction and socialization. The process of human exchange aims at knowledge sharing to foster product-, process- and business model innovation. There are many reasons why supplier engineers will not share their ideas freely in innovation projects. One is that they need to overcome conflict of interest, as they need to take care of their own company's interest as well the manufacturer's innovations interest. This conflict is not easily solved. Next, they need to be valued in their area of expertise as an engineer and as a human being. A hostile, 'not invented here' culture, arrogance, or downright ignorance at the manufacturer, could jeopardize a supplier engineers' motivation to contribute.
- Contractual governance. Our research shows that collaborative innovation cannot do without contractual guidance. Here, R&D managers should differentiate between low and high degrees of formality. High degree of formality would create clear guidance to all stakeholders. High degree of formality should be accompanied with a fair sharing of the pains and gains of the innovation project. When R&D managers opt for low formality in terms of contractual governance, they should do more. They should make sure that the supplier engineers feel at home, and feel respected as a valued member of the team. Investing in onboarding practices is inevitable. Being inconsistent i.e. changing between low and high formality in dealing with supplier engineers, will make them uncertain and uneasy and unwilling to share their knowledge and insight. Wrong contractual incentives may exacerbate the problem.
- Relational governance. As collaborative innovation seems to be predominantly a human interaction process, the value of investing in the relationship with suppliers can hardly be overstated. Suppliers should be considered by the manufacturer as is an important asset to the company. More particularly, manufacturers should aim for constantly improving the quality of the relationship with business-critical suppliers. Suppliers who have outstanding performance, should be rewarded with more business and deeper engagement in new product development projects. When they do, suppliers should have a fair return on their investments. In case of project failure, the consequences for all parties involved should be clear upfront and remedies in-line with contributions. Professional project management, risk management and relationship management would be necessary in order to create a climate in which a supplier can contribute.

8 Heading for Early Supplier Involvement: Are You Ready for It?

Based upon our previous discussion, it is now clear why extant research on the effects of early supplier involvement has produced such contrasting results. Given the many challenges that need to be overcome, early supplier involvement may easily result in disappointment. Joint collaborative new product innovation represents a trajectory which is full of problems, risks and disappointments. Is early supplier involvement a mission impossible or not? It is for those who think that early supplier involvement can be managed as a systematic structured process. It is not for those who think that early supplier involvement primarily is all about fostering human interaction in conflicting business settings. However, then engaging supplier engineers in new product development is far from easy. When embarking on such a journey, we recommend manufacturers and supply partners to start with the beginning. Which is: to start with the human side of the enterprise. To establish a cross functional, cross organizational team of capable engineers and specialists that are well prepared and equipped for their tasks and which is supported by an adequate (though sparse) governance and project structure. Next, the team should be equipped with sufficient resources. This seems more valuable than to try to structure all 17 processes around the three areas as this will, apart from the huge effort, only provide for limited control and certainty. In reality, every collaborative innovation will develop differently than originally anticipated. The joint project team should be allowed, based upon the manufacturer's initial feasibility studies, to develop their own project mission, restate the project objectives and prepare a global project and work plan. These may serve as the input for an initial, flexible development contract, which stipulates how parties will deal with intellectual property and how investments made will be recovered, i.e. how financial losses will be spread. Rather than working sequentially, parties need to prepare for iterative loops which allows parties based on the progress made, to regularly review and mitigate incumbent project and work plans. Next, the quality of the relationship should be reviewed and discussed regularly to secure that everyone is still committed and contributes. Only then early supplier involvement may turn into a mission possible.

Acknowledgments The author wants to express his gratitude to Dr. Jeroen Schepers, Dr. Jelle de Vries and Prof. Dr. J.Y.F. Wynstra for their comments on earlier versions of this article.

References

Argyris C (1990) Overcoming organizational defenses. Pearson Education, USA, p 358

Argyris M, Schön D (1974) Theory in practice. Increasing professional effectiveness. Jossey-Bass, San Francisco

Axelsson B, Hakansson H (1984) Inköp for Konkurrenskraft. Liber, Stockholm

Bell SJ, Whitwell GJ, Lukas BA (2002) Schools of thought in organizational learning. J Acad Mark Sci 30(1):70–86

Birou LM (1994) Ter role of the buyer-supplier linkage in an integrated product development environment. Doctoral thesis, Michigan State University

Clark KB (1989) Project scope and project performance. The effects of parts strategy and supplier involvement on product development. Manag Sci 35(10):1247–1263

Cowley S, Larson S (2005) IBM, Compuware reach $400M settlement. Computerworld. Available via http://www.computerworld.com/article/2556607/technology-law-regulation/ibm--compuware-reach--400m-settlement.html. Accessed 13 Jul 2016

Davis JH, Schoorman FD, Donaldson (1999) Toward a stewardship theory of management. Acad Manage Rev 22(1):20–47
de Vries JJAP (2017) Release external expertise in collaborative innovation. Doctoral thesis, Eindhoven University of Technology, The Netherlands
Dowlathahi S (1992) Purchasing's role in a concurrent engineering environment. Int J Purch Mater Manag Winter:21–25
Eisenhardt K (1989) Agency theory: an assessment and review. Acad Manage Rev 14(1):57–74
Eisenhardt KM, Tabrizi BN (1995) Accelerating adaptive processes: product innovation in the global computer industry. Adm Sci Q 40:84–110
Fehr E, Schmidt KM (1999) A theory of fairness, competition and cooperation. Q J Econ 114 (3):817–868
Golden TD, Raghuram S (2010) Teleworker knowledge sharing and the role of altered relational and technological interactions. J Organ Behav 31(8):1061–1085
Hakansson H (1989) Corporate technological behavior: co-operation in neworks. Routledge, London
Hartley JL, Zirger BJ, Kamath RR (1997) Managing the buyer-supplier interface for on time performance in product development. J Oper Manag 15(1):57–70
Primo MA, Amundson SD (2002) An exploratory study of the effects of supplier relationships on new product development outcomes. J Oper Manag 20(1):33
Ragatz GL, Handfield RB, Scannell TV (1997) Success factors for integrating suppliers into product development. J Prod Innov Manag 14(3) and 55(5):389
Thomas E (2003) Jury awards Lexar over $380 million for Toshiba's theft of trade secrets and breach of fiduciary duty|Lexar. Lexar Media. Available via http://www.lexar.com/about/news room/press-releases/jury-awards-lexar-over-380-million-toshibas-theft-trade-secrets-and-br. Accessed 13 Jul 2016
Van Echtelt FEA (2004) New product development: shifting suppliers into gear. Doctoral thesis, Eindhoven University of Technology, The Netherlands
Williamson OE (1979) Transaction-cost economics: the governance of contractual relations. J Law Econ 22(2):233–261
Womack JP, Jones DT, Ross D (1990) The machine that changed the world. Rawson Associates, New York, p 323
Wynstra JYF (1998) Purchasing involvement in product development. Doctoral thesis, Eindhoven University of Technology, The Netherlands
Wynstra JYF, Ten Pierick E (2000) Managing supplier involvement in new product development: a portfolio approach. J Purch Supply Manag 6(1):49–57

Arjan J. van Weele (PhD, Eindhoven University of Technology) holds the NEVI Chair in Purchasing and Supply Management at the School of Industrial Engineering at Eindhoven University of Technology, The Netherlands. Dr. VanWeele's research interests lie in managing upstream global value chains, where companies have to rely on interorganizational networks and have to deal with interorganizational conflict of interests. In this respect, international contracting and contract management are subjects of ongoing research. Dr. Van Weele is the author of several leading textbooks in the field, including *Purchasing and Supply Chain Management*, which has been published in several languages. He has coauthored *International Contracting: Contract Management for Complex Projects*, which explains how to contract for large projects. His articles have been published in, among others, the *Journal of Purchasing and Supply Management*, *Journal of Supply Chain Management*, *Industrial Marketing Management*, *Journal of Product Innovation Management*, and *European Management Journal.*

Part III
Strategies and Implications for Innovation

Purchasing Involvement in Discontinuous Innovation: An Emerging Research Agenda

Richard Calvi, Thomas Johnsen, and Katia Picaud Bello

Abstract Building on a systematic review of the literature, we define and discuss why and how purchasing needs to be involved in the discontinuous innovation process. We argue that purchasing involvement in NPD should be considered mainly when the customer firm faces discontinuous innovation. Seeking to promote this emerging research agenda, we present three propositions to focus future studies and inspire practices: (a) technology sourcing and scanning out of the boundary of the supply base is an important stake to support discontinuous innovation as (b) to form an ambidextrous purchasing organization and (c) to develop absorptive capacity within purchasing function. The paper concludes by summarizing the conceptual implications of the paper, outlining some initial managerial recommendations.

1 Introduction

A plethora of research has evolved over the last 25 years concerning Early Supplier Involvement (ESI) in New Product Development (NPD) (e.g. Petersen et al. 2005; Cousins et al. 2006; van Echtelt et al. 2008; Johnsen 2009). This body of research has demonstrated that suppliers are critical sources of innovation and that collaborating with suppliers as part of the NPD process enables innovating companies to capitalize on suppliers' complementary capabilities, thereby improving innovation and NPD performance (Brem and Tidd 2012).

R. Calvi (✉)
IREGE Laboratory, Université Savoie Mont Blanc, Chambery, France
e-mail: richard.calvi@univ-savoie.fr

T. Johnsen
Politecnico di Milano School of Management, Milan, Italy
e-mail: thomaserik.johnsen@polimi.it

K. Picaud Bello
ESSCA School of Management, Angers, France
e-mail: katia.picaud-bello@essca.fr

A. C. Moreira et al. (eds.), *Innovation and Supply Chain Management*, Contributions to Management Science, https://doi.org/10.1007/978-3-319-74304-2_8

Various functions interact with suppliers as part of NPD projects but Purchasing performs as an important go-between function facilitating ESI processes (Wynstra et al. 2000; Lakemond et al. 2001). However, although more than 30 years have passed since Farmer (1981) argued the need for Purchasing to be involved in NPD, relatively little progress has been made on research on this challenge. In fact, despite the upsurge of research on ESI, most of the literature overlooks the role of purchasing in this process, suggesting little interest in, for example, the role of the organizational structure of the purchasing department (Schiele 2010). Researches on Early Purchasing Involvement (EPI) are still at the infancy and this chapter tries to fulfill this gap.

A complementary gap exists in Purchasing research on product innovation involving radical or discontinuous change—an increasingly significant problem as companies (and their competitors) develop new products and technologies that represent not only incremental improvements, but break with existing technological paradigms. Consider for example the challenge of the automotive industry in developing electric cars: radically different technological solutions are required, rendering existing competences and technologies obsolete. This suggests an important research gap because recent research has questioned the relevance of ESI in NPD projects characterized by high technological uncertainty (Song and Parry 1999; Primo and Amundson 2002; Song and Di Benedetto 2008). This challenge should be logically more important facing discontinuous innovation which seeks to provide a radically new product on a specific market but often combining technologies already used in some other sector. By scouting the market we can legitimately except that the Purchasing function should have an impact of the efficiency of the discontinuous innovation process.

Addressing this gap in current research, we seek to instigate a new research agenda in purchasing around the question: How do purchasing involvement in NPD change when faced with discontinuous innovation? The paper begins by presenting a systematic literature review focusing on definitions of discontinuous, disruptive and radical innovations and research on these types of innovation within the purchasing literature. On the basis of the systematic literature review we argue that developing and managing discontinuous innovation require an open, networked approach but a question arise whether a collaborative approach is necessarily appropriate or not. We briefly provide an overview of the literature on ESI that focuses specifically on purchasing involvement, before we set out some propositions for purchasing involvement in discontinuous innovation to focus and guide future research into this emerging field. The paper concludes by summarizing the conceptual implications and outlining some managerial contributions.

2 Systematic Literature Review on Discontinuous/Disruptive/Radical Innovation and Purchasing

This paper provides a rigorous starting point for discussing the role of purchasing in discontinuous innovation by providing a systematic review of the literature on purchasing and the three types of innovation that clearly concern a high degree of technological uncertainty: "discontinuous", "disruptive" and "radical". The main purpose of this is to provide a conceptual clarity over a field which is replete with different terminology covering seemingly very similar concepts. For instance, some authors argue that radical, disruptive and discontinuous innovations differ in, effectively, the degree of newness. For example, Linton (2002) argues that: "disruptive technologies are discontinuous, but discontinuous technologies are not necessarily disruptive". This suggests that disruptive innovations are more significant game-changers. However, the terms technological punctuation, shift, or breakthrough are also associated with discontinuous innovation, making it problematic to suggest that discontinuous involves less change than disruptive innovation (Rice et al. 2000; Magnusson et al. 2003; Phillips et al. 2006a).

This paper presents a systematic literature review method focusing on the three types of innovations: discontinuous, disruptive and radical innovation in the purchasing literature. We followed the steps recommended by Tranfield et al. (2003) and Brereton et al. (2007) for the search, filtering and review process, adapting their recommendations to fit the context and purpose of our research.

2.1 *The Search Process*

A systematic search was performed for journal articles dealing with the subject of discontinuous innovations and purchasing. We used the well-established databases EBSCO Business Source Complete, Science Direct and Wiley. The following keywords were used: "discontinuous innovation", "radical innovation", "and disruptive innovation" in combination with "purchasing", "sourcing", "supply", "procurement" and "buying". Using truncated keywords to capture different variations of these terms, we used these to search across titles and abstracts to ensure that we initially captured as many relevant articles as possible. We did not limit the search to any time period. Finding that the term disruptive innovation appeared relatively late from around in 2000 following on from Christensen and Rosenbloom's (1995) work on disruptive technology, we also decided to use the term "disruptive technology" following the same process. The initial search generated a list of 46 articles using the term "discontinuous innovation", 43 for "disruptive innovation" and 198 for "radical innovation": a total of 287 articles (Table 1).

Table 1 Paper searching and filtering process

Keyword hits	Search process Truncated search terms	Filter process 1 Inclusion/ Exclusion criteria based on titles and abstracts	Filter process 2 Further relevance based on journal quality (ABS)	Papers analysed Exclude papers based on consumer purchasing or supply chain management issues
Discontinuous innovation *AND* Purchasing[a]	46	36	26	10
Disruptive innovation *AND* Purchasing	43	24	22	2
Radical innovation *AND* Purchasing	198	55	36	10
Total	287	→ 11	→ 84	→ 22

2.2 *Relevance and Quality Filtering Process*

The titles and abstracts of the papers within this initial sample of 287 papers were first checked for relevance removing those that were obviously out of scope, such as those that dealt with public administration and those concerning finance. Then, we filtered these papers according to the quality of the journal. To ensure that only high quality research was considered, only articles published in major English language North American and European journals were included, excluding those that were not ranked in the Association of Business schools (ABS) journal quality list (Harvey et al. 2010). The ABS ranking draws from several other highly regarded journal quality rankings; although deselecting articles on the basis of any journal ranking is inevitable contentious, the ABS ranking is widely viewed as providing a reliable measure of research rigor and quality. We decided on two exceptions to this requirement, the Journal of High Technology Management Research and Research Technology Management, which were found to be significant in the volume of relevant papers that publish innovation studies and are widely recognized as quality journals.

The first filtering process narrowed the number of papers to 115. Having used three different databases for paper searching, we double-checked all papers across the entire database to remove duplicate articles. For example, some articles discussed two types of innovation and were therefore initially counted twice. This second manual filtering process thus reduced the list to 84 papers: these papers provided the basis for the review of definitions of the terms discontinuous innovation, disruptive innovation and radical innovation.

The sample of 84 papers was evaluated to identify articles that focused specifically on the role of the purchasing function in discontinuous/disruptive/radical innovation. One researcher performed an initial exclusion of those that were clearly irrelevant and where there was any doubt about whether or not to include a paper, another two researchers independently reviewed it for relevance: the final decision was a consensus amongst the three researchers. This filtering process excluded, for example, papers focused on consumer purchasing or wider supply chain management issues, finally resulting in a list of 22 articles that provided the basis for identifying themes specifically related to purchasing. Then, we categorized each paper into one of the three types of innovation. This was done by reading each paper in its entirety and completing a database. The database included information on the following variables: study (authors and year), method, definitions, focus, and main findings (Table 2).

2.3 *Findings*

In general, the literature review demonstrated that the use of the terms “discontinuous innovation”, “disruptive innovation” and “radical innovation” have been adopted by scholars based on their research objectives. As can be seen in Table 2, researches have used the three types of innovation to position the product system (discontinuous versus radical), market acceptance of the product (discontinuous versus disruptive), product objective (disruptive versus radical), or product newness (incremental, discontinuous and radical).

Based on our literature review, we argue that the term discontinuous innovation refers to the development of a product with high technological and market newness. Viewing the product innovation as a system may require the application or adoption of specific technology knowledge that may not have existed at all or may already exist in another marketplace, which can change the product configuration (Bergek et al. 2013; Magnusson et al. 2003). Viewing discontinuous innovation in this light, suggests a need to move towards networked or open innovation (Chesbrough 2003). In order to deal with today’s rapid technological changes, the innovation process has become collective and combinatorial in character, the emphasis shifting towards the firm’s external network of relationships as a means of accessing and acquiring new capabilities and combining these in novel ways (Bergek et al. 2013).

The themes highlighted in Table 2 indicate that the purchasing-related literature in this emerging field focuses on, in particular, the potential implications for technology sourcing and supplier relationships. In fact, this theme is not limited to papers on discontinuous innovation (Phillips et al. 2006a; Rohrbeck 2010) but is also evident in papers that focus on disruptive (e.g. Tomaselli and Di Serio 2013) and radical innovations (e.g. Perrons et al. 2004; Bunduchi 2013).

Figure 1 provides an overview of the overall findings from the Table 2. One observation is that one overlapping theme of the papers, which use the terms

Table 2 Discontinuous, radical and disruptive innovation studies with purchasing themes

Study	Method	Focus	Main findings	Journal
Discontinuous innovation				
Lambe and Spekman (1997)	Literature review	Provide understanding of interrelated dynamics of discontinuous technology, technology-sourcing alliances and new product development	Emergence of dominant design enhances importance of technology provided by a partner. Relative bargaining power can change over time as partners look to increase their "slice of the pie". Technology is critical to long-term strategic interest and ensures that absorption from alliance	Journal of Product Innovation Management
Theoharakis and Wong (2002)	Quantitative analysis of literature using socio-cognitive approach	Analyze and categorize market stories related to Local Area Network (LAN) technologies	Discontinuous innovations require active scanning of environment through internal and external information sources to recognize opportunities and enjoy higher performance. Product availability in external environment reflects increased interest of suppliers' product development and overall marketing efforts	Journal of Product Innovation Management
Rice et al. (2002)	Case study of 12 discontinuous innovation projects	Organizational challenges and resource uncertainty of managing transition from R&D project to operating unit	Managing transition from R&D project to operating unit is complicated due to: (1) Key external partners (significant contributors during development) may come up short during final phase (2) R&D unprepared to	IEE Transactions on Engineering Management

(continued)

Table 2 (continued)

Study	Method	Focus	Main findings	Journal
			cover business development costs associated with transition due to funding uncertainty	
Magnusson et al. (2003)	Two case studies based on two product development projects	Adapting organisation in order to manage architectural and modular innovations	Suppliers engaged at early stages of innovation process to ensure manufacturing considerations	International Journal of Innovation
Reid and De Brentani (2004)	Conceptual	Fuzzy front-end of discontinuous innovation, examining three perspectives: environment, individual role and organization	Fuzzy front-end model involves information gathering and adoption from environment based on assumption that external sources are primary source of new ideas and that even in-house ideas have input from external sources	Journal of Product Innovation Management
Phillips et al. (2006a)	7 case studies of different industries in a participative action research	Identify and develop specific tools and techniques for management of discontinuous innovation projects	In seeking a discontinuous technology, the customer must search for a supplier that may exist "below the radar" in dark and unfamiliar selection environments. Supplier relationships as strategic dalliances: short-lived but results in novel learning for both parties	R&D Management
Rohrbeck (2010)	Case study of three telecommunication operators	Using technology foresight (TF) to identify, anticipate and assess discontinuous technological change	Technology scouting can support sourcing of technologies by identifying opportunities and threats arising from developments at early stage	R&D Management

(continued)

Table 2 (continued)

Study	Method	Focus	Main findings	Journal
Athaide and Zhang (2011)	Structural equation model through field study interviews and questionnaire	Test a conceptual model of buyer-seller interactions during NPD	Product co-development relationships are most appropriate when targeting knowledgeable buyers with whom seller has enjoyed extensive relationships. Development of customized innovations calls for co-development relationship while discontinuous innovation requires emphasis on unilateral education based relationships	Journal of Product Innovation Management
Buffington et al. (2012)	Conceptual	Use of computer technology and creative human cognition to improve capabilities of seeking and achieving discontinuous innovation in product design	Development of discontinuous innovation via generative customization occurs through dynamic interaction between product and non-product parameters (new suppliers, consumers and market conditions), developed/interacted through generative and agent-based modeling simulation process	International Journal of Production Research
Bergek et al. (2013)	Case study in gas turbine and car industries	Analyze discontinuous innovation in two industries to test theories of competence-destroying, disruptive and creative accumulation	Technological discontinuities seldom lead to creative destruction, neither disruptive innovation, nor competence-destroying. Concept of creative accumulation proposed for explaining dynamic	Research Policy

(continued)

Table 2 (continued)

Study	Method	Focus	Main findings	Journal
			developments in industries supplying complex products	
Disruptive innovation				
Golicic and Sebastiao (2011)	Exploratory case study of multiple supply chains	Examines supply chain strategy and activities during mini-launch/early commercialization stage for disruptive product	Supply chain strategy for disruptive innovations should be based on market legitimacy; building supply chain capabilities i.e. align or affiliate with suppliers who possess key complementary skills or assets and common objective in targeting both (willing) suppliers and customers	Journal of Business Logistics
Tomaselli and Di Serio (2013)	Multiple case studies	Supply chain strategies in video game industry	Video game industry is dominated by companies that search to increase partnerships for development of consoles adopting a strategy that is horizontal and modular e.g. Sony's value chain control allowed it to reduce costs and gain market share	Journal of Technology Management & Innovation
Radical innovation				
McDermott and Handfield (2000)	Exploratory case studies	Exploring processes associated with development of radical new products	Suppliers of key components selected by cross-functional team early in planning stage. Project managers must involve purchasing personnel to identify potential suppliers with demonstrated record that offer technological solutions to meet market needs	Journal of High Technology Management Research

(continued)

Table 2 (continued)

Study	Method	Focus	Main findings	Journal
Perrons et al. (2004)	Survey	Measure how business unit dealt with principal suppliers during radical technology jumps	Maintaining strong ties with suppliers yields no significant long-term benefit for firms contending with radical new technologies	International Journal of Innovation Management
Song and Benedetto (2008)	Survey of 173 radical innovation projects	Supplier involvement in radical innovation ventures	Supplier involvements in development of radical innovation by new ventures depend on attracting right suppliers and depending on them for continued financial support	Journal of Operations Management
Dell'Era et al. (2010)	Two exploratory case studies	New Product Development in Italian design-driven companies	Radical innovations require experimentations, involving external technologies. Design-driven companies need supply network of varied technical capabilities, flexible and willing to experiment	Research Technology Management
Van de Vrande et al. (2011)	Patent counts from database (financial data, patent data, grants)	Test hypothesis with sample of 153 firms in pharmaceutical industry with large number of patents	Small investments in supply relationship with high level of flexibility appear to be most appropriate way to invest in breakthrough technologies. Investing in distant or unfamiliar technologies appears not to be favorable to successful development of pioneering technologies	Journal of Product Innovation Management

(continued)

Table 2 (continued)

Study	Method	Focus	Main findings	Journal
Ritala and Hurmelinna-Laukkanen (2013)	Cross-industry survey	Examine why some firms are better able than others to reap benefits from collaborating with competitors in innovation	Firm's ability to scan, evaluate and acquire knowledge from external sources (potential absorptive capacity) and to protect its innovations and core knowledge against imitation (appropriability regime) are relevant in increasing radical innovation outcomes of collaborating with competitors	Journal of Product Innovation Management
Xu et al. (2013)	64 pharmaceutical firms	Test hypothesized effects between alliance participation and innovation outcomes	Managers must carefully consider a firm's internal capabilities along with potential benefits gained from external alliances in developing radical innovations	Journal of Product Innovation Management
Datta and Jessup (2013)	Structural equation modeling: sample of patents in IT industry	Predict radicalness of innovation by exterior sourcing and technology distinctness	Selection capabilities of partners, especially when technologies are markedly different from firm's existing ones are crucial not only to develop radical innovations but also for long-term performance and strategic renewal	Technovation
Bunduchi (2013)	Case studies	The role of trust in supplier selection for collaborative new product development	Selecting suppliers for new product development by overreliance on trust hampers radical innovation as it encourages firms to explore information and competencies only within their supply base	Production, Planning and Control

(continued)

Table 2 (continued)

Study	Method	Focus	Main findings	Journal
Chiang and Wu (2016)	Modeling scenario	The role of contract design in radical process innovations	The manufacturer should engage key suppliers early under a contingent contract to fully exploit ESI when there are promising leads for radical improvement but insufficient in-house expertise	IEE Transactions on Engineering Management

[a]Purchasing = indicates that the following search terms were used: purchasing OR sourcing OR supply OR buying OR procurement

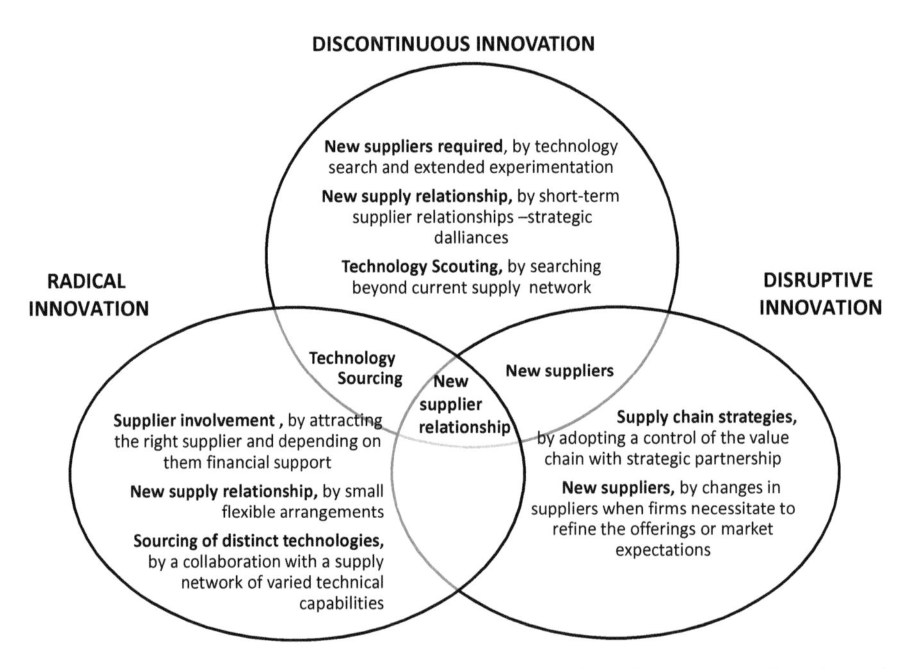

Fig. 1 Distinct and overlapping purchasing themes in research on discontinuous, disruptive and radical innovation

discontinuous and radical innovation, concerns challenges of technology sourcing. This includes research on technology scouting or searching and the need for flexibility supplier arrangements such as the need to balance partnership type relationships with short-term dalliances. In contrast, the purchasing-related research, which uses the term disruptive innovation, most notably Golicic and Sebastiao (2011), concentrates on issues to do with new forms of supply chain strategy, which also relate to the need to develop new suppliers but focusing more widely on issues concerning supply chain design. A common theme across the research on all three types of innovation is the need for changing supplier relationships (e.g. Phillips et al. 2006a, b; Perrons et al. 2004; Tomaselli and di Serio 2013), yet very little research focuses on the role of the purchasing function in relation to discontinuous innovation.

3 Discussion and Propositions

3.1 How Do Purchasing Involvement in NPD Change When Faced with Discontinuous Innovation?

Our systematic literature review revealed that an emerging field of research is beginning to question if purchasing involvement in NPD and sourcing practices that are suitable for incremental, or continuous, innovation is also suitable for discontinuous innovation (e.g. Perrons et al. 2004; Rohrbeck 2010; Van de Vrande et al. 2011).

An early contribution to research on purchasing involvement in NPD was made by Burt and Soukup (1985) who identified six points in the design process where purchasing should provide information and advice to engineering. They found that purchasing can act as a facilitator between NPD projects and suppliers' capabilities. Furthermore, purchasing can provide information about cost, performance, supply market availability, quality, and reliability of components. R&D or engineering teams would not normally have by their own this knowledge necessary to avoid supply problems stemming in NPD process. Thus, the specific knowledge of the supply market and a high level of interaction with other functions involved in NPD provides purchasing with a unique opportunity to facilitate the transfer of needs to supplier network. In the language of the French Sociologist Crozier and Friedberg (1977), such a boundary-spanning role can be described as "marginal-secant"—"the position of an actor that is the stakeholder of different systems of action, playing the role of a go-between and interpreter" (p. 86). They suggest that in a steady state of the environment, the importance of this role is low; however, when faced with uncertainty and discontinuity, such actors gain importance and power within the organization.

3.2 Three Propositions for Purchasing in Discontinuous Innovation Context

Based on our literature review we present three propositions, which we intend to be used to frame future research on this emerging topic.

Proposition 1 *Purchasing must search outside the existing supply chain.*

As highlighted earlier, a number of research (Perrons et al.'s 2004; Swink 1999; Primo and Amundson 2002; Knudsen 2007) have found that deep collaborative relationships with existing suppliers are insufficient in the face of innovations representing high technological uncertainty.

Discontinuous innovation changes the "rules of the game", creating the need to look in "dark" areas and develop relationships with organizations from unfamiliar zones (Bessant et al. 2005; Phillips et al. 2006a). Phillips et al. (2006a) propose that innovating firms seek to develop short-term "dalliances" with suppliers located on the periphery or even outside the firm's usual supply chain boundary. Rather than referring to short-term relationships with existing suppliers, dalliances are essentially 'flirtations' with new potential suppliers outside the firm's usual supply network. This idea is in conflicts with conventional wisdom of ESI whereby firms should collaborate with existing supply partners (Bonaccorsi and Lipparini 1994) with whom they have accrued experience and trust (Cousins et al. 2006).

This thought fit perfectly when you are facing incremental innovation but less when you ask for some new need who require suppliers able to think "out of the box". We can find also some other theoretical explanation of this mechanism looking at the "dark side" of buyer-supplier relationships (Villena et al. 2011). These authors advocate that there is an inverted curvilinear relationship between the "structural social capital" develop over time with a partner and the performance of the relationship for the buyer firm. They explain this fact by the information overloading developed in the interaction who can induce a lack of learning and some difficulties in decision making process. Bidault and Castello (2010) find the same inverted curvilinear relationship between the amount of trust between the participants of a NPD project and the level of innovation in the final product. It means that the NDP teams need a minimum of trust to innovate but also that with too much trust due to a long working experience together the partners are no longing thinking differently that could impede the innovation intensity. Our proposition here is not that purchasing should replace their partnership strategies with opportunistic behavior but that each context of co innovation requires different organizational responses.

As discontinuous innovation represents a highly uncertain situation, long-term stable supplier relationships may be inappropriate and 'dancing' with new potential business partners may be a better strategy than becoming entrenched in a long-term partnership arrangement. It is akin to the search-transfer problem (Hansen 1999) and Granovetter's classic concept of the strength of weak ties (1973): distant and infrequent relationships are efficient for knowledge sharing because they provide access to novel information by bridging otherwise disconnected actors.

More particularly, discontinuous innovation requires a different form of sourcing or scanning for new technology and a key strategic issue for firms is the delicate balancing of long-term collaborative relationships "with the ability to remain flexible and constantly scan for new "breakthrough" technologies from different partners" (Cousins et al. 2011, p. 932). Like Rohrbeck (2010) they do not, however, mention how this affects purchasing or sourcing departments but we propose purchasing can play a vital role in such *scouting* activities. We develop some advices for that in the following proposition.

Proposition 2 *Through ambidexterity Purchasing function can manage both exploitation and exploration activities.*

Our second proposition for future research continues the theme of developing the purchasing function to better handle the challenges of discontinuous innovation. The proposition focuses specifically on the need to develop an organizational model that can respond to the dual need for continuous and discontinuous innovation.

One challenge concerning ambidexterity is to develop alternative skills and competences for discontinuous innovation, which exist in parallel with those that are predominantly designed for continuous innovation, and to develop organizational behaviors and routines for each type of innovation (Bessant et al. 2005). In contrast with normal processes for continuous innovation situations, skills for discontinuous innovation are said to be "non-linear, stochastic, highly explorative and experimental, involving probing and learning, rather than targeting and developing" (Rice et al. 1998; Phillips et al. 2006a). Little is known about how such skills can be developed within purchasing but this might require recruitment of a specific group of new people as well as skills and competence development through training (see e.g. Nicholas et al. 2013). To support these changes new modes of organization may be required.

There is a distinct lack of research on how to organize the purchasing department so that it can take a leading role in the sourcing of new technologies and capabilities necessary for discontinuous innovation. Schiele (2010) stressed the dual role of purchasing in NPD: supporting the process of innovation while at the same time maintaining cost and integration responsibility over the entire product life cycle. This duality suggests a classic exploration—exploitation paradox of organization originally proposed by March (1991), who suggested that the effort for excellence in exploration and exploitation are naturally competing for scarce resources, they tend to crowd each other out.

The challenge of ambidexterity is difficult for a function as intimately connected to the exploitation process as Purchasing: many of its activities are concerned with current operations and continuous improvements, not least ongoing cost reduction. Early ambidexterity models suggested that a structural separation of exploration and exploitation activities enables firms to pursue both simultaneously. Structural separation is necessary because individuals, who have operational responsibilities, cannot explore and exploit simultaneously, as dealing with such contradictory frames creates operational inconsistencies and implementation conflicts (Gilbert 2006). Concretely a lot of firms as opted for a structural distinction between a department called 'advanced (or forward) sourcing' and another department called 'strategic sourcing' (Calvi 2000). The advanced sourcing team is integrated into all NPD projects while the strategic sourcing team has a stronger commercial focus and a connection with internal customers.

We can even imagine some ambidexterity arrangements inside the "advanced sourcing team" depending from the nature of innovation. The idea that discontinuous innovation requires a different organizational arrangement than incremental innovation is supported by O'Reilly and Tushman (2004), who suggest that the effectiveness of organization depends on the nature of the innovation effort sustained by the exploration process. Indeed, Schiele (2010) reported that BMW has divided its "advanced sourcing" department in two: one dedicated to support the NPD process and another dedicated to the scanning of their supply market for innovations. Such reports in the literature, however, remain very rare and more research is needed to investigate organizational responses to manage the sourcing requirements for both continuous and discontinuous innovation.

An alternative solution to the problem of ambidexterity has been offered by other Gibson and Birkinshaw (2004). They suggest that ambidexterity is something that should be present in the mind of each employee rather than being incorporated into the structure of the organization. They assert that ambidexterity is achieved by building an organizational context at the business unit level that emphasizes both performance management and social support. They assert that structural separation between exploration and exploitation units can lead to harmful isolation, and frameworks that are based exclusively on organizational structure are top-down by nature. This is in line with the work of Phillips et al. (2006b), which highlights the difficulties of ambidexterity including the problem of building entrepreneurial activities within established firms where activities need to be integrated, to some degree, with the rest of the organization.

Ambidexterity is a construct that is not directly observable, but like Godfrey and Hill (1995) demonstrated, unobservable constructs lie at the core of a number of influential theories in strategic management and by extension organization theory. Despite ambidexterity being a rather indiscernible concept, the problem of how to organize the purchasing department for new technology sourcing, the question of structural separation, and what forms of new skills and competences to develop, all these topics warrant further research.

Proposition 3 *Purchasing must develop absorptive capacity.*

Whilst one purchasing challenge is the sourcing of new technologies and capabilities from outside the existing supply chain, another challenge is to develop the ability to scout disparate marketplaces and then develop that market intelligence, code it, and absorb it within the firm (Cousins et al. 2011). In other words, firms and their purchasing departments need to develop *absorptive capacity* (Cohen and Levinthal 1990) so that they are able to appropriate external knowledge and transform it into new product and service offerings.

In a large empirical study, Rothaermel and Alexandre (2009) found that firms with greater levels of absorptive capacity obtain commensurately greater benefits from ambidexterity in technology sourcing. In short, absorptive capacity is the fulcrum that allows firms to leverage ambidexterity by better combining internal and external technology sourcing. This result is aligned with Tsai's (2009) suggestion that a firm with low absorptive capacity may not only find it very difficult to recognize the value of new ideas generated from supplier relationships, but may lack adequate ability to assimilate ideas into product innovation. Thus, external searching and scouting does not work effectively without extensive internal effort. In order to gain competitive advantage from external resources, managers should “Ask not what your suppliers can do for you; ask what you can do with your suppliers” (Takeishi 2001, p. 419).

Zahra and George (2002) define absorptive capacity as a set of organizational routines and processes, by which firms *acquire*, *assimilate*, *transform*, and *exploit* external knowledge to produce a new offer. Absorptive capacity focuses specifically on internal functional requirements for managing external collaboration and depends on individuals who stand at the crossroad of the firm and the external environment. Tu et al. (2006) argue that absorptive capacity requires organizational mechanisms that can deal with both internal and external knowledge, information and technology. They defined absorptive capacity as “the organizational mechanisms that help to identify, communicate, and assimilate relevant external and internal knowledge. The elements of absorptive capacity are considered to be the firm’s existing knowledge base, the effectiveness of systems that scan the environment, and the efficacy of the firm’s communication processes”. We believe that if purchasing can develop absorptive capacity it is well-positioned to play a key role in acting as a facilitator, for example, in relation to R&D or Engineering that are usually closely involved in, or responsible for, boundary-spanning activities such as technology scouting (Rohrbeck 2010). The role of purchasing in the absorptive capacity process has not been studied in any detail to date. Furthermore, it remains unclear whether these NPD-focused roles are also appropriate when innovation challenges switch from continuous to discontinuous innovation.

4 Contributions and Conclusion

The systematic review of the literature clearly highlighted the critical role of an open, networked, approach to innovation. We subsequently discussed how purchasing can facilitate the transfer of knowledge and technology from suppliers by sourcing practices. We argued that purchasing needs to play a key role in this process and presented three propositions for future research. The three propositions build on emerging research can also provide guidelines for purchasing practices.

As an **academic contribution** this chapter has sought to advance a new research agenda in an important new field of research within purchasing and innovation management theory: how purchasing can—and should—be involved in facilitating discontinuous innovation. This paper has contributed to this gap by identifying a set of articles that specifically address with this issue and analyzing and discussing these studies, identifying that this literature suggests important challenges for purchasing function wishing to be contributing in discontinuous innovation context (Fig. 1). Thus, we have formulated three propositions issue from the literature review to guide future research into purchasing involvement in discontinuous innovation.

There is clearly a largely untapped need for further conceptual and empirical research on changes in the role of purchasing in discontinuous innovation. The paper has formulated and explained propositions to frame how this research could be taken forward. We suggest that researchers initially employ qualitative research designs to provide rich in-depth case studies of the process of managing purchasing involvement in discontinuous innovation. Once rich insights have been created through qualitative research, we suggest that quantitative research be conducted to validate findings at a large scale in a variety of context.

The propositions put forward during the discussion can also serve as a guide a contribution for purchasing practitioners who are anxious to contribute to the deployment of discontinuous innovation in their firm. The first message is to be not satisfied with the company's historical supplier base if one wants to "think out of the box". This requires adapting the supplier selection processes which, in large companies, are so formalized and constraining that they can hamper business with start-ups for example. Thus, to by-pass the classical process of the firm or to create adapted process should be the prerequisite to the customer firms not to lose business with start-ups capable of contributing actively to their offer creation process. Our second proposition encourages the CPOs to build ambidextrous organizations, in particular in the specialization of buyers in "advanced sourcing team" connected with NPD projects and even in activities purely linked to innovation as we can already observe in some firms with the creation of innovation buyer position (Servajean and Calvi 2018). The third one highlights the necessary evolution of the purchasing practices in order to create a real absorptive capacity for external resources. The challenge is here in the evolution of sourcing (acquisition) and internal collaboration (assimilation) practices.

In conclusion this paper has investigated the question: *How do purchasing involvement in NPD change when faced with discontinuous innovation?* We addressed this question first by defining discontinuous innovation, in relation to radical and disruptive innovations, through a systematic literature review. Based on the existing literature we found little support for a meaningful separation of these types of innovation, merely some subtle differences. Our use of the term discontinuous innovation is closely related to Rice et al. (2000) in focusing on innovations that transform the relationship between customers and suppliers, restructure marketplace economies, displace current products, create entirely new product categories, and provide a platform for the long-term growth sought by corporate leaders.

References

Athaide GA, Zhang JQ (2011) The determinants of seller-buyer interactions during new product development in technology-based industrial markets. J Prod Innov Manag 28(1):146–158

Bergek A, Berggren C, Magnusson T, Hobday M (2013) Technological discontinuities and the challenge for incumbent firms: destruction, disruption or creative accumulation? Res Policy 42 (6):1210–1224

Bessant JR, Lamming RC, Noke H, Phillips WE (2005) Managing innovation beyond the steady state. Technovation 25(12):1366–1376
Bidault F, Castello A (2010) Why too much trust is death to innovation. MIT Sloan Manag Rev 51 (4):33–39
Bonaccorsi A, Lipparini A (1994) Strategic partnerships in new product development: an Italian case study. J Prod Innov Manag 11:134–145
Brem A, Tidd J (2012) Perspectives on supplier innovation. Imperial College Press, London
Brereton P, Kitchenham BA, Budgen D, Turner M, Khalil M (2007) Lessons from applying the systematic literature review process within the software engineering domain. J Syst Softw 80 (4):571–583
Buffington J, Amini M, Keskinturk T (2012) Development of a product design and supply-chain fulfillment system for discontinuous innovation. Int J Prod Res 50(14):3776–3785
Bunduchi R (2013) Trust, partner selection and innovation outcome in collaborative new product development. Prod Plan Control 24(2–3):145–157
Burt DN, Soukup WR (1985) Purchasing's role in new product development. Harv Bus Rev 63 (3):90–97
Calvi R (2000) Le role des services achats dans le developpement des produits nouveaux: une approche organisationnelle [The role of purchasing in new product development: an organisational approach]. Finance Contrôle Stratégie 3(2):31–55
Chesbrough H (2003) Open innovation: the new imperative for creating and profiting from technology. Harvard Business School Press, Boston, MA
Chiang IR, Wu SJ (2016) Supplier involvement and contract design during new product development. IEEE Trans Eng Manag 63(2):248–258
Christensen C, Rosenbloom RS (1995) The attacker's advantage: technological paradigms, organizational dynamics and the value network. Res Policy 24:233–257
Cohen WM, Levinthal DA (1990) Absorptive capacity: a new perspective on learning and innovation. Adm Sci Q 35(1):128–152
Cousins P, Handfield R, Lawson B, Petersen K (2006) Creating supply chain relational capital: the impact of formal and informal socialization processes. J Oper Manag 24(6):851–863
Cousins PD, Lawson B, Petersen KJ, Handfield RB (2011) Breakthrough scanning, supplier knowledge exchange, and new product development performance. J Prod Innov Manag 28 (6):930–942
Crozier M, Friedberg F (1977) L'Acteur et le Système. Les Contraintes de l'action Collective. Éditions du Seuil, Paris
Datta A, Jessup LM (2013) Looking beyond the focal industry and existing technologies for radical innovations. Technovation 33(10/11):355–367
Dell'Era C, Marchesi A, Verganti R (2010) Mastering technologies in design-driven innovation. Res Technol Manag 53(2):12–23
Farmer D (1981) The role of procurement in new product development. Int J Phys Distrib Mater Manag 11(2/3):46–54
Gibson CB, Birkinshaw J (2004) The antecedents, consequences, and mediating role of organizational ambidexterity. Acad Manag J 47(2):209–226
Gilbert CG (2006) Change in the presence of residual fit: can competing frames coexist? Organ Sci 17(1):150–167
Godfrey PC, Hill CWL (1995) The problem of unobservable in strategic management research. Strateg Manag J 16:519–533
Golicic SL, Sebastiao HJ (2011) Supply chain strategy in nascent markets: the role of supply chain development in the commercialization process. J Bus Logist 32(3):254–273
Granovetter MS (1973) The strength of weak ties. Am J Sociol 78(6):1360–1380
Hansen MT (1999) The search-transfer problem: the role of weak ties in sharing knowledge across subunits. Adm Sci Q 44(1):82–111
Harvey C, Kelly A, Morris H, Rowlinson M (2010) Academic journal quality guide version, 4. The Association of Business Schools, London

Johnsen TE (2009) Supplier involvement in product development and innovation – taking stock and looking to the future. J Purch Supply Manag 15(3):187–197
Knudsen MP (2007) The relative importance of interfirm relationships and knowledge transfer for new product development success. J Prod Innov Manag 24(2):117–138
Lakemond N, van Echtelt F, Wynstra F (2001) A configuration typology for involving purchasing specialists in product development. J Supply Chain Manag 37(4):11–27
Lambe CJ, Spekman RE (1997) Alliances, external technology acquisition, and discontinuous technological change. J Prod Innov Manag 14(2):102–116
Linton JD (2002) Forecasting the market diffusion of disruptive and discontinuous innovation. IEEE Trans Eng Manag 49(4):365–374
Magnusson T, Lindström G, Berggren C (2003) Architectural or modular innovation? Managing discontinuous product development in response to challenging environmental performance targets. Int J Innov Manag 7(1):1–26
March JG (1991) Exploration and exploitation in organizational learning. Organ Sci 2(1):71–87
McDermott C, Handfield R (2000) Concurrent development and strategic outsourcing: do the rules change in breakthrough innovation? J High Technol Manag Res 11(1):35–57
Nicholas J, Ledwith A, Bessant J (2013) Reframing the search space for radical innovation. Res Technol Manag 56:27–35
O'Reilly CA, Tushman ML (2004) The ambidextrous organization. Harv Bus Rev 82(4):74–81
Perrons RK, Richards MG, Platts K (2004) The effect of industry clockspeed on make-buy decisions in the face of radical innovations: an empirical test. Int J Innov Manag 8(04):431–454
Petersen KJ, Handfield RB, Ragatz GL (2005) Supplier integration into new product development: coordinating product, process and supply chain design. J Oper Manag 23(3–4):371–388
Phillips WE, Lamming RC, Bessant JR, Noke H (2006a) Discontinuous innovation and supply relationships: strategic dalliances. R&D Manag 36(4):451–461
Phillips WE, Noke H, Bessant J, Lamming R (2006b) Beyond the steady state: managing discontinuous product and process innovation. Int J Innov Manag 10(2):1–23
Primo MAM, Amundson SD (2002) An exploratory study of the effects of supplier relationships on new product development outcomes. J Oper Manag 20(1):33–52
Reid SE, De Brentani U (2004) The fuzzy front end of new product development for discontinuous innovations: a theoretical model. J Prod Innov Manag 21(3):170–184
Rice MP, Leifer R, O'Connor G (2000) Managing the transition of a discontinuous innovation project to operational status. IEEE Trans Eng Manag 15:586–590
Rice MP, Leifer R, O'Connor G (2002) Commercializing discontinuous innovations: bridging the gap from discontinuous innovation project to operations. IEEE Trans Eng Manag 49(4):330–340
Rice MP, O'Connor GC, Peters LS, Morone JG (1998) Managing discontinuous innovation. Res Technol Manag 41(3):52
Ritala P, Hurmelinna-Laukkanen P (2013) Incremental and radical innovation in coopetition—the role of absorptive capacity and appropriability. J Prod Innov Manag 30(1):154–169
Rohrbeck R (2010) Harnessing a network of experts for competitive advantage: technology scouting in the ICT industry. R&D Manag 40(2):169–180
Rothaermel FT, Alexandre MT (2009) Ambidexterity in technology sourcing: the moderating role of absorptive capacity. Organ Sci 20(4):759–780
Schiele H (2010) Early supplier integration: the dual role of purchasing in new product development. R&D Manag 40(2):138–153
Schiele H, Calvi R, Gibbert M (2012) Customer attractiveness, supplier satisfaction and preferred customer status: introduction, definitions and an overarching framework. Ind Mark Manag 41 (8):1178–1185
Servajean R, Calvi R (2018) Shades of the innovation-purchasing function – the missing link of open innovation. Int J Innov Mgt 22(1):1850008
Song M, Di Benedetto A (2008) Supplier's involvement and success of radical new product development in new ventures. J Oper Manag 26(1):1–22
Song M, Parry ME (1999) Challenges of managing the development of breakthrough products in Japan. J Oper Manag 17:665–688

Swink ML (1999) Threats to new product manufacturability and the effects of development team integration processes. J Oper Manag 17:691–709
Takeishi A (2001) Bridging inter- and intra-firm boundaries: management of supplier involvement in automobile product development. Strateg Manag J 22(5):403–433
Theoharakis V, Wong V (2002) Marking high-technology market evolution through the foci of market stories: the case of local area networks. J Prod Innov Manag 19(6):400–411
Tomaselli FC, Di Serio LC (2013) Supply networks and value creation in high innovation and strong network externalities industry. J Technol Manag Innov 8(4):177–185
Tranfield D, Denyer D, Smart P (2003) Towards a methodology for developing evidence-informed management knowledge by means of systematic review. Br J Manag 14(3):207–222
Tsai KH (2009) Collaborative networks and product innovation performance: toward a contingency perspective. Res Policy 38(5):765–778
Tu Q, Vonderembse MA, Ragu-Nathan TS, Sharkey TW (2006) Absorptive capacity: enhancing the assimilation of time-based manufacturing practices. J Oper Manag 24(5):692–710
Van de Vrande V, Vanhaverbeke W, Duysters G (2011) Technology in-sourcing and the creation of pioneering technologies. J Prod Innov Manag 28(6):974–987
Van Echtelt FEA, Wynstra F, van Weele AJ, Duyesters G (2008) Managing supplier involvement in npd: a multiple-case study. J Prod Innov Manag 25(2):180–201
Villena VH, Revilla E, Choi TY (2011) The dark side of buyer–supplier relationships: a social capital perspective. J Oper Manag 29(6):561–576
Wynstra JYF, Axelsson B, van Weele A (2000) Driving and enabling factors for purchasing involvement in product development. Eur J Purch Supply Manag 6(2):129–141
Xu S, Wu F, Cavusgil E (2013) Complements or substitutes? Internal technological strength, competitor alliance participation, and innovation development. J Prod Innov Manag 30 (4):750–762
Zahra SA, George G (2002) Absorptive capacity: a review, reconceptualization and extension. Acad Manage Rev 27(2):185–203

Richard Calvi PhD, HDR, from Grenoble University is full Professor at Savoie University (France). He works on the topic of purchasing and supply management (PSM) for more than 20 years, and he is an active researcher in the IPSERA network (International Purchasing and Supply Education and Research Association). He's directing a master's degree specialized in PSM, and he had written more than 30 articles on this topic in various academic journals such as *R&D Management*, *Industrial Marketing Management*, *International Journal of Purchasing Management*, and *International Journal of Innovation Management*. He is also the scientific director of the French Purchasing National Council (CNA) and coeditor of the "Excellence HA" journal.

Thomas Johnsen PhD, is the Gianluca Spina Professor of Supply Chain Management at Politecnico di Milano School of Management. Prior to this role, he was Professor of Purchasing and Supply Management at Audencia Nantes School of Management and, most recently, at ESC-Rennes School of Business. He is currently associate editor of the *Journal of Purchasing & Supply Management*. He has been executive board member of the International Purchasing & Supply Education & Research Association (IPSERA) and was Chair of the IPSERA 2013 conference at Audencia. His book (with M. Howard and J. Miemczyk) *Purchasing and Supply Chain Management: A Sustainability Perspective* was published by Routledge in April 2014 and awarded the ACA-Bruel coup de coeur prize.

Katia Picaud has a PhD degree in management from the University of Nantes, France, and today, she works as a lecturer in the field of purchasing and supply management, strategic management, and international business management at Audencia Business School. Her research interests include new product development, innovation management, purchasing involvement in innovation, and organizational learning theories. She has a BA in International Relations and an International Master in Management. She received the Eiffel Scholarship for Academic Excellence by the French Ministry of Foreign Affairs to pursue her PhD in 2013. She has published conference papers in the European Academy of Management (EURAM), International Purchasing and Supply Education and Research Association (IPSERA), and Association International de Management Stratégique (AIMS).

National Culture as an Antecedent for Information Sharing in Supply Chains: A Study of Manufacturing Companies in OECD Countries

Ruggero Golini, Andrea Mazzoleni, and Matteo Kalchschmidt

Abstract The chapter investigates the importance that differences in national culture characteristics have in explaining the investment that companies do in collaboration with their supply chain partners. Empirical analysis is based on the fifth round of the International Manufacturing Strategy Survey (IMSS V), through which have been gathered data among 392 companies belonging to 16 Organization for Economic Cooperation and Development (OECD) economies. Two specific cultural traits have been considered: individualism-collectivism and power distance. Results prove evidence of significant and complex relationship between the mentioned cultural characteristics and the amount of investment that a focal company is willing to develop in information sharing with its supply chain partners.

1 Introduction

The twentieth century has brought profound changes in the economic and social environment. The borders of countries have been blurring and market competition has consequently increased both from/in developed and developing countries. Companies have started to deal with increasingly more demanding customers, globalized supply chains and uncertainties. This has led traditional approaches such as arm-length relationship to become no longer effective to support competition; thus, manufacturers have started to strategically interact with their supply chain partners to effectively meet market requirements, leading to integrated supply chains (e.g., Pagell et al. 2005; Flynn et al. 2010; Frohlich and Westbrook 2001; Zhao et al. 2008) where integration can be seen as the evolution of approaches such as customer-supplier partnership (e.g., Lamming 1993). In this strand, several studies have argued the relevance of supply chain integration to build a competitive advantage (e.g., Flynn et al. 2010; Sanders 2008) based on the contribution of each supply

R. Golini (✉) · A. Mazzoleni · M. Kalchschmidt
Department of Management, Information and Production Engineering, University of Bergamo, Dalmine, BG, Italy
e-mail: ruggero.golini@unibg.it; andrea.mazzoleni@unibg.it; matteo.kalchschmidt@unibg.it

A. C. Moreira et al. (eds.), *Innovation and Supply Chain Management*, Contributions to Management Science, https://doi.org/10.1007/978-3-319-74304-2_9

chain partner to enhance the overall supply chain performance (Yeung et al. 2009). The counterproof is that a lack of integration can be detrimental on performance (Frohlich and Westbrook 2001; Wiengarten et al. 2014) especially in global supply chains (Golini and Kalchschmidt 2011).

However, some authors (e.g., Zhao et al. 2008; Yeung et al. 2009) have suggested how the understanding of supply chain integration enablers and antecedents is still scarcely investigated. In particular, country-related factors have been seldom considered (Wiengarten et al. 2014). Among these, national culture is one of the most neglected factors, despite many scholars acknowledge its importance in the operations and supply chain management fields (e.g., Flynn et al. 2010). Indeed, culture affects the way through which people act when engaging in a business relationship (e.g., Griffith and Myers 2005). It follows that supply chain decisions cannot be undertaken disregarding the culture of the country in which a company is operating (Prasad and Babbar 2000; Pagell et al. 2005). This issue becomes even more relevant in global supply chains where companies belonging to different countries (i.e. cultures) need to interact.

As a consequence, in this work we aim to provide some evidence on the effect of national culture on a specific aspect of supply chain integration, that is supply chain information sharing (SCIS) with suppliers and customers, thus we refer to external integration (e.g., Flynn et al. 2010). Although previous research (e.g., Flynn et al. 2010) acknowledges the relatedness between internal integration and external integration, our chapter is limited to external integration and, in particular, to external supply chain information sharing.

As previous scholars have noted (e.g., Frohlich and Westbrook 2001; Yeung et al. 2009; Cagliano et al. 2003) external integration can be achieved through investments in information sharing and physical system coupling. SCIS reflects the exchange of information about production plans, inventories level, order tracking and tracing between a manufacturer and its supply chain partners (e.g. Cagliano et al. 2003) whilst system coupling represents the joint investments that customers and suppliers carried out to coordinate physical activities (e.g., just in time, vendor managed inventory, continuous replenishment) and achieve, as a result, a faster flow of products with less inventory levels along the supply chain (Power 2005).

Here we d focus on information sharing mainly for two reasons. First of all, information sharing can be performed with a wider array of suppliers and customers both at the local and global scale, whilst system coupling requires some degree of physical proximity and greater investments (Narasimhan and Nair 2005). Second, information sharing is considered an enabler for system coupling, therefore the study of information sharing (Zhou and Benton 2007) is the first step to be undertaken towards the understanding of national culture as an antecedent for supply chain integration in broader terms.

Acknowledging the abovementioned gaps, this work addresses the following research question:

What is the relationship between national culture and the willingness of a company to invest in information sharing both with their suppliers and customers?

Specifically, we use multi-country data at the plant level (International Manufacturing Strategy Survey) to assess SCIS while the cultural characteristics are assessed using the Hofstede (1980) taxonomy. In particular, the cultural characteristics of individualism-collectivism and power distance will be considered as we identified those as the most theoretically related to SCIS. Each of them, reflects a relevant cultural trait for SCIS adoption as these aspects are related to how people engage in a relationship and how people perceive power and authority (Zhao et al. 2008).

In doing so, we focused only on Organization for Economic Cooperation and Development (OECD) countries. Firstly, this choice allows us to deeply focus on a specific environment by considering the OECD market economies. Second, this might help to reduce issues related to differences in economic development, enhancing the role of the cultural characteristics in explaining differences in the investment in SCIS.

The results show that both individualism-collectivism and power distance are significant in explaining variations in SCIS integration (with power distance being stronger). In other terms, a manufacturer located in a country characterized by a low individualism-collectivism and high power distance shows a higher propensity to invest in SCIS integration with its supply chain partner. Most interestingly, the effect of culture is stronger towards suppliers, witnessing that national culture particularly affects upstream supply chain integration.

The work is structured as follows: initially, a detailed literature review regarding supply chain integration, information sharing and national culture will allow us to establish the research hypotheses. Then, we turn our attention on the methodology. Afterwards, empirical results are provided as well as the theoretical and managerial implications. Lastly, conclusions and opportunities for future developments will be drawn.

2 Theoretical Background and Hypotheses Development

2.1 Supply Chain Integration and National Culture

Although supply chain integration is still an emerging area of research (Flynn et al. 2010), several studies have suggested its importance to achieve a competitive advantage and to improve performance (Frohlich and Westbrook 2001; Vickery et al. 2003; Zhao et al. 2008) as well as its relevance in preventing issues such as the well-known bullwhip effect (Lee et al. 1997). It is therefore worth understanding which are the antecedents for supply chain integration (Zhao et al. 2008; Yeung et al. 2009), such as country related factors (Wiengarten et al. 2014). Responding to this call, recent research on supply chain integration has begun to pay attention to the role of cultural values, either as an enabler of integration as well as in relation with supply chain performance (e.g., Zhao et al. 2006, 2008; Cannon et al. 2010; Cai et al. 2010). Considering a global supply chain perspective, Griffith and Myers (2005) have argued how cultural values might address people's behavior that in turn might affect how practices are applied and businesses are managed, a suggestion that is traceable in other authors (e.g., Cannon et al. 2010; Naor et al. 2010). In a similar vein, Chen et al. (1998) as well as Huff and Kelley (2003) have identified how cultural traits might influence the way through which people tend to cooperate with each other.

From a supply chain point of view, by considering buyer-supplier relationship, Cannon et al. (2010) have noted how cultural differences might undermine aspects such as trust, commitment and long-term orientation and how "cultural differences may well present challenges to the health of these relationship" (Cannon et al. 2010, p. 506). The same authors have also assessed how cultural differences might change the level of buyer's long-term orientation, when buyer's trust and supplier performance have been considered as antecedents. In a similar vein, Zhao et al. (2008) have considered culture as an antecedent of supply chain integration considering the Chinese's cultural environment. They found that there are two aspects strictly tied to supply chain integration: power, defined according to Yeung et al. (2009, p. 69) as "the member's ability to influence the behavior and decision of other members", and relationship commitment, i.e. the willingness of members to invest resources in a long-term relationship. Furthermore, Griffith and Myers (2005) have compared a sample of US and Japanese companies with the aim to understand if the fit between the mentioned governance strategies and the expectation of the several supply chain members, arising from their cultural values, improves companies' performance.

Despite a quite rich background of articles, several gaps still exist. First of all, the extant literature has not assessed culture in a direct way: specifically, cultural values have been assessed according to trust, power and commitment (e.g., Zhao et al. 2008). While this point of view offers a detailed and company or relation-specific perspective, it lacks in offering a broader and country-level perspective. Moreover, supply chain integration is often analyzed as a whole, including information sharing, collaboration, risk/revenue sharing and system coupling while each of these elements has peculiar characteristics in terms, for instance, of level of investment.

As a consequence, we try to fill this gap by assessing culture at country level and focusing on a specific aspect of supply chain integration that is information sharing.

In particular, we rely on the Hofstede (1980)'s dimensions of individualism-collectivism and power distance, two cultural traits that might reflect, respectively, how people engage in a relationship and how people perceive power and authority. This suggestion is traceable in Zhao et al. (2008). Therefore, in the next sections the meaning of the mentioned cultural traits is revised, and research hypothesis established.

2.2 Supply Chain Information Sharing

Supply chain integration is a holistic concept that encompasses ongoing collaboration, commitment, mutual trust, sharing of risks, information, money and rewards as well as administrative tasks along a manufacturer and its customers and suppliers (Flynn et al. 2010).

A manufacturer can integrate its activities either internally, externally or both, reflecting the multidimensionality of supply chain integration (see, Flynn et al. 2010; Wong et al. 2011). Internal integration is achieved when initiatives and programs are carried out internally within the manufacturer (i.e. among functional units), whilst external integration when a manufacturer coordinates activities with either its customers, suppliers or both; supplier integration is built when a focal company collaborates with its supplier by sharing with them information about inventories and production plans (Wong et al. 2011); similarly customer integration encompasses a joint activities between a focal companies and its customers, with the aim to anticipate market requirements and "matching supply with demand" (Wong et al. 2011, p. 605).

In this regard, literature has considered supply chain integration through two main areas of application named *technological collaboration*, that reflects joint efforts made in product development (Dowlatshahi 1998; Hartley et al. 1997; Petersen et al. 2006; He et al. 2014) and *operational collaboration*, that mirrors the integration in the production-logistics processes (Cagliano et al. 2006; Frohlich and Westbrook 2001).

Focusing on external operational collaboration, two dimensions have been identified: information sharing and physical system coupling (e.g., Frohlich and Westbrook 2001; Yeung et al. 2009; Cagliano et al. 2003). As already mentioned in the introduction, in this paper we focus on information sharing that is considered the first stage and an enabler for system coupling (Narasimhan and Nair 2005; White et al. 2005).

Information sharing, alongside collaborative planning (Cai et al. 2010), has been recognized as an activity through which a focal company can achieve integration about information with either its customers and suppliers. From a supply chain integration perspective, information sharing is a theme through which integration can arise (Cagliano et al. 2006; Yeung et al. 2009; Frohlich and Westbrook 2001); more specifically, through information sharing, the coordination of the supply chain, as well as its planning, can be enhanced (e.g., Welker et al. 2008).

In detail, information sharing refers to the exchange of information between a focal company and its supply chain partners, concerning issues such as production plans, inventories level and market demand. According to Yeung et al. (2009, p. 67), information sharing can be defined as "the degree to which a firm can coordinate the activities of information sharing, and combines core elements from heterogeneous data management systems, content management systems, data warehouses, and other enterprise applications into a common platform, in order to substantiate integrative supply chain strategies"; a definition that underlines either the technological and managerial issues that might hinder an effective implementation of the considered activity. Information sharing requires, besides the willingness to exchange, receive and manage data (Van der Vaart et al. 2012), standardized supply chain practices (Zhou and Benton 2007) as well as the ICT system integration between the focal company and its supply chain partners, resulting beneficial in reducing issue such as the bullwhip effect (e.g., Lee and Whang 1997). In the context of global supply chains, however, such benefits, even if existing (Gunasekaran and Ngai 2005; Lee and Whang 1997), can be more difficult to be achieved since coordination of all decisions and activities is always needed (Yeung et al. 2009). Moreover, the "trust" that should exist among supply chain partners can be hampered if supply chain partners belong to different cultures (e.g., Zhao et al. 2008; Cai et al. 2010; Yeung et al. 2009). As a consequence, we focused on the role that national culture can have on SCIS external integration.

2.3 Individualism-Collectivism and Information Sharing

In the field of manufacturing and supply chain management, individualism-collectivism has been seen as a construct able to explain differences in cross-cultural behavior (e.g., Power et al. 2010; Cannon et al. 2010). In his research, Hofstede and Hofstede (1991) defines individualism (IDV) as "a preference for loosely-knit social framework in which individuals are expected to take care of themselves and their immediate families only" whilst individualism-collectivism as "a preference for a tightly-knit framework in society in which individuals can expect their relatives or members of a particular in-group to look after them in exchange for unquestioning loyalty". More specifically, people in individualistic culture tend to act accordingly to their own interests rather than the society's goals. Conversely, people in a more collectivistic culture place the society's interest above their own. According to Power et al. (2010), the mentioned attitudes underline the people's goals orientation in performing tasks. The same suggestion is traceable in Doney et al. (1998, p. 608), according to which "individualism-collectivism reflects the way people interact, such as the importance of unilateral versus group goals, the strength of interpersonal ties, respect for individual accomplishment, and tolerance for individual opinion". Therefore, according to Cannon et al. (2010), people in individualistic culture will be more autonomous and more self-confident; conversely, people in a collectivistic culture will be more interdependent reflecting how the mentioned cultural trait might affect the way through which people establish and sustain their relationships, an aspect that can be crucial in the light of information sharing integration activities. According to Zhao et al. (2008), a collectivistic culture such as China, could be more suitable to engage in a type of relationship that underline mutual trust and long-term orientation (i.e. normative relationship), such as SCIS requires. Therefore, we posit:

Hp1a: A manufacturer located in an individualistic culture will be less willing to share information with its suppliers.

Hp1b: A manufacturer located in an individualistic culture will be less willing to share information with its customers.

2.4 Power Distance and Information Sharing

Strictly tied with individualism-collectivism, the cultural trait of power distance (PDI) "expresses the degree to which the less powerful members of a society accept and expect that power is distributed unequally" (Hofstede and Hofstede 1991). More specifically, power distance reflects how people is comfortable with decisions taken from the most powerful members in a society (e.g., Zhao et al. 2008). It follows, that a manufacturer located in a high PDI culture might be more amenable with regards to the sources of power, such as a customer. In fact, customers have two sources of power: the first one is the traditional bargaining power (Porter 1980) the second one is related to knowledge, reputation and legitimacy (i.e. expert power, referent power and legitimate power) (Zhao et al. 2008). Similarly, Yeung et al. (2009) have point out how, in a high PDI culture, a powerful supplier might affect a buyer in implementing supply chain integration. Furthermore, SCIS carries some specific risks (e.g., misuse of information, opportunistic behavior by customers) (e.g., Handfield and Bechtel 2002), thus there can be resistance to its implementation. However, if the manufacturer supplier acknowledges the power of the customer or a supplier it will share information more easily. Furthermore, supply chain integration is based on the idea to have a better coordination of the supply chain partners at the different stages to enhance the overall supply chain performance (Yeung et al. 2009). This coordination, however, requires that some decisional autonomy at the company level is left in favor of the decisions at the supply chain level. Since in a high PDI culture, authority is centralized, companies in high PDI countries will be more willing to leave decisions affecting their operations to those parties that coordinate the entire supply chain. In turn, companies characterized by high PDI will be more willing to integrate their supply chains and, by consequence, to share information with suppliers and customers.

Therefore, we posit:

Hp2a: A manufacturer located in a high power distance culture will be more willing to share information with its suppliers.

Hp2b: A manufacturer located in a high power distance culture will be more willing to share information with its customers.

3 Research Methodology

3.1 Sample Description and Data Collection

The research hypotheses are tested relying on data gathered throughout the fifth edition of the International Manufacturing Strategy Survey (IMSS V), carried out in 2009. Originally launched by London Business School and Chalmers University of Technology, this project studies manufacturing and supply chain strategies within assembly industry (ISIC 28-35 classification), through a detailed questionnaire administered simultaneously in different countries by local research groups. Responses are gathered in a unique database (Lindberg et al. 1998), which is available only to those who have actively participated in the data collection process.

The basic structure of the questionnaire is as follows: the first section of the questionnaire pertains to the business unit, in order to gather general information (e.g., company size, industry, production network configuration, competitive strategy and business performance) on the context in which manufacturing takes place, whereas the other sections refer to the plant's dominant activity, focusing on manufacturing strategies, practices and performance. Dominant activity is defined as the most important activity, which best represents the plant. The plant is chosen as the unit of analysis in order to avoid problems related to business units with multiple plants operating in different ways.

In each edition, the questionnaire is partially redesigned in order to ensure alignment with the most recent research goals. This update is carried out by a design team composed of a pool of international researchers and, thus, avoids the researchers' country-biases (Van de Vijver and Leung 1997). Data in each country are gathered in the country's native language and the questionnaire is translated and back-translated to check for consistency (Behling and Law 2000). Companies are selected from a convenience sample or randomly selected from economic datasets and then the operations, production or plant managers are contacted and asked to assist in the research. If the respondent agrees, the questionnaire is sent and, where appropriate, a reminder is sent after a few weeks. Questionnaires that are sent back are controlled for missing data, typically handled on a case-by-case basis by directly contacting the company again. Every country then controls the gathered data for late respondent bias by company size and industry. The overall response rate is 18.3% of the questionnaires sent (10.6% of the contacted companies).

The sample is limited to those companies whose answers were valid for our analysis and to those countries for which the Hofstede's indexes are available. Further, we drop cases declaring to have <20 employees or more than 16,000 and cases not providing the ISIC code classification. Lastly, our analysis is limited to those countries belonging to an OECD economy. Data concerning Ireland are dropped since the limited number of observations. Therefore, 392 companies (from the 729 of the overall dataset) belonging to 16 countries were used in the analysis. Table 1 shows the sample description, whilst Table 2 the Hofstede's scores of individualism-collectivism and power distance. Coherently with Kull and Wacker (2010), the Hofstede's scores are normalized. Each index has been divided by its

Table 1 Sample description

Country	N	%	Country	N	%	ISIC[b] code (Rev. 3.1)	N	%	Size[a]	N	%
Belgium	27	6.89	Korea	15	3.83	28	150	38.27	Small	221	56.38
Canada	15	3.83	Mexico	11	2.81	29	103	26.28	Medium	72	18.37
Denmark	13	3.32	Netherlands	39	9.95	30	2	0.51	Large	99	25.26
Estonia	25	6.38	Portugal	8	2.04	31	53	13.52			
Germany	27	6.89	Spain	32	8.16	32	14	3.57			
Hungary	61	15.56	Switzerland	26	6.63	33	20	5.10			
Italy	41	10.46	UK	9	2.30	34	21	5.36			
Japan	13	3.32	USA	30	7.65	35	29	7.40			

Sample size: 392
[a]Size: Small: <250 employees, Medium: 251–500 employees, Large: over 501 employees
[b]ISIC Code (Rev. 3.1)
28: Manufacture of fabricated metal products, except machinery and equipment; 29: Manufacture of machinery and equipment not classified elsewhere; 30: Manufacture of office, accounting and computing machinery; 31: Manufacture of electrical machinery and apparatus not classified elsewhere; 32: Manufacture of radio, television and communication equipment and apparatus; 33 Manufacture of medical, precision and optical instruments, watches and clocks; 34: Manufacture of motor vehicles, trailers and semi-trailers; 35: Manufacture of other transport equipment

Table 2 Hofstede's score of individualism-collectivism and power distance

Country	IDV	PDI	Country	IDV	PDI
Belgium	75	65	Korea	18	60
Canada	80	39	Mexico	30	81
Denmark	74	18	Netherlands	80	38
Estonia	60	40	Portugal	27	63
Germany	67	35	Spain	51	57
Hungary	80	46	Switzerland	68	34
Italy	76	50	UK	89	35
Japan	46	54	USA	91	40

Table 3 Adjusted Hofstede cultural dimensions

Country	IDV	PDI	Country	IDV	PDI
Belgium	109.06	140.95	Korea	26.17	130.11
Canada	116.33	84.57	Mexico	43.62	175.64
Denmark	107.60	39.03	Netherland	116.33	82.40
Estonia	87.25	86.74	Portugal	39.26	136.61
Germany	97.42	75.89	Spain	74.16	123.60
Hungary	116.33	99.75	Switzerland	98.88	73.72
Italy	110.51	108.42	UK	129.42	75.89
Japan	66.89	117.09	USA	132.32	86.74

overall mean and then multiplied by 100, in order to enhance its interpretability in comparison with the overall mean (Table 3). For the sake of clarity, the normalized score of individualism-collectivism in Belgium shows how this country is the 9.06% above the average score and, similarly, how Japan is the 33.11% below it, reflecting a more collectivistic cultural orientation.

3.2 Measures

Consistently with previous literature (Cagliano et al. 2003; Frohlich and Westbrook 2001), the degree of SCIS integration is assessed through a set of supply chain management practices that a focal company has adopted with its supply chain partners. Specifically, we have asked to respondents to indicate the extent through which production planning decisions, as well as flows of goods, are coordinated both with their customers and suppliers. Measuring are assessed through a Likert-scale ranging from "none" (value = 1) to high (value = 5).

Table 4 Explanatory factor analysis: extraction method: principal component factors

SCIS-suppliers	Loading	CITC
Share inventory level	0.7516	0.5264
Share production planning/forecast	0.7931	0.5794
Order tracking/tracing	0.7076	0.4808
Agreement on delivery frequency	0.7182	0.4900
Kaiser–Meier–Olkin adequacy	0.724	
Bartlett test	Chi-Squared 315.809	
Cronbach's alpha	0.727	
SCIS-customers	Loading	CITC
Share inventory level	0.7732	0.5856
Share production planning/forecast	0.8247	0.6588
Order tracking/tracing	0.7752	0.5933
Agreement on delivery frequency	0.7820	0.5969
Kaiser–Meier–Olkin adequacy	0.731	
Bartlett test	Chi-Squared 488.095	
Cronbach's alpha	0.796	

Rotation method: promax with Kaiser Normalization

Content validity, constructs validity as well as reliability are discussed coherently with previous researches (e.g., Zhou and Benton 2007). An exploratory factor analysis (EFA) has been carried out in order to build a multi-item measure, which reflects the extent through which a focal company has invested in SCIS activities with their supply chain partners.

First of all, according to previous researches (e.g., Frohlich and Westbrook 2001; Wiengarten et al. 2014), content validity is assured through the IMSS survey protocol, that encompasses the involvement of several managers and academics. Moreover, as Table 4 shows, factor loading are all above the 0.4 threshold and Cronbach's alpha is higher than 0.7 for each construct, indicating unidimensionality and reliability of the scales (Nunnally and Bernstein 1994). Kaiser–Meyer–Olkin and Corrected-Item Total Correlation reliability tests are also provided in order to further validate the scale reliability. In this regard, literature recommends a minimum Kaiser–Meyer–Olkin of 0.5 (e.g., Frohlich and Westbrook 2001): as shown in Table 4, Kaiser–Meyer–Olkin is 0.724 for what concern the scale underlying the SCIS investments that a focal company has applied in collaboration with its suppliers and 0.731 for what concern the scale underlying the SCIS investments that a focal company has applied with its customers. Similarly, literature (e.g., Zhou and Benton 2007) recommends a minimum value of Corrected-Item Total Correlation of 0.3; as indicated in Table 4, all items satisfy these criteria.

Therefore, two latent factors named *"SCIS-suppliers"* and *"SCIS-customers"* have been built by averaging the items that constitute them. Lastly, as shown in Appendix, the latent factors exhibit a strong correlation with the items they intend to measure and a low correlation with the items they do not intend to measure; item-to-factor correlations are all above the 0.5 threshold and correlations of within-factor items are higher than correlations with non-factor items.

3.3 *Hofstede's Measure of Culture*

The Hofstede's (1980) model has been used in order to assess the country's cultural peculiarities. This taxonomy, through a survey conducted among more than 100,000 IBM employees around the world, classifies countries through fifth cultural values: power distance, individualism-collectivism, uncertainty avoidance, masculinity-femininity and long-term orientation. Each index is measured through a score and due to its extensive use (see, Kirkman et al. 2006 for a review) as well as to its validity for management research (Merrit 2000; Magnusson et al. 2008) the Hofstede's model is been chosen in this chapter. In addition, this framework has been used in several manufacturing studies (e.g., Wacker and Sprague 1998; Flynn and Saladin 2006). As a result, the Hofstede's indexes of power distance and individualism-collectivism are used to test the research hypothesis.

3.4 *Control Variables*

A set of control variables has been added in order to guarantee the generalizability of the empirical results: consistently with previous literature (Power et al. 2010; Boyer and Pagell 2000) we controlled for company's size (measured through the logarithm of the total number of employees) and for company's process choice. Specifically, company's process choice is assessed by asking to respondents to indicate the percentage of customer orders processed as made to stock. According to Welker et al. (2008), company's process choice reflects a proxy of the business condition in which a manufacturer operates: a propensity to be "engineered to order" might underline a more uncertain environment whilst a propensity to be "made to stock" a more stable environment. With the aim to capture the degree of internationalization of the focal company, the percentage of suppliers and customers locate inside the same country of the plant has been added. The first variable is named *"local sourcing"* and is added for what concern the model considering the SCIS investments between a focal company and its suppliers. The second variable is named *"local sales"* and enters as a control in the model considering the investments in SCIS applied by a focal company in collaboration with its customers. Moreover, the position of the focal company in the supply chain is considered by asking to respondents to indicate the percentage of sales to end-users (Wiengarten et al. 2014). Lastly, one more control variable has been considered as a proxy of market uncertainties: demand variability. Specifically, we asked to respondents to indicate on a Likert-scale (1–5) their weekly degree of demand fluctuations.

3.5 Hierarchical Linear Modeling (HLM)

The abovementioned research hypotheses are tested using a Hierarchical Linear Modeling (Raudenbush and Bryk 2002) treating each research hypothesis separately.

The latent factors, reflecting the extent of the investments in information sharing applied by a focal company with its customers and suppliers, are the dependent variables. Consistently with previous studies (e.g., Kull and Wacker 2010; Naor et al. 2010), HLM has been applied in order to take into account the nested structure of the data. In the detail, our sample contains variables describing either the plants characteristics as well as the country cultural peculiarities. Therefore, the empirical analysis combines plant level data ($n_1 = 392$) with country level data ($n_2 = 16$). Responses concerning size, company's process choice, supply chain position, percentage of local sourcing and percentage of local sales are the plant data (level-1), whilst the country data are the Hofstede's cultural scores of power distance and individualism-collectivism (level-2). Coherently with Naor et al. (2010), all the dependent variables were standardized and grand mean centered in order to avoid multicollinearity. Correlation matrix is shown in Appendix. Further, we controlled each step of the procedure by evaluating the variance inflation factor and the condition indexes. The highest mean variance inflation factor is 1.03 on a cut-off point between 5 and 10 (Hair et al. 1998; Menard 2002; Neter et al. 1989) whereas the highest condition index is 7.97 (Belsley et al. 2004). Therefore, multicollinearity is not an issue for any model.

Initially, a variance components model is analyzed (Snijders and Bosker 2012). This model allows the intercept to vary across countries and is equivalent to a one-way random effect ANOVA in which no explanatory variables are included (see, Kull and Wacker 2010). The model divides the variance of the investments in SCIS integration carried out in collaboration with suppliers and customers, into within-group variance σ^2 (level-1 data) and between group variance τ^2_0 (level-2) where γ_{00} is the grand mean. The deviance value D is the lack of fit measure and is used in order to compare the several empirical models. Intra-class correlation (ICC), i.e. the ratio between the variance of the country random effect τ^2_0 and the overall variance, is an index indicating the proportion of the total variance in the investments in SCIS that is due to differences across countries. Lastly, the likelihood ratio test (LR-test) indicates if there is a significant evidence of differences across countries for what concern the investments in SCIS either in collaboration with suppliers as well as customers. If the null hypothesis ($\tau^2_0 = 0$) is rejected, the use of the variance-components model in comparison with an ordinary least square regression is justify (Rabe-Hesketh and Skrondal 2008, p. 69).

The second step of the analysis considers a random intercept model. Specifically, we have added to the variance-components model the level-1 variables, describing the plant's characteristics. In this model, the intercept is allowed to vary across countries whilst the effects of the level-1 variables are treated as fixed. If the LR-test is significant, the intercepts differ across countries. Further, for what concern the investments in SCIS applied between a focal company with its supply chain partners, the amount of variance that the level-1 (level-1 R^2) variables explain is computed and the Wald test is reported in order to test if the covariates are jointly significant.

Lastly, in order to account for the variations in the intercepts, the third step of the analysis adds to the random intercept model the country's cultural peculiarities (level-2 data). Yet, the intercept is allowed to vary across countries whilst the effects of the level-1 and level-2 covariates are treated as fixed. Similarly, the level-2 R^2, indicating the proportion of the variance in the intercepts explained by the country's cultural characteristics, is computed.

4 Results

4.1 SCIS-Suppliers

When the investments in SCIS activities carried out by a focal company in collaboration with its suppliers are considered, the variance components model (Table 5) shows how the grand mean γ_{00} is significant. It means that, averaging across plants and countries, the extent of the level in the investments in SCIS is 3.275 on a scale ranging from none (value = 1) to high (value = 5). Further, the variance of the country random effect τ^2_0 is 0.067 whilst the variance of the plant random effect σ^2 is 0.653. As previously mentioned, the intra-class correlation is the amount of variance that is due to differences across countries; in the case of suppliers ICC is 0.093, indicating that around the 9.3% of the variance in the investments in SCIS with suppliers is due to differences across countries. The likelihood ratio test (LR-test) is significant (χ^2 = 16.38) providing support for the use of HLM rather than an ordinary least square regression.

Therefore, a random intercept model is considered. In this model, the intercept is allowed to vary between countries and the level-1 variables, describing the plant's

Table 5 HLM results—variance component model

SCIS-suppliers	Empty model (std. error) [z-value]	SCIS-customers	Empty model (std. error) [z-value]
Grand mean γ_{00}	**3.275** (0.079) [41.38]***	Grand mean γ_{00}	**3.129** (0.106) [29.36]***
Deviance (D)	964.717	Deviance (D)	1110.361
τ^2_0	0.067	τ^2_0	0.133
σ^2	0.653	σ^2	0.939
ICC	0.093	ICC	0.124
LR test (χ^2)	16.38***	LR test (χ^2)	26.78***

***Sig < 0.01

Table 6 HLM results—random intercept model

SCIS investments suppliers	Random intercept model (std. error) [z-value]	SCIS investments customers	Random intercept model (std. error) [z-value]
Grand mean γ_{00}	**3.263** (0.083) [39.02]***	Grand mean γ_{00}	**3.128** (0.105) [29.63]***
Level-1 variables		Level-1 variables	
Size	**0.117** (0.044) [2.62]***	Size	0.037 (0.053) [0.71]
MTS	−0.017 (0.045) [−0.39]	MTS	0.023 (0.054) [0.43]
SC position	−0.065 (0.041) [−1.59]	SC position	**−0.261** (0.048) [−5.37] ***
Demand variability	**0.128** (0.040) [3.18]***	Demand variability	**0.118** (0.048) [2.47]**
Local sourcing	**−0.127** (0.045) [−2.80] ***	Local sourcing	
Local sales		Local sales	−0.002 (0.054) [−0.04]
Deviance (D)	934.969	Deviance (D)	1072.586
τ^2_0	0.080	τ^2_0	0.133
σ^2	0.599	σ^2	0.850
Level-1 R^2	0.056	Level-1 R^2	0.082
Wald test (χ^2)	32.13***	Wald test (χ^2)	39.70***
LR test (χ^2)	25.03***	LR test (χ^2)	31.78***

**Sig < 0.05
***Sig < 0.01

characteristics, are entered in the model as fixed effect. The LR-test is significant at conventional level, indicating how intercepts differ across countries. More in the detail, Table 6 shows how there was a reduction in the deviance value (D = 934.969). Further, the fixed effects of size and demand variability are positively related to the extent of the investments in SCIS applied by a focal company with its suppliers. Conversely, the fixed effect of local sourcing is negatively related to it. The level-1 covariates explain the 5.6% of the total variance in the investments in SCIS applied in collaboration with company's suppliers.

With the aim to account for the variances in the intercepts, a random intercept model with level-2 covariates is considered. Tables 7 and 8 show how the deviance value is further reduced (D = 930.066) and how both the fixed effects of power distance and individualism-collectivism are significant at conventional level. Specifically, power distance is positively related to the extent of investments in SCIS applied by a focal company with its suppliers whilst individualism-collectivism negatively. The level-1 covariates are still statistically significant. Moreover, the level-2 R^2 is 0.2243 for what concern power distance and 0.2521 for what concern individualism-collectivism. It means, that power distance accounts the 22.43% of the variance in the intercepts, whilst individualism-collectivism the 25.21%.

Table 7 HLM results—random intercept model with level-2 covariates (PDI)

SCIS investments suppliers	Random intercept model with level-2 variables (std. error) [z-value]	SCIS investments customers	Random intercept model with level-2 variables (std. error) [z-value]
Grand mean γ_{00}	**3.255** (0.076) [42.61]***	Grand mean γ_{00}	**3.121** (0.096) [32.37]***
Level-1 variables		Level-1 variables	
Size	**0.114** (0.044) [2.59]***	Size	0.034 (0.052) [0.65]
MTS	−0.019 (0.045) [−0.44]	MTS	0.018 (0.054) [0.35]
SC position	−0.062 (0.041) [−1.53]	SC position	−**0.259** (0.048) [−5.35]***
Demand variability	**0.125** (0.040) [3.13]***	Demand variability	**0.115** (0.047) [2.41]**
Local sourcing	−**0.121** (0.045) [−2.68]***	Local sourcing	
Local sales		Local sales	0.001 (0.054) [0.03]
Level-2 variable		Level-2 variable	
Power distance	**0.135** (0.065) [2.07]**	Power distance	**0.174** (0.081) [2.13]**
Deviance (D)	930.066	Deviance (D)	1068.328
τ^2_0	0.062	τ^2_0	0.104
σ^2	0.598	σ^2	0.847
Level-2 R^2	0.2243	Level-2 R^2	0.2181
Wald test (χ^2)	36.23***	Wald test (χ^2)	44.46***
LR test (χ^2)	21.31***	LR test (χ^2)	29.50***

**Sig < 0.05
***Sig < 0.01

4.2 SCIS-Customers

When the investments in SCIS activities carried out by a focal company in collaboration with its customers are considered, the variance components model (Table 5) shows how, averaging across plants and countries, the extent of the level in the investments in SCIS is 3.129. The variance of the country random effect τ^2_0 is 0.133 whilst the variance of the plant random effect σ^2 is 0.939. ICC is 0.124 indicating that around the 12.4% of the variance in the investments in SCIS with customers is due to differences across countries. The likelihood ratio test (LR) is significant ($\chi^2 = 26.78$), proving support for the use of HLM rather than an ordinary least square regression.

The random intercept model (Table 6) with level-1 variables, describing the plant's characteristics, shows how the extent of the investments in SCIS applied by a focal company in collaboration with its customers is positively related to the fixed effect of demand variability and negatively related to the fixed effect of supply

Table 8 HLM results—random intercept model with level-2 covariates (IDV)

SCIS investments suppliers	Random intercept model with level-2 variables (std. error) [z-value]	SCIS investments customers	Random intercept model with level-2 variables (std. error) [z-value]
Grand mean γ_{00}	**3.233** (0.076) [42.19]***	Grand mean γ_{00}	**3.099** (0.102) [30.29]***
Level-1 variables		Level-1 variables	
Size	**0.116** (0.044) [2.64]***	Size	0.0373 (0.052) [0.71]
MTS	−0.019 (0.045) [−0.43]	MTS	0.0208 (0.054) [0.39]
SC position	−0.061 (0.041) [−1.50]	SC position	**−0.260** (0.048) [−5.34]***
Demand variability	**0.125** (0.040) [3.13]***	Demand variability	**0.116** (0.047) [2.43]**
Local sourcing	**−0.126** (0.045) [−2.79]***	Local sourcing	
Local sales		Local sales	−0.000 (0.054) [−0.00]
Level-2 variable		Level-2 variable	
Individualism-collectivism	**−0.138** (0.065) [−2.13]**	Individualism-collectivism	−0.136 [−1.59]
Deviance (D)	929.9008	Deviance (D)	1070.1278
τ^2_0	0.0601206	τ^2_0	0.1168479
σ^2	0.5985151	σ^2	0.8482555
Level-2 R^2	0.2521	Level-2 R^2	0.1237
Wald test (χ^2)	36.41***	Wald test (χ^2)	42.38***
LR test (χ^2)	0.0000	LR test (χ^2)	32.34***

**Sig < 0.05
***Sig < 0.01

chain position (i.e. the percentage of sales to end-user). The level-1 R^2 is equal to 0.082, indicating that the plant's characteristics explain around the 8.2% of the total variance in the investments in SCIS between a focal company and its customers. The deviance value D is equal to 1072.586 and the LR test is significant at conventional level, providing support for the variation in the intercepts.

When the level-2 variables, reflecting the country's cultural characteristics, are added to the random intercept model, Tables 7 and 8 show how the deviance value is reduced (D = 1068.328) for what concern the model involving power distance, as well as for what concern the model involving the cultural trait of individualism-collectivism (D = 1070.127). However, the extent of the investments in SCIS in collaboration with customers is positively related to the fixed effect of power distance whilst is not related to the fixed effect of individualism-collectivism, which is not statistically significant at conventional level.

5 Discussion

Although previous research has highlighted the relationship between SCI and company's operational and business performance (e.g., Flynn et al. 2010; Frohlich and Westbrook 2001; Vickery et al. 2003), some authors have suggested how the understanding of its enablers and antecedents, such as competitive environment, relationship commitment and national culture is still scarcely investigated (e.g., Flynn et al. 2010; Zhao et al. 2008). Therefore, starting from this premise, this research addresses the question if cultural values might influence the willingness of a manufacturer in choosing the level of external integration either with its customers and suppliers. In particular, due to their "relational" characteristics, that might be amenable to the supply chain members behavior and, as a mirror, of cultural values, the investments in supply chain integration involving the mutual exchange of information between a manufacturer and its supply chain partners are considered. In particular, the chapter focuses on the specific role of the country's cultural peculiarities of power distance and individualism-collectivism in influencing the degree of external SCIS integration between a manufacturer and its customers and suppliers. The mentioned cultural traits are considered since previous research has recognized how individualism-collectivism can explain differences in cross-cultural behavior (e.g., Power et al. 2010; Cannon et al. 2010) as well as its relatedness, due to the national wealth, with the cultural trait of power distance (e.g., Hofstede 1983; Flynn and Saladin 2006). In this sense, we have advanced a set of four research hypotheses of which three are empirically supported while one is rejected (Table 9).

As noted throughout the chapter, the mutual exchange of information about production planning, scheduling and deliveries is fundamental in order to achieve an effective supply chain integration. These premises, constitute the rationale upon which our findings about the role of cultural values can be interpreted. Indeed, there are some factors that might hamper or foster the extent of integration and, among these, the country level ones can play an important role (Wiengarten et al. 2014). In this study, we focused on the country's cultural characteristics as an enable of the willingness of a focal company to invest in SCIS with its supply chain partners.

Previous research has addressed the role of power relationship in managing supply chain (e.g., Zhao et al. 2008; Yeung et al. 2009). In line with this, power has been classified according to fifth typologies: expert, referent, legitimate, reward and coercive (French and Raven 1959). In particular, Zhao et al. (2008) have traced how expert, referent and legitimate power might reflect the extent through which a manufacturer decides to be influenced by its customers and how, conversely, reward and coercive power could be seen as the "weapon" that a customer has in order to influence the manufacturer's behavior. Yet, Zhao et al. (2008) have suggested how in a high PDI culture, people might be more willing to accept the use of coercion and

Table 9 Summary of statistical results

National culture	SCIS-suppliers	SCIS-customers
Individualism-collectivism	Negative effect (Hp1a confirmed)	No effect (Hp1b not confirmed)
Power distance	Positive effect (Hp2a confirmed)	Positive effect (Hp2b confirmed)

might be more amenable to the perception of power driven by identification of values, skills and knowledge.

Within this scope, we found that the cultural trait of PDI is directly related to the level of SCIS integration activities that a manufacturer has applied either with its suppliers and customers; thus, Hp2a and Hp2b are supported. Combining these two empirical evidences, a manufacturer located in a cultural environment in which people perceive hierarchical levels and is comfortable with decisions taken from the most powerful members in society (customers, suppliers or supply chain coordinators), seem to be more willing to achieve external integration in terms of information sharing activities on both sides of its supply chain (or being more "outward facing" as in Frohlich and Westbrook 2001).

In addition, as previous research has noted (e.g., Yeung et al. 2009), relationships within a supply chain are managed through two main mechanisms: power and trust. This latter, in particular, might be helpful in order to shed a light on the role that IDV exerts on the degree of SCIS. In particular, our findings show how IDV is negatively related to the degree of SCIS that a manufacturer applies in with its suppliers, but not with customers. Thus, our results show that IDV pushes companies to be more "customer-facing" in Frohlich and Westbrook (2001) terms. When exchanging information in the supply chain, the threat of opportunistic behavior might be perceived higher when looking at the upstream part of the supply chain. SCIS deals with exchange of critical data (e.g., market demand, production planning) among a manufacturer and its supply chain partners. Moreover, integration with suppliers can increase the switching costs thus reducing the bargaining power. It follows that a manufacturer located in an individualistic culture could be less willing to share critical data with its suppliers since it might be less confident in a trustworthy collaboration. Conversely, this aspect could become secondary when sharing information with customer or even being a marketing leverage to consolidate the relationship with customers. In this sense, this suggestion could reflect, as outlined in Flynn et al. (2010), the primary objective of SCI, i.e. the customer-orientation.

6 Conclusions

The chapter, considering the investments in external SCIS, sheds a further light on the relevant role of the country's cultural characteristics in influencing the extent of the investments that a manufacturer has developed in collaboration with its customers and suppliers. In this regard, critical suggestions are advanced: a manufacturer who perceives more differences in power status might be more favorable to exchange its critical data about production planning, market demand and deliveries with its supply chain partner achieving, as a result, a higher level of SCIS. The empirical evidence is argued considering how, in a high PDI culture, a manufacturer might be more amenable to the sources of power. In this regard, a manufacturer could be more willing to exchange data with its supply chain partner in the light of the recognition of bargaining power, skills, knowledge and expertise available within a supplier as well as within a customer. The managerial implication is that companies engaging in relationships with new suppliers or customers in high PDI countries can expect a higher willingness or they can even be asked to share information.

Another finding is that IDV exerts a negative effect, but only on the supplier-side. We interpreted this result using the concepts of opportunistic behavior and switching costs: a manufacturer located in an individualistic country may be more reluctant to implement SCIS integration with its suppliers since it might perceive a higher the threat of leakages due to an opportunistic behavior or to increase its dependency on that supplier. However, this does not apply with customers. The managerial implication is that a company willing to become a supplier of another company in a high IDV country will find resistances when asking for sharing information. On the other side, a company will find suppliers willing to share information even in high IDV countries.

In both the case of PDI and IDV, culture can act as an enabler or barrier for SCIS. Moreover, it can create clashes, especially in the case of a buyer with high PDI (high propensity for SCIS) and a supplier with low PDI and high IDV (very low propensity to SCIS).

These findings, support us to deem that our chapter provides an interesting contribute either to theory and practice. Theoretically, the research might extend the debate on supply chain integration at global level and, from a managerial point of view, might help managers to recognize the cultural implications of cross-cultural collaboration. This is due, not only to cultural conflicts, as literature has already addressed, but also to the willingness of other companies of investing in collaboration means.

Lastly, we would like to highlight that our work is far from being free of limitations. First of all, previous research has advanced the importance to consider the role of internal integration (e.g., Flynn et al. 2010) alongside external integration.

The lack of data did not allow us to address this issue. However, it would be advisable that future research fills this gap. In addition, this study suffers the limits already advanced in similar studies using country level data (e.g., Wiengarten et al. 2014). The cultural values of power distance and individualism-collectivism are the same for all the plants located in a country. In this regard, future studies should to shed a light concerning the role of each facility's cultural peculiarities in influencing the degree of supply chain information sharing integration as well as of supply chain integration as whole. We also acknowledge that other cultural models, such as the GLOBE project (House et al. 2004) could lead to different results. Yet, we have deliberately decided to focus our attention on those countries located in a OECD economy, with the aim to avoid issues related to differences in economic development across countries. As a result, we have considered only 16 countries; though it provides a good representation of different cultural archetypes and it is many more compared to what done in previous studies addressing supply chain integration, currently available in literature. To conclude a last limitation is advanced: in this study, we have considered only those companies belonging to assembly industries and no other kinds of businesses (e.g., process industries).

Appendix

Table 10 Factor loadings

	Factors	
	Info-sharing supplier	Info-sharing customer
Share inventory level-Supplier	**0.7620**	0.3686
Share production planning-Supplier	**0.7773**	0.3416
Order tracking/tracing-Supplier	**0.7256**	0.3944
Agreement on delivery frequency-Supplier	**0.7048**	0.4136
Share inventory level-Customer	0.3495	**0.7821**
Share production planning-Customer	0.3735	**0.8140**
Order tracking/tracing-Customer	0.4472	**0.7803**
Agreement on delivery frequency-Customer	0.4390	**0.7787**

Table 11 Correlation matrix

Variables	(1)	(2)	(3)	(4)	(5)	(6)	(7)	(8)	(9)	(10)
SCIS investments-suppliers (1)	1									
SCIS investments-customers (2)	0.5097*	1								
Size (3)	0.1510*	0.0359	1							
MTS (4)	0.0071	0.0236	0.1238	1						
SC position (5)	−0.091	−0.2646*	−0.055	−0.1134	1					
Demand variability (6)	0.1308*	0.1031	−0.007	0.0253	−0.0694	1				
Local sales (7)	−0.1163	−0.1349*	−0.1440*	−0.0473	0.1537*	0.0446	1			
Local sourcing (8)	−0.0369	0.0372	−0.1081	−0.0221	0.0599	0.0805	0.3666*	1		
Power distance (9)	0.1591*	0.1359*	0.0707	0.0452	−0.0365	0.051	−0.0698	−0.0413	1	
Individualism-collectivism (10)	0.1414*	−0.0763	−0.0248	−0.031	0.0292	−0.0327	−0.0302	0.0156	−0.5445*	1

*Sig < 0.01

References

Behling O, Law KS (2000) Translating questionnaires and other research instruments: problems and solutions. Sage, Thousand Oaks, CA
Belsley DA, Kuh E, Welsch RE (2004) Regression diagnostics: identifying influential data and sources of collinearity. Wiley-IEEE, New York, NY
Boyer KK, Pagell M (2000) Measurement issues in empirical research: improving measures of operations strategy and advanced manufacturing technology. J Oper Manag 18(3):361–374
Cagliano R, Caniato F, Spina G (2003) E-business strategy: how companies are shaping their supply chain through the internet. Int J Oper Prod Manag 23(10):1142–1162
Cagliano R, Caniato F, Spina G (2006) The linkage between supply chain integration and manufacturing improvement programmes. Int J Oper Prod Manag 26(3):282–299
Cai S, Minjoon J, Yang Z (2010) Implementing supply chain integration in China: the role of institutional force and trust. J Oper Manag 28(3):257–268
Cannon JP, Doney PM, Mullen MR, Petersen KJ (2010) Building long-term orientation in buyers-suppliers relationship: the moderating role of culture. J Oper Manag 28(6):506–521
Chen CC, Chen XP, Meindl JR (1998) How can cooperation be fostered? The cultural effects of individualism-collectivism. Acad Manage Rev 23(2):285–304
Doney PM, Cannon JP, Mullen MR (1998) Understanding the influence of national culture on the development of trust. Acad Manage Rev 23(3):601–620
Dowlatshahi S (1998) Implementing early supplier involvement: a conceptual framework. Int J Oper Prod Manag 18(2):143–167
Flynn BB, Saladin B (2006) Relevance of Baldrige constructs in an international context: a study of national culture. J Oper Manag 24(5):583–603
Flynn BB, Huo B, Zhao X (2010) The impact of supply chain integration on performance: a contingent and configurational approach. J Oper Manag 28(1):58–71
French RP, Raven BH (1959) The bases of social power. In: Cartwright D (ed) Studies in social power. University of Michigan Press, Ann Arbor, MI
Frohlich MT, Westbrook R (2001) Arcs of integration: an empirical study of supply chain strategies. J Oper Manag 19(2):185–200
Golini R, Kalchschmidt M (2011) Moderating the impact of global sourcing on inventories through supply chain management. Int J Prod Econ 133(1):86–94
Griffith DA, Myers MB (2005) The performance implications of strategic fit of relational norm governance strategies in global supply chain relationships. J Int Bus Stud 36(3):254–269
Gunasekaran A, Ngai EWT (2005) Build-to-order supply chain management: a literature review and framework for development. J Oper Manag 23(5):423–451
Hair JF, Black WC, Babin BJ, Anderson RE, Tatham RL (1998) Multivariate data analysis. Prentice Hall, Upper Saddle River, NJ
Handfield RB, Bechtel C (2002) The role of trust and relationship structure in improving supply chain responsiveness. Ind Mark Manag 31(4):367–382
Hartley JL, Zirger BJ, Kamath RR (1997) Managing the buyer-supplier interface for on-time performance in product development. J Oper Manag 15(1):57–70
He Y et al (2014) The impact of supplier integration on customer integration and new product performance. Int J Prod Econ 147:260–270
Hofstede G (1980) Culture's consequences: international difference in work-related values. Sage, Beverly Hills, CA
Hofstede G (1983) National cultures in four dimensions. Int Stud Manag Organ 8(1–2):46–74
Hofstede G, Hofstede GJ (1991) Culture and organizations: Software of the mind: intercultural cooperation and its importance for survival. HarperCollins, New York, NY
House R, Hanges P, Javidan M, Dorfman P, Gupta V (2004) Culture, leadership, and organizations: the globe study of 62 societies. Sage, London
Huff L, Kelley L (2003) Levels of organizational trust in individualist versus collectivist societies: a seven-nation study. Organ Sci 14(1):81–90

Kirkman BL, Lowe KB, Gibson C (2006) A quarter century of Culture's Consequences: a review of the empirical research incorporating Hofstede's cultural value framework. J Int Bus Stud 36 (3):285–320
Kull TJ, Wacker JG (2010) Quality management effectiveness in Asia: the influence of culture. J Oper Manag 28(3):223–239
Lamming R (1993) Beyond partnership: strategies for innovation and lean supply. Prentice Hall, Hempstead
Lee HL, Whang S (1997) Information sharing in a supply chain. Int J Manuf Technol Manag 1 (1):79–93
Lee HL, Padmanabhan V, Whang S (1997) Information distortion in a supply chain: the bullwhip effect. Manag Sci 43(4):546–558
Lindberg P, Voss CA, Blackmon (1998) International manufacturing strategies: context, content and change. Kluwer Academic Publishers, Dordrecht
Magnusson P, Wilson RT, Zdravkovic S, Zhou JX, Westjohn SA (2008) Breaking through the cultural clutter; a comparative assessment of multiple cultural and institutional frameworks. Int Mark Rev 25(2):183–201
Menard SW (2002) Applied logistic regression analysis. Sage, Thousand Oaks, CA
Merrit A (2000) Culture in the cockpit: do Hofstede's dimension replicate? J Cross Cult Psychol 31 (3):283–301
Naor M, Linderman K, Schroeder R (2010) The globalization of operations in Eastern and Western countries: unpacking the relationship between national and organizational culture and its impact on manufacturing performance. J Oper Manag 28(3):194–205
Narasimhan R, Nair A (2005) The antecedent role of quality, information sharing and supply chain proximity on strategic alliance formation and performance. Int J Prod Econ 96(3):301–313
Neter J, Wasserman W, Kutner MH (1989) Applied linear regression models. Irwin, Homewood, IL
Nunnally JC, Bernstein IH (1994) Psychometric theory. McGraw-Hill, New York, NY
Pagell M, Katz JP, Sheu C (2005) The importance of national culture in operations management research. Int J Oper Prod Manag 25(4):371–394
Petersen KJ, Prayer DJ, Scannell TV (2006) An empirical investigation of global sourcing strategy effectiveness. J Supply Chain Manag 36(2):29–38
Porter ME (1980) Competitive strategy: techniques for analyzing industries and competitors: with a new introduction. Free Press, New York, NY
Power D (2005) Supply chain management integration and implementation: a literature review. Supply Chain Manag Int J 10(4):252–263
Power D, Schoenherr T, Samson D (2010) The cultural characteristic of individualism/collectivism: a comparative study of implications for investment in operations between emerging Asian and industrialized Western countries. J Oper Manag 28(3):206–222
Prasad S, Babbar S (2000) International operations management research. J Oper Manag 18 (2):209–247
Rabe-Hesketh S, Skrondal A (2008) Multilevel and longitudinal modeling using STATA, 2nd edn. STATA Press, College Station, TX
Raudenbush SW, Bryk AS (2002) Hierarchical linear models: applications and data analysis methods (advanced quantitative techniques in social sciences), 2nd edn. Sage, Thousand Oaks, CA
Sanders NR (2008) Pattern of information technology use: the impact of buyer–supplier coordination and performance. J Oper Manag 26(3):349–367
Snijders T, Bosker R (2012) Multilevel analysis. Sage, London
Van de Vijver FJR, Leung K (1997) Methods and data analysis for cross-cultural research. Sage, Thousand Oaks, CA
Van der Vaart T, Van Donk DP, Gimenez C, Sierra V (2012) Modelling the integration-performance relationship. Int J Oper Prod Manag 32(9):1043–1074

Vickery SK, Jayaram J, Doge C, Calantine R (2003) The effects of an integrative supply chain strategy on customer service and financial performance: an analysis of direct versus indirect relationship. J Oper Manag 21(5):523–539
Wacker JG, Sprague LG (1998) Forecasting accuracy: comparing the relative effectiveness of practices between seven developed countries. J Oper Manag 16(2–3):271–290
Welker GA, van der Vaart T, van Donk DP (2008) The influence of business conditions on supply chain information-sharing mechanisms: a study among supply chain links of SMEs. Int J Prod Econ 113(2):706–720
White A, Daniel E, Mohdzain M (2005) The role of emergent information technologies and systems in enabling supply chain agility. Int J Inf Manag 25:396–410
Wiengarten F, Pagell M, Ahmed MU, Gimenez C (2014) Do a country's logistical capabilities moderate the external integration performance relationship? J Oper Manag 32(1):51–63
Wong CY, Boon-itt S, Wong CWY (2011) The contingency effects of environmental uncertainty on the relationship between supply chain integration and operational performance. J Oper Manag 29(6):604–615
Yeung JH, Selen W, Zhang M, Huo B (2009) The effect of trust and coercive power on supplier integration. Int J Prod Econ 120(1):66–78
Zhao X, Flynn BB, Roth AV (2006) Decision sciences research in China: a critical review and research agenda. Decis Sci 37(4):451–496
Zhao X, Huo B, Flynn BB, Yeung JHY (2008) The impact of power and relationship commitment on the integration between manufacturers and customers in a supply chain. J Oper Manag 26 (3):368–388
Zhou H, Benton WC (2007) Supply chain practice and information sharing. J Oper Manag 25 (6):1348–1365

Ruggero Golini is Associate Professor of General Management and Supply Chain Management at the University of Bergamo. In 2011, he received his PhD in economics and technology management with the thesis "Global Supply Chain Management in the Manufacturing Industry—Configurations, Improvement Programs and Performance." His research interests are focused on global supply chain management and global value chains. Since 2012, he is one of the international coordinators of the International Manufacturing Strategy Survey (global network of universities from more than 20 countries). He is author of more than 20 papers published in international peer-reviewed journals. He is member of several international associations such as the European Association of Operations Management (EurOMA) and Production and Operations Management Society (POMS).

Andrea Mazzoleni has received his PhD in economics and management of technology at the University of Bergamo in 2014 with a thesis entitled "Is the world really flat: cultural issues in manufacturing industries." His research interests have been focused on the influence of cultural traits on operations and supply chain management.

Matteo Kalchschmidt has received his PhD in management engineering at Politecnico di Milano and is currently Full Professor of Project and Innovation Management at the University of Bergamo. His research interests have been focused on operations management with specific attention to demand management and forecasting, global supply chain management, and sustainable supply chain management. He is author of more than 100 publications among which several are in different international journals. He is member of several international associations such as the European Association of Operations Management (EurOMA) and Production and Operations Management Society (POMS).

Risk Allocation, Supplier Development and Product Innovation in Automotive Supply Chains: A Study of Nissan Europe

Arnaldo Camuffo

Abstract As new technologies and globalization change the vertical contracting structure of the auto industry, risk allocation in OEM-supplier relationships remain critical to ensure innovation and competitiveness. Developing previous, agency theory based research on the levels and the determinants of risk sharing, this study of Nissan Europe's supply chain shows that the OEM absorbs more risk (a) the greater the supplier's environmental uncertainty, (b) the more risk averse the supplier, and (c) the less severe the supplier's moral hazard. The study also shows that Nissan, though still absorbing risk from their suppliers to a nonnegligible degree, has moved to a more market-based approach to supplier selection and development as a consequence of technological change, the industry globalization and the merger with Renault.

1 Introduction

As the auto industry is disrupted by new technologies and globalization (Jacobides et al. 2016), conventional wisdom on automakers supply chain configurations is challenged. Conventional wisdom suggests that, whenever the complexity of vertical relationships is non-negligible—and especially in presence of technological change—, collaboration between buyers and suppliers can improve the ordering, logistics, inventory management and new product development processes, facilitating the understanding of supply chain dynamics and fostering cross-firm learning and problem solving (Lamming 1993). Collaborative relationships based on information exchange, trust and risk sharing facilitate global supply chain coordination in various ways. They allow to envision and incorporate new technologies into new products as OEMs design new architectures and suppliers develop incremental or modular innovations. They favor the reduction of product design and process-related

A. Camuffo (✉)
Department of Management and Technology and ICRIOS-Invernizzi Center for Research in Innovation, Organization, Strategy and Entrepreneurship, Bocconi University, Milan, Italy
e-mail: arnaldo.camuffo@unibocconi.it

A. C. Moreira et al. (eds.), *Innovation and Supply Chain Management*, Contributions to Management Science, https://doi.org/10.1007/978-3-319-74304-2_10

errors, thereby enhancing quality, time, and customer responsiveness (Henke et al. 2009). They also facilitate problem solving, improving the novelty and quality of component design, shortening customer response time, reducing the costs of protecting against opportunistic behavior, and increasing cost savings through product design reviews and operational efficiencies.

In collaborative "voice" relationships (Helper and Sako 1995, 1998), buyers and suppliers engage in collaboration through which they continuously improve their joint products and processes (Helper et al. 2000), control opportunism, and share risk, thus nurturing the formation of "relational rents" (Asanuma 1989; Dyer and Singh 1998; Baker et al. 2002).

The first evidence of the peculiarities of collaborative supplier relations in the auto industry was provided by Cusumano (1985) and Asanuma (1985a, b), who found that Toyota, Nissan, Honda and Mazda, although with differences, sought to develop close and longstanding relationships with first-tier suppliers (Nishiguchi 1994). More specifically, the Japanese automakers undertook specific procedures to help suppliers improve their capabilities (Fujimoto 2001; Sako 2004). They sent their own engineers to the supplier's product design department to co-design a given component or to the supplier's shop floor to help solve a problem with a specific component in order to meet the product launch date. They may provide consulting or training to suppliers' engineers and employees and ask suppliers to work on the improvement of a specific component or production line for an extended period. OEMs and suppliers would work together with a view to learning heuristics to achieve faster innovation introduction, cost and inventory reduction or quality improvement. Moreover, according to most of the studies (Asanuma 1989; Smitka 1989; Sako 1996), collaborative supply relations normally include mechanisms geared towards: (a) curbing potential negative knowledge spillovers to other OEMs (via suppliers); (b) mitigating the effect of uncertainty and risk on suppliers.

During the last two decades, most of the auto producers around the world have benchmarked and imitated such collaborative practices (Bensaou and Venkatraman 1995; Camuffo and Volpato 1997; Dyer 2000; Dyer and Chu 2003) interpreting and adapting the original features of the relationships between Japanese assemblers and suppliers (Smitka 1989; Womack et al. 1990; Cusumano and Takeishi 1991; Nishiguchi 1994).

Till recently, the "superior quality" of Japanese-style supply relationship management practices has somewhat been taken for granted, and their nature considered unchanged over time. Indeed, some studies have focused on their evolution, but they put prevalent emphasis on how these practices were adapted in transplants in Europe and North America (Florida and Kenney 1993; Liker et al. 1999), not on how competitive pressure brought about by globalization, new technologies and M&As have impacted on them (Ahmadjian and Lincoln 2001).

However, global competition, the recent crisis and the transition towards new propulsion systems (hybrids, EVs, fuel cells, hydrogen) has tremendously increased uncertainty, exacerbated competitive pressure around the globe, and challenged the idea that assemblers and suppliers would eventually converge towards collaborative supply relationship management practices (Jacobides et al. 2016). In this new competitive landscape, even Japanese makers and suppliers seem to have moved towards a more market based approach in some of the aspects of the supply relationships (MacDuffie and Helper 2006; Aoki and Lennerfors 2013) either as a temporary reaction to financial pressure or as a more permanent response to threats and opportunities created by the globalization of the industry (Henke et al. 2008), sustainability related regulations and technological uncertainty (Lee et al. 2010).

This chapter contributes to fill this research gap analyzing post-merge (with Renault) Nissan supply relationship management practices in Europe (Stevens 2008; Aoki and Lennerfors 2013). More specifically, the chapter analyzes vertical interfirm relationships at the Nissan Europe Barcelona plant and explore: (a) to what extent Nissan shares risk with its suppliers; and (b) whether and how the degree of risk sharing relates to suppliers' financial, structural, location and technological characteristics.

2 Theory

Theoretically, we frame our research question as located at the crossroads of organizational economics, agency theory and the theory of relational contracts applied to supply chain management.

Repeated games theory provides a rationale for cooperation and risk-sharing in buyer-supplier relations, since reputational considerations can induce self-interested economic actors to give up the gains from squeezing the last dollar of profit out of the current transaction if long-run losses coming from destroying one's reputation for dealing honorably are relevant (McAfee and McMillan 1986; McMillan 1990). Relational contract theory suggests that, in supply chains characterized by complex transactions and long-run, hand-in-glove buyer-supplier relationships, the parties may reach accommodations when unforeseen and uncontracted-for events occur by means of agreements 'based on outcomes that are observed by only the contracting parties ex post, and also on outcomes that are prohibitively costly to specify ex ante' (Baker et al. 2002, p. 39). Relational contracts, therefore, allow the parties to utilize their detailed knowledge of their specific situation and adapt to new information as it becomes available. Vertical inter-firm relationships can also be modeled as agency relationships in which a manufacturer/buyer is the principal who delegates to suppliers (agents) the task to design and/or produce different parts or components (Aoki 1988). Supplier relations are conceptualized as contracts or payment schemes through which the buyer seeks to align the supplier behavior and efforts, curbing potential opportunism and allocating risk efficiently (Levinthal 1988). Assuming the actors are self-interested, have a conflict of interest and there is information asymmetry between them, different contracts may have different comparative efficiency under different conditions (Eisenhardt 1989).

This chapter focuses on a specific aspect of supply chain management, i.e. how buyers and suppliers accommodate for the risk resulting from unpredictable cost fluctuations. Contingent on the characteristics of suppliers and transactions, buyers may opt for different risk allocation strategies. These can be conceptualized as lying within a continuum defined by two opposing strategies: risk shifting and risk absorption (Kawasaki and McMillan 1987; Aoki 1988).

Under the risk-shifting hypothesis, buyers transfer the risk involved in their business onto their suppliers. Buyers wish to keep control of suppliers, try to exploit them to drive costs down and use them as a buffer against business fluctuations. However, as the information relative to suppliers' behaviors, technology and costs may be limited, suppliers may take advantage of this private knowledge. That is, there is potential for moral hazard and hold up problems. In order to decrease this potential, buyers wish to gather as much detailed information as possible on suppliers (source of business, cost structure, product and process technologies, innovation capability, manufacturing capacity, inventories and financial position) and monitor, on the basis of this information, their behaviors and results. Because of their conflict of interests, buyers and suppliers determine their own course of action independent of the impact of their decision on other parties.

Under the risk absorption hypothesis, buyers are concerned not only with short-term reductions of purchasing costs, to be obtained by squeezing suppliers' profit margins no matter what the source of cost fluctuations or the cause of volume variability are, but also by building and maintaining long-term relationships with reliable, innovative and capable suppliers. Providing support to suppliers and sharing information with them on business and technological issues help establishing stable relationships which eventually improves the overall business performance also of the buyer (Dyer 2000). Within this framework, buyers have an interest in absorbing at least part of the risk deriving from unpredictable cost or demand fluctuations. If they do not provide suppliers with some kind of "insurance" against unexpected cost fluctuations, suppliers' commitment and performance are likely to worsen and, eventually, negatively affect also buyers' bottom line.

The risk shifting and the risk absorption hypotheses underlie different logics in the design of supply contracts and in the management of supplier relations.

Elaborating on seminal work by McAfee and McMillan (1986), Holmstrom and Milgrom (1987), and Kawasaki and McMillan (1987), and on the developments proposed by Asanuma and Kikutani (1992), Tabeta and Rahman (1999), Yun (1999), Okamuro (2001), and Camuffo et al. (2007), this study assesses Nissan Europe's risk allocation strategy in supplier relations and tests, through regression analysis, an agency model of the determinants of Nissan Europe's supply chain risk allocation.

3 Nissan's Supplier Relationships

Sako (2004) provides an in-depth account of Nissan' supplier development practices up to the merge with Renault (1999). She analyzes the early stages of Nissan's supplier development in the 1950s and 1960s, illustrating how it evolved broadening and deepening its scope from developing 'maintenance capability' on the shop floor to developing comprehensive suppliers' management capabilities (including Total

Quality Control). She also describes the evolution of Nissan's supplier network (*Takarakai*). Since the mid-1990s, Nissan has developed a whole series of measures for suppliers (Nissan's Capability Enhancement Activity program) concerning (a) financial performance, (b) quality, cost and delivery, and (c) systems governing components, factories, and companies. Sako (2004) briefly describes also the early changes in Nissan's supplier relations introduced by Carlos Ghosn after the alliance with Renault. His 2001 Nissan Revival Plan to return Nissan to profitable growth included the divestment of equity owned in affiliated suppliers, the sharing of ten vehicle platforms accounting for 90% of combined production volume, and the establishment of Renault Nissan Purchasing Organization (RNPO) to manage approximately half of the combined global annual purchasing spending of the Renault Nissan Group. The new purchasing organization and the associated new sourcing strategy resulted in pressures to change supplier relationship management, switching to a more market-based approach to supplier selection and management. Overall, these activities contributed to reducing Nissan's purchasing costs by 20% in the early 2000s. Nissan got to the shakeout of the second half of 2008 as a globally competitive automobile manufacturer, and the merge with Renault made it part of the fourth largest auto manufacturer worldwide. After achieving outstanding performance in the 2002–2006 period, its financial performance worsened after 2006, but navigated the great recession remaining a key player in the industry. The purchasing function of Nissan is governed by Renault Nissan Purchasing Organization (RNPO) and Nissan Europe Purchasing. RNPO was established in 2001. It ensures that Renault and Nissan coordinate their purchasing activities and leverage economies of scale. Joint purchasing through RNPO has made it possible to generate approximately 0.5% of yearly cost savings on each component since 2002. In addition, having common Renault and Nissan suppliers makes it possible to design shared components, which generates design cost savings of approximately 2–5% a year on each component. RNPO helps Renault and Nissan to select the best suppliers with respect to quality. This organization is responsible for the determination of the supplier network/base for the two car makers. Continuous information exchange between Renault and Nissan favors the adoption and diffusion of best practices. Renault has adopted the strict quality procedures of Nissan and Nissan has adopted the cost analysis and price target-setting approaches of Renault. Nissan Europe Purchasing organization manages the supplier relationships of the European plants. It employs approximately 200 people. Buyers are grouped by type of commodity and not according to the vehicle project or supplier. The main reason to organize buyers in this way is to make the communication with the engineering/design department easier, since the latter is organized in the same way. Besides, this purchasing organizational structure allows component specialization and the acquisition of component specific knowledge, which facilitates cross-supplier comparisons and improves the effectiveness of cross-supplier evaluations as regards quotations and prices. Finally, buyers grouped by component category can fully leverage the purchasing volumes to pressure suppliers for price reductions. Nissan Europe's purchasing organization includes also people fully dedicated to vendor tooling. Stevens (2008) describes in detail how Nissan's keiretsu system changed with the Renault–Nissan Purchasing Organization (RNPO). Starting with the initial aim of 30% of commonly purchased parts, RNPO accounted for 75% of parts purchased for both carmakers in 2006. Our interviews with Nissan Europe's design and purchasing managers also highlighted an important distinction between Nissan's supplier development teams' behavior before and after the merge with Renault. Before the alliance with Renault, Nissan's supplier development teams provided general support to suppliers helping them to become more efficient across the board. Since the alliance with Renault, however, Nissan's supplier development teams and activities have focused more narrowly on the improvement of Nissan components' production cost, promoting and activating joint improvement projects. Nissan provides technical and managerial support to certain suppliers on how to improve engineering and production and when the project is implemented and targeted results achieved, Nissan buyers ask for purchasing price reductions. The more critical the component, the more likely the supplier will be involved in joint improvement activities.

4 An Agency Model for Risk Sharing in Automotive Supplier Relations

Kawasaki and McMillan (1987) pioneered the study of risk allocation in supplier relationships in the auto industry. Their original model was an attempt to understand Japanese supply relationship management practices as the outcome of rational and self-interested behaviors and to ground it on the theory of repeated games and organizational economics.

Kawasaki and McMillan (1987) consider the manufacturer/buyer as a principal who delegates to suppliers (agents) the task to produce different parts or components. The work that the buyer delegates to suppliers consists in the design and production of a good or a service that is part of a more complex product the buyer designs and assembles. Supplier relationships are conceptualized as contracts through which the buyer decides if and how to share the risk arising from unpredictable fluctuations of suppliers' production costs. The model's aim is to explain rationally risk allocation in buyer-supplier relations and therefore assumes that both parties are selfish and their objective is to maximize their own profits. This represents a conflict of interest between the buyer and the supplier firms, as the purchasing costs of the former represent the revenues of the latter. In addition, there is information asymmetry between the parties who cannot observe reciprocally their behaviors. More specifically, the supplier has information (about cost structure, cost reduction initiatives, etc.) the OEM does not have. Also, buyers and suppliers are assumed to have different tolerance towards risk. The buyer firm is assumed risk neutral and the supplier-risk averse or risk neutral. This assumption is based on the belief that OEMs are normally more capable to diversify its investment portfolio and, hence, risk. In addition, the fluctuations associated with a single contract, i.e. a single supplier, can be small relative to the OEM's total profit.

In the model, the contract is the payment scheme that the buyer offers to the supplier Kawasaki and McMillan (1987) base their model on the Holmstrom and Milgrom's (1987) result that the optimal contract between principal and agent is linear in the end-of-period accumulated production costs. That is, even if the supplier produces and its costs take place in continuous time, the buyer pays the supplier only at discrete points in time and therefore the payment is based on the accumulated production costs up to the time of the payment. Therefore, the contract between buyer and supplier can be represented through the following payment function:

$$p = b + a\,(c - b),$$

where p is the price paid, c is the accumulated production cost. The parameters a and b are chosen in advance by the buyer. The parameter b reflects the target price. The actual cost, c, can be higher or lower with respect to the target. The difference between target and actual cost depends on agent's cost reduction efforts and on the environmental conditions. The parameter a is the risk sharing parameter. It determines how the difference between target and actual costs has to be shared.

The payment is a sum of two components. The first (b) does not vary with cost fluctuations and represents the insurance part of the payment. The second ($a(c - b)$) is the variable part of the payment and represents the incentive part of the payment. It depends on supplier's actual cost value and therefore on the level of effort it puts to reduce costs.

Depending on the value of a there are fundamentally three types of contracts:

- If $a = 0$, the contract is fixed price. The principal pays always the target. In this way, he shifts all the risk of cost fluctuations to the supplier. The principal gives to the agent incentives to reduce his costs. Supplier's cost reduction efforts are paid by the result they have generated.
- If $a = 1$, the contract is cost plus. The principal pays to the supplier all his costs (these include the profit). Therefore, we can say he absorbs all the risk of supplier's cost fluctuations. The principal insures the suppler, by providing him the same profits. These are independent with respect to supplier's cost evolution.
- If $0 < a < 1$, the risk is shared between the buyer and supplier firms. The contract balances the incentive and the insurance part of the payment.

The supplier's accumulated production cost can be represented as the sum of three components:

$$c = c^* + w - \xi,$$

where $c*$ represents the ex-ante (before signing the contract) expected cost that is common knowledge; w is a random variable that represents the unpredictable cost fluctuations observed by the supplier in the course of doing the work. The buyer does not know its value but knows its distribution, which is assumed to be normal with mean 0, and variance σ^2; ξ is the cost reduction achieved as a result of the supplier's cost-reduction effort. This effort represents an additional cost for the supplier, that Kawasaki and McMillan (1987) model as a quadratic function:

$$h(\xi) = \xi^2/2\delta.$$

The function is quadratic in order to represent the situation where the cost reducing effort has diminishing marginal returns. The buyer does not know the level of this effort and is not able to estimate it (or the estimation is too costly). This information asymmetry generates the potential for supplier's opportunistic behaviors or *moral hazard*. That is, the buyer/principal does not observe the supplier's/agent's actions after the parties signed the contract. Therefore, the buyer cannot pay the supplier on the basis of the effort dedicated to cost reduction. The cost reduction activities can include for example: searching lower priced inputs, carefully managing raw-material or final goods inventories, diminishing waste, improving labor methods, etc.

The buyer's optimal contract (and the corresponding optimal choice of α), implies the minimization of the expected value of its payment (purchasing price) to the supplier, subject to two constraints:

(a) The supplier optimizes its expected utility function by choosing the optimal cost-reducing effort (individual rationality constraint);
(b) The supplier accepts the contract only if (the expected utility of) profit is at least as large as that it could gain from the best alternative option/buyer it has (this is taken to be given exogenously) (incentive compatibility constraint).

To solve for the optimal contract, the first step is to find the second mover (agent/supplier) optimal answer as a function of the offered contract. The principal (buyer) anticipates how the supplier will answer in terms of effort exerted, as a function of the risk sharing parameter (a). So he should decide which is the optimal payment scheme. The choice of the contract is essentially the choice of the risk sharing parameter, since once a is determined, the other contract parameter (b) is also determined.

As the agent is self-interested and will maximize its own profit function, the optimal level of effort is:

$$\xi = \delta(1 - a).$$

The optimal level of supplier's cost reduction effort decreases as a increases. The larger is a, the less the supplier is responsible for its own costs and therefore the weaker is the incentive to undertake cost reduction activities (efforts that are costly). The difference between production under cost plus contracts ($a = 1$) and production cost under fixed contracts ($a = 0$) is equal to δ. Hence this provides a natural measure of the extent of *moral hazard*.

The quantities here are normalized to one. That is, this economic model ignores uncertainty and risk deriving from demand fluctuations (Okamuro 2001).

The solution of this constrained minimization problem[1] results in the following first-order condition for the buyer's optimal choice of α:

$$\alpha = \lambda\sigma^2 / \left(\delta + \lambda\sigma^2\right),$$

where λ represents the supplier's risk aversion ($\lambda \geq 0$ is the Arrow-Pratt measure of absolute risk aversion) and σ^2, the uncertainty.

This result relates the risk-sharing parameter α, to three variables:

- The supplier's environmental uncertainty (cost fluctuations σ^2),
- The supplier's risk aversion (λ) and
- The supplier's moral hazard (ease to drive down cost for given levels of cost-reducing effort -δ).

Our research hypotheses follow.

The issue of risk arises because outcomes only partly depend on actors' behaviors. The business cycle, technological change, government policies, financial markets dynamics, economic climate, competitors' actions cause unpredictable variation in outcomes. Given our assumption that buyers are less risk averse than suppliers, bearing risk should be less costly for them than for the suppliers. When uncertainty decreases, the cost for the buyer of shifting risk to the supplier is low and therefore the risk sharing parameter should increase. However, as uncertainty increases, it becomes increasingly expensive for the buyer to shift risk onto the supplier. As a proxy of supplier's environmental uncertainty, we use supplier's cost fluctuations.

Hypotheses 1A *Suppliers' cost fluctuation is positively related to risk absorption.*

The more risk averse the supplier is, the higher is the cost for the buyer to shift risk. Indeed, the more risk averse the supplier is, the larger the risk-premium it requires to take on risk. Therefore, in this case, the buyer shifts less risk. Suppliers' risk aversion is stronger the smaller it is (because it is more difficult and expensive for a smaller supplier to diversify its portfolio of customers) and the weaker, in terms of financial structure, it is. As we proxy supplier's risk aversion with its size and financial stability, we formulate the following two hypotheses:

Hypothesis 2A *Suppliers' size is negatively related to risk absorption.*

Hypothesis 2B *Suppliers' financial stability is negatively related to risk absorption.*

The more capable the supplier is to improve design and reduce costs, the less willing the buyer is to absorb risk, since it fears supplier's potential opportunism in terms of both potential knowledge spillovers to competitors and ability to extract rents form the relation.

We proxy supplier's moral hazard with two variables.

[1]For proof, see Kawasaki and McMillan (1987, pp. 330–332). Using Holmstrom and Milgrom's (1987) theorem about the linearity of the optimal contract in end-of-period accumulated production costs, they solve a dynamic principal-agent problem as if it were a static problem, with the only additional restriction that the principal's payment function is linear.

The first is the extent to which the supplier takes on responsibility in the design and technological development of the component it produces for the customer/car maker. The higher is this responsibility the higher is the supplier's technological capability and, consequently, the higher the potential for moral hazard. Therefore:

Hypothesis 3A *Suppliers' technological capability is negatively related to risk absorption.*

The second is participation to joint improvement projects in which the buyer and the supplier team up to solve problems and improve the design, quality or cost of supply at various stages (design, production logistics, etc.). These activities reduce information asymmetries between the parties and can be considered suppliers' development initiatives undertaken by the buyer to improve the suppliers' capability. The car maker's supplier development team works jointly with the supplier's management and engineering team, often in the supplier's plant and premises, providing assistance, for example through resident engineers, in the form of training and consulting. In this situation, both the car maker and the supplier exercise costly effort aimed at improvement and share the risks and the benefits produced by it. Through these projects the car maker reduces the information asymmetry on suppliers' engineering, operations, costs and way of working. In this way there is less room for opportunistic behavior and therefore the suppliers' moral hazard is reduced. Consequently, the buyer has incentives to absorb the risk deriving from unanticipated fluctuations in the supplier's production cost. The hypothesis follows:

Hypothesis 3B *Supplier's participation in joint improvement projects with the buyer is positively related to risk absorption.*

5 Data and Research Method

5.1 Data

We constructed a unique dataset that includes information on the supplier relations of Nissan in Spain. Our sample includes 113 companies that supply about 80% of the total purchasing volumes for the car models produced there. For these suppliers, Nissan represents a significant share of their business, up to 60% of their revenues.

The data were collected from a variety of sources and consisted in: (a) purchasing turnover from each supplier in the sample, on a 5-year period (2002–2007); (b) type(s) of component(s) bought from each supplier in the sample over the same time-frame; (c) product structure for vehicles; (d) manufacturing cost breakdown (including the proportion of each component cost to the total).

Moreover, we conducted several hours of structured interviews with OEMs' purchasing managers and engineers in order to complement the set of information. More specifically, Nissan Europe provided information with regard to the component's technological content and to suppliers' technological capabilities.

Finally, we gathered financial information for each supplier in the sample from the SABI (INFORMA D&B) and Amadeus (Bureau Van Dijk) databases. These two databases provide the financial statements to access revenues and cost data of the 113 analyzed suppliers in the sample for the 5-year period 2002–2007.

5.2 *Measures*

5.2.1 Dependent Variable

The dependent variable in our model is the risk sharing parameter α, calculated according to the Kawasaki and McMillan's (1987) methodology. They argue that the effects of the choice of sharing parameter α can be estimated without detailed information on individual contracts. The idea is that if the principal (buyer) absorbs part of the fluctuations of supplier's production costs, the variance of supplier's profit (s^2) will be lower than the variance of its costs (σ^2).

Kawasaki and McMillan (1987, p. 332) derive the expression for the risk-sharing parameter from the following:

$$s^2 = (1 - \alpha)^2 \sigma^2,$$

and therefore the risk sharing parameter can be represented as:

$$\alpha = 1 - s/\sigma,$$

where α refers to a specific buyer-supplier contract or relation, s is the standard deviation of supplier's profit, and σ is the standard deviation of supplier's costs. α is close to 1 when the standard deviation of supplier's profit is low relative to the standard deviation of supplier's cost. In this case the buyer absorbs risk. α is close to 0 when the standard deviation of supplier's profit is high relative to the standard deviation of supplier's cost. In this case, the buyer shifts risk. Therefore s, σ, and, consequently, α could conceivably be measured using contract or relation-specific profit and cost data for each buyer-supplier contract/relation. Unfortunately, this data is not only unavailable in our data set but it would also be almost impossible to collect using current cost accounting techniques.[2]

[2]For each buyer-supplier contract/relation it would be necessary to identify revenues and direct design and production costs. However, especially in small and medium companies such as the analyzed suppliers (although this also applies to larger firms), cost accounting is not carried out for different contract/relations, not even with regard to direct design and manufacturing costs. Furthermore, all indirect manufacturing costs, as well as most sales and administrative expenses, are shared across products, customers and contracts. None of the analyzed firms (whether buyers or suppliers) allocate these costs to obtain a 'full contract/relation' cost figure. However, we argue that, even if indirect and general costs were allocated to each contract/relation on a conventional basis (e.g., contract revenues) following standard cost accounting techniques, this would not provide a fair picture of the costs and profits of each contract/relation. Therefore, an assessment of contract/

However, we know that Nissan is the key customer for most of the analyzed suppliers, and that the proportion of the supplier's sales to Nissan to total sales is rather large (up to a maximum of 60%). Furthermore, we also know that Nissan tends to use a single sourcing policy. In this situation, it is reasonable to assume that the variation of supplier's profit relative to the variation of supplier's cost in the analyzed buyer-supplier relationships is similar to the variation of supplier's overall profit relative to the variation of supplier's overall cost, even without specific information about the proportion of the supplier's sales to Nissan to the supplier's total sales.

5.2.2 Independent Variables

Supplier's environmental uncertainty Following Kawasaki and McMillan (1987) and Asanuma and Kikutani (1992), we used the variance of the suppliers' operating costs as the measure for the suppliers' cost fluctuation (*VARCOST*). We follow Camuffo et al. (2007) to avoid endogeneity and apply two-stage least squares using as instrumental variable for the endogenous regressor an alternative measure of the supplier's cost fluctuations: the variance of raw, subsidiary and expendable materials (*VARMP*).

Suppliers' risk aversion We used two proxies:

1. Size. Due to scale economies and risk-pooling effects, a supplier tends to be less risk averse toward any particular relationship, the smaller this is relative to its overall operations. The larger the supplier, the smaller is the impact of a single customer's variance on its profit, and the smaller its risk aversion. We used the supplier's number of employees (annual average) as the measure of the supplier's size[3] (*NUM*).
2. Financial stability. The supplier's capability of absorbing financial turbulence is a proxy for (the inverse of) its risk aversion (Okamuro 2001). The more financially stable the supplier, the lower its risk aversion, since it is better able to face unpredictable financial turbulence (e.g., interest rate and exchange rate volatility, exogenous changes in credit availability, variation in the cost of capital, etc.) without support from external entities. We used the proportion of supplier's

relation specific profits and costs on the basis of state-of-the-art cost accounting techniques would not be reliable.

[3]Following Asanuma and Kikutani (1992, p. 15), we used the number of employees, not total net sales, to measure suppliers' size. Indeed, risk aversion has to be constant and not depend on profit (or, indirectly, on variables correlated to profit such as total net sales).

equity to supplier's total assets as the measure of financial stability (*STAB*). The larger this ratio, the more stable, from a financial standpoint, is the supplier.

Suppliers' moral hazard We used suppliers' technological capability and suppliers' participation in joint improvement projects with the buyer as proxies for suppliers' moral hazard.

As regards suppliers' technological capability we follow Eisenhardt, who defines task programmability as 'the degree to which appropriate behavior of the agent can be specified in advance by the principal' (Eisenhardt 1989, p. 62). In our model, the more programmable the supplier's task, the easier it becomes for the buyer to control the supplier's behavior, namely, its improvement/cost reduction effort. A routine task (e.g., the mere production of a simple component designed entirely by the buyer) is more easily observed because information concerning the supplier's behavior is already, or more readily, available. If the buyer carries out the entire design process and the supplier just manufactures, the supplier has little technological capability (Yun 1999). In this case, the buyer probably has a fairly detailed knowledge not only of the overall final product architecture, but also of the components the supplier manufactures. This implies full knowledge of the supplier's processes and cost structure and more willingness to absorb risk. Conversely, if the supplier plays an innovative role in new product development, its task is not routine, and therefore its technological capability is high.[4] The supplier's technological capability is the highest in the case of proprietary technology and/or of in-house developed components, i.e., those designed and built by the supplier without any knowledge contribution from the buyer (so-called black-box parts). In this case, the supplier's task is not observable and the buyer has little knowledge of the supplier's processes and cost structure; consequently, it is more willing to shift risk. On the basis of the information gathered during our interviews with Nissan Europe's design department, we classified each component (and corresponding supplier) into the following typology, which entails an increasing level of the component's technological complexity and the supplier's design responsibility: (a) the buyer completely designs the component and the supplier just manufactures it (Drawings Supplied—DS), (b) the buyer defines the product concept domain and the functional parameter domain while the supplier works out the design details and manufactures the component (Drawings Approved—DA), and (c) the buyer purchases the component that has been fully designed and manufactured by the supplier (Marketed Goods—MG). We classified suppliers into one of these three categories,[5]

[4]It could be argued that only large suppliers can afford investment in technology and to develop valuable know-how. We checked our data for collinearity and found no statistical correlation between the supplier's size (proxy for risk aversion) and the supplier's technological capability (proxy for moral hazard).

[5]In the case of suppliers selling more than one component, we picked the one that was the most technologically complex.

and transformed the variable into two dummies: DA and MG. DS suppliers are coded by DA = 0 and MG = 0, DA suppliers are coded by DA = 1 and MG = 0, and MG suppliers are coded by DA = 0 and MG = 1. Thus, DA and MG have, respectively, additional effects on the risk-sharing parameter as the supplier's technological capability increases from DS (the supplier's state taken as a floor) to DA and MG suppliers.

The second proxy we used to measure supplier's moral hazard is whether this is engaged in joint improvement projects. The participation in such projects reduces the informational asymmetry between the parties and curbs potential opportunism, thus facilitating risk sharing. Through joint improvement initiatives, Nissan Europe and its suppliers share information and reciprocally obtain knowledge on several aspects of the supply, including cost structures and production processes for the exchanged item. Lower informational asymmetries facilitate integrative negotiations, help trust building, reduce moral hazard and, hence, favor risk sharing.

On the basis of data gathered during our interviews, we measure the extent to which suppliers participate in joint improvement programs through a dummy variable (*CPR*) that takes value one when the supplier does, and value zero otherwise.

Supplier's location As mainstream literature in supplier relations in the auto industry considers geographical proximity or site specificity (Dyer 1997, 2000) as a potential driver of OEM-supplier relations configuration and performance, we control for supplier's distance from the analyzed plant. Other things equal, supplier co-location should improve inter-firm coordination, reduce information asymmetry, enhance trust and risk sharing.

6 Findings

6.1 Suppliers' Risk Aversion

As a preliminary step, we verified two basic underlying assumptions of the agency model we apply: (a) suppliers are risk averse and (b) large suppliers are less risk averse.

Following Kawasaki and McMillan (1987, p. 338), we tested suppliers' risk aversion assuming the existence of a linear relationship between the mean (μ) and the variance (s^2) of suppliers' profits:

$$\mu = (½\, \lambda)\, s^2 + k$$

where $(½\, \lambda)\, s^2$ is the risk premium and k is the residual profit. If the mean and the variation of the supplier's profits are positively and significantly related, λ is positive and we can be sure that suppliers are risk averse. We estimate the above mentioned equation by ordinary least squares (OLS). Results are reported in Table 1.

Table 1 Estimates for suppliers' risk aversion (standard errors in parentheses)

Sample data	Number of observations	½ λ (s.e.)	k (s.e.)	R^2
Total sample	113	0.7700743**	1055820*	0.2333
		(0.1325047)	(500599.2)	
Sub sample "small suppliers" (Suppliers with less than 394 employees (total sample median)	57	0.9610825**	718309.3**	0.3287
		(0.1869168)	(278866.4)	
Sub sample "large suppliers" (Suppliers with more than 394 employees (total sample median)	56	0.7413373**	1329170	0.2124
		(0.1942938)	(1001607)	

+$p < 0.10$; *$p < 0.05$; **$p < 0.01$

The OLS results confirm a positive and significant value for λ. Furthermore, in order to verify that larger suppliers are less risk averse, we estimate the same model for two subsamples, obtained dividing the total sample into two subsamples: the first including "small" suppliers, i.e. suppliers with a number of employees lower than the median number of employees of the total sample; the second including "large" suppliers, i.e. suppliers with a number of employees larger than the median number of employees. The results of the corresponding OLS models, also included in Table 1, shows that the estimated value for λ is higher for the "small" suppliers' subsample. This confirms that larger suppliers are less risk averse.

6.2 *Risk Sharing Parameter Values*

The risk sharing parameter α is calculated by using the following formula:

$$\alpha = 1 - s/\sigma,$$

as derived in the previous section. The sample mean value of α is 0.72, while the median is about 0.80. Although it is difficult to assess the degree of risk-sharing, these values seem to suggest that Nissan absorbs supplier's cost uncertainty to nonnegligible degree. However, comparing these data with similar data from earlier studies in the auto industry (Kawasaki and McMillan 1987; Asanuma and Kikutani 1992; Camuffo and Volpato 1997; Yun 1999), the mean values of α for Nissan Europe's suppliers during the period 2002–2007 are lower than those of other automakers during the 1980s and 1990s, and, even more interestingly, are lower than those for Nissan Japan in the 1980s. Table 2 reports the sample mean and variance values of α for selected studies in the auto industry using the same methodology.

Please note that this comparison is somewhat speculative and that the data should be interpreted very cautiously.

Table 2 Descriptive statistics for the risk-sharing parameter α and international comparison with earlier similar studies

α	Nissan Europe (current)	Nissan Japan (Asanuma and Kikutani 1992)	Toyota (Asanuma and Kikutani 1992)	Mazda (Asanuma and Kikutani 1992)	Mitsubishi (Asanuma and Kikutani 1992)	Fiat (Camuffo and Volpato 1997)	Korean suppliers (Yun 1999)
Number of observations	113	75	96	87	97	92	93
Mean α	0.7261	0.9133	0.9061	0.9081	0.9031	0.7273	0.85
Variance α	0.0567	0.0043	0.0056	0.0057	0.0052	0.0794	0.04

6.3 Determinants of Risk Sharing

We then tested the agency model and the related research hypotheses. We recall the buyer's optimal choice of α:

$$\alpha = \lambda\sigma^2 / \left(\delta + \lambda\sigma^2\right).$$

We linearize the expression by rearranging and taking logarithms on both sides of the equation, which gives us:

$$\ln\left(1/\alpha - 1\right) = \ln\left(1/\sigma^2\right) + \ln\left(1/\lambda\right) + \ln\ \delta,$$

where σ^2 is the supplier's cost variance, λ is the supplier's constant absolute risk aversion and δ represents the supplier's moral hazard. In order to conduct regression analysis on our sample to understand the determinants of risk sharing, we used the proxies defined in the Method section for α, λ, σ^2 and, δ, obtaining the following model:

$$\ln\left(1/\alpha - 1\right) = a_0 + a_1 \ln\left(1/\sigma^2\right) + a_2 NUM + a_3 STAB + a_4.DA + a_5 MG + a_6 CRP + a_7 LOC + \varepsilon$$

Table 3 reports the descriptive statistics and the correlation coefficients for the variables in the model.

Then we tested our model through OLS analysis (results in Table 4). Plain OLS analysis, however, suffers from potential endogeneity because the variance of the supplier's operating costs enters in the definition of both an independent and the dependent variable. As a consequence, any error in the measurement of this variance induces a correlation between the explanatory variable and the disturbance term of the regression.

Following Camuffo et al. (2007), we apply two-stage least squares (TSLS) using as an instrumental variable for the endogenous regressor an alternative measure of the supplier's cost fluctuation: the variance of the cost of raw, subsidiary, and expendable materials (VARMP). The possible measurement errors of these costs are likely to be uncorrelated, or at least less correlated, to the measurement errors of the dependent variable. The results of the TSLS analysis are also reported in Table 4.

The regression models in Table 4 shows an adjusted R^2 of 0.52. This value is high compared with that of earlier, similar studies. Besides, the F-statistic is significant at 1%, indicating that the current variables together significantly explain the variation in the dependent variable.

We obtained the same significant coefficients when we estimate robust OLS and TSLS models. We also tested for homoskedasticity (Breusch-Pagan/Cook-Weisberg test) and omitted variables (Ramsey RESET test) getting full support for our model.

Both the OLS and TSLS models support our hypotheses. All the coefficients are significant, although their impact on risk sharing is modest and lower than that of previous studies.

Table 3 Descriptive statistics and correlation matrix (N = 113)

	Mean (S.E.)	$\ln(1/\alpha - 1)$	ln(1/VARCOST)	NUM	STAB	DA	MG	CPR	LOC
$\ln(1/\alpha - 1)$	1.295783 (1.431414)	1							
ln(1/VARCOST)	−13.129 (3.501822)	0.2779**	1						
NUM	945.3628 (1512.221)	0.5796**	0.1712^{+}	1					
STAB	35.5677 (24.0185)	0.2835**	0.1270	0.1801^{+}	1				
DA	0.4247788 (0.4965112)	0.0869	−0.1936*	−0.0255	−0.0423	1			
MG	0.2035398 (0.404424)	0.4009**	0.1798^{+}	0.2646**	0.0526	−0.4344**	1		
CPR	0.3893805 (0.4897818)	−0.3850**	−0.2231**	−0.2343**	−0.0829	−0.0621	−0.0882	1	
LOC	0.1858407 (0.3907107)	−0.4288**	−0.1484	−0.1090	−0.2019*	−0.1804	−0.0720	0.5516**	1

+$p < 0.10$; *$p < 0.05$; **$p < 0.01$

Table 4 OLS and TSLS results

	OLS model	TSLS model
Constant	−1.105583**	−1.137178**
	(0.3508862)	(0.3504826)
ln(1/VARCOST)	0.0328424**	0.0315868**
	(0.0110764)	(0.0111918)
NUM	0.000222*	0.000222*
	(0.0000905)	(0.0000906)
STAB	0.0069541*	0.0069872*
	(0.0030359)	(0.0030272)
DA	0.8177391**	0.8145983**
	(0.2093896)	(0.2094757)
MG	1.19229**	1.194018**
	(0.363065)	(0.3618567)
CPR	−0.6338093**	−0.6372378**
	(0.1909391)	(0.1912732)
LOC	−0.5359667*	−0.5371313*
	(0.2604751)	(0.2607696)
R^2	0.5643	0.5643

Dependent variable is $\log(1/\alpha - 1)$. For the TSLS model, LOG (1/VARMP) instrument for LOG(1/VARCOST). N = 113
+$p < 0.10$; *$p < 0.05$; **$p < 0.01$

In addition, the control variable related to supplier's site proximity is has the expected negative and significant ($p < 0.05$) coefficient in our regression. This confirms what was expected, that is risk sharing increases when the supplier is located near the Car Maker.

7 Discussion

Our microeconometric analysis of Nissan Europe's supplier relations provides some insights into how buyers and suppliers share the risk deriving from unpredictable cost fluctuations. It shows that Nissan Europe absorbs risk to a non-negligible degree, but that this level of risk sharing is lower than in the past and similar to that of other European and Korean automakers.

This result is consistent with the anecdotal evidence we gathered during our interviews and with the results of other qualitative studies (Stevens 2008; Aoki and Lennerfors 2013), suggesting that, due to the merge with Renault and to competitive pressure related to the globalization of components' sourcing in the auto industry, Nissan Europe has adapted the original Nissan supplier development practices, moving towards a more competitive configuration characterized by a lower, or at least more selective, degree of risk absorption.

On the one hand, Nissan Europe's operations and purchasing activities are largely grounded on and determined by the use of the structures and systems of its local based partner Renault, whose supplier relations management practices remain closer to a market-based approach (multiple sourcing for the same component, competitive bidding, short term contracts, little support and development, etc.). On the other hand, the establishment of RNPO as a centralized global purchasing unit generated pressures to switch to a more market-based approach to supplier selection and management, within the context of sharing a global supplier base with Renault whose supplier relations.

With regard to the determinants of risk sharing, our study confirms agency theory predictions that buyers absorb more risk:

(a) The greater the supplier's environmental uncertainty;
(b) The more risk averse the supplier; and
(c) The less severe the supplier's moral hazard.

We measured supplier's moral hazard not only as supplier's technological capability, but also as supplier's participation to joint improvement projects promoted by Nissan. This variable measures the extent to which the buyer and the suppliers share information and collaborate for a common goal which curbs the potential for moral hazard. Alternatively, these joint improvement projects can be interpreted as the means through which the parties build relational contracts and implement routines like benchmarking, co-design, and 'root cause' error detection and correction, i.e. the pragmatist mechanisms that constitute 'learning by monitoring'—a relationship in which buyers and suppliers (a) continuously improve their joint products and processes; and (b) control opportunism and share risk (Helper et al. 2000).

8 Conclusion

Our microeconometric analysis shows that Nissan still absorbs risk from their supplier to a nonnegligible degree, but that global pressure to reduce cost, technological changes and organizational changes related to the alliance with Renault moved towards a more competitive configuration.

This study also confirms agency theory predictions and its findings are consistent with the theory of repeated games and the theory of relational contracts that provide a rationale for cooperation and risk-sharing in buyer-supplier relations (McAfee and McMillan 1986; McMillan 1990; Dyer and Singh 1998; Baker et al. 2002). Besides, our analysis bridges organizational economics with the resource based view that, in vertical inter-firm relationships, collaborative routines represent shared capabilities.

From a research perspective, this chapter contributes to the understanding of risk allocation in vertical inter-firm relationships in the auto industry by: (a) allowing for longitudinal comparisons in order to assess the dynamics of sourcing policies; (b) proposing new proxies for moral hazard, i.e. capturing under which conditions (capability endowment) suppliers may not have incentives to collaborate; (c) constructing original firm-level databases, based mostly on primary and certified data sources, which provides a more reliable ground for statistical analysis; and (d) solving a problem of endogeneity which affected previous studies.

This chapter also offers to practitioners some insights as regards the design of supply contracts, the optimal allocation of risk across supply chains and the management of supply networks.

First, risk sharing could be included as a conceptual milestone in the design and management of supply networks. For example, a somehow refined, customer-specific version of the risk sharing parameter α, calculated applying activity based costing methodologies, could complement rating techniques and become integrative part of suppliers' rating systems.

Second, some of the findings of this study could be used as a basis for supply chain policy making. For example, buyers who wish to engage in the technological development of small, rapidly evolving suppliers, should be prepared to measure and maneuver risk sharing as the supplier and the corresponding relation evolve. Buyers should be willing to absorb more risk in the early stage of the supplier relation life-cycle, nurturing and protecting the supplier against environmental uncertainty by stabilizing its profits. As the relation evolves and suppliers grow and become more technologically capable, however, since the supplier's potential moral hazard increases, they should become more cautious in risk sharing and invest more heavily to curb opportunism *via* informational asymmetry reduction and trust building. Similarly, the availability of data on how a buyer has shared the risk with suppliers provides a more solid basis for the ex ante design of "smart" supply contracts. For example, performing sensitivity analysis on risk sharing parameters can lead to more effective price and quantity negotiations in buy back and revenue sharing contracts.

Further research along three directions (largely corresponding to three limitations this study shares with similar, previous ones) would improve the scientific rigor and managerial relevance of this stream of investigation.

First, the estimation of the risk sharing parameter α remains somewhat problematic (Okamuro 2001). The fluctuation of supplier's costs and profits depends on changes not only in unit costs and prices, but also in quantities. If reliable data on volume variability (and on inventory variations) were available (but in our case neither buyers nor suppliers were willing or ready to provide them), it would become possible to distinguish between the volume-related and the cost-related components of risk, leading to a more articulated estimate and understanding of risk allocation in supply chain contracting.

A second conceptual limit also relates to the nature of the risk sharing parameter α. Since α is calculated using the supplier's operating costs and income, it is a comprehensive measure which refers to all the supplier's clients and not to a specific customer. Therefore, α is a characteristic of the supplier and not of a specific buyer-supplier relation. Given this assumption, the models used so far remain oversimplified, especially when the supplier's portfolio of customers is diversified. Further research should address this issue breaking down the analysis by customer, for example calculating, using state-of-the-art cost accounting methodologies, customer specific α values, and then modeling risk allocation at this more disaggregated level of analysis.

Third, the agency model we applied does not allow clear identification of the causal mechanism underlying the relationship between risk allocation and the nature of the supplier relation. Further evidence of a direct link between cooperative, stable supplier relations and risk sharing should be sought complementing the model with variables able to capture relational aspects like trust or the degree of customer-supplier integration and technological aspects like the product architecture and the specification of product performance parameters.

References

Ahmadjian CL, Lincoln JR (2001) Keiretsu, governance, and learning: case studies in change from the Japanese automotive industry. Org Sci 12(6):683–701

Aoki M (1988) Information incentives and bargaining structure in the Japanese economy. Cambridge University Press, Cambridge

Aoki K, Lennerfors TT (2013) Whither Japanese keiretsu? The transformation of vertical keiretsu in Toyota, Nissan and Honda 1991–2011. Asia Pac Bus Rev 19(1):13–27

Asanuma B (1985a) The contractual framework of parts purchases in the Japanese automotive industry. Jpn Econ Stud 15:32–78

Asanuma B (1985b) The organization of parts supply in the Japanese automotive industry. Jpn Econ Stud 13:33–51

Asanuma B (1989) Manufacturer–supplier relationships in Japan and the concept of relation-specific skill. J Jpn Int Econ 3:1–30

Asanuma B, Kikutani T (1992) Risk absorption in Japanese subcontracting: a microeconometric study of the automobile industry. J Jpn Int Econ 6:1–29

Baker G, Gibbons R, Murphy KJ (2002) Relational contracts and the theory of the firm. Q J Econ 117:39–84

Bensaou M, Venkatraman N (1995) Configurations of interorganizational relationships: a comparison between U.S. and Japanese automakers. Manage Sci 41(9):1471–1492

Camuffo A, Volpato G (1997) Nuove forme di integrazione operativa: Il caso della componentistica automobilistica. Franco Angeli, Milan

Camuffo A, Furlan A, Rettore E (2007) Risk sharing in supplier relations: an agency model for the Italian air-conditioning industry. Strateg Manage J 28:1257–1266

Cusumano MA (1985) The Japanese automobile industry: technology and management at Nissan and Toyota. Harvard University Press, Cambridge

Cusumano MA, Takeishi A (1991) Supplier relations and management: a survey of Japanese, Japanese-transplant, and US auto plants. Strateg Manage J 12:563–588

Dyer JH (1997) Effective interfirm collaboration: how firms minimize transaction costs and maximize transaction value. Strateg Manage J 18(7):535–556

Dyer JH (2000) The determinants of trust in supplier-automaker relationships in the U.S., Japan, and Korea. J Int Bus Stud 31(2):259–285

Dyer JH, Chu W (2003) The role of trustworthiness in reducing transaction costs and improving performance: empirical evidence from the United States, Japan, and Korea. Org Sci 14(1):57–68

Dyer JH, Singh H (1998) The relational view: cooperative strategy and sources of interorganizational competitive advantage. Acad Manage Rev 23:660–679

Eisenhardt KM (1989) Agency theory: an assessment and review. Acad Manage Rev 14:57–74

Florida R, Kenney M (1993) Beyond mass production: the Japanese system and its transfer to the US. Oxford University Press, Oxford

Fujimoto T (2001) The Japanese automobile parts supplier system: the triplet of effective inter-firm routines. Int J Automot Technol Manage 1(1):1–34

Helper S, Sako M (1995) Supplier relations in Japan and the United States: are they converging? Sloan Manage Rev 36(3):77–84

Helper S, Sako M (1998) Determinants of trust in supplier relations: evidence from the automotive industry in Japan and the United States. J Econ Behav Org 34(3):387–418

Helper S, MacDuffie JP, Sabel C (2000) Pragmatic collaborations: advancing knowledge while controlling opportunism. Ind Corp Change 9:443–483

Henke J Jr, Parameswaran W, Pisharodi R, Mohan R (2008) Manufacturer price reduction pressure and supplier relations. J Bus Ind Mark 23(5):287–300

Henke jr J, Zhang C, Griffith DA (2009) Do buyer cooperative actions matter under conditions of relational stress? Evidence from Japanese and U.S. assemblers in the U.S. automotive industry. J Oper Manage 27(6):479–494

Holmstrom B, Milgrom P (1987) Aggregation and linearity in the provision of intertemporal incentives. Econometrica 55:303–328

Jacobides MG, MacDuffie JP, Tae CJ (2016) Agency, structure, and the dominance of OEMs: change and stability in the automotive sector. Strateg Manage J 37(9):1942–1967

Kawasaki S, McMillan J (1987) The design of contracts: evidence from Japanese subcontracting. J Jpn Int Econ 1(3):327–349

Lamming RC (1993) Beyond partnership: strategies for innovation and lean supply. Prentice Hall, Hemel Hempstead

Lee J, Veloso FM, Hounshell DA, Rubin ES (2010) Forcing technological change: a case of automobile emissions control technology development in the US. Technovation 30(4):249–264

Levinthal D (1988) A survey of agency models of organization. J Econ Behav Org 9:34–45

Liker J, Fruin M, Adler PS (eds) (1999) Remade in America: transplanting and transforming Japanese management systems. Oxford University Press, New York

MacDuffie JP, Helper RS (2006) Collaboration in supply chains: with and without trust. In: Heckscher C, Adler PS (eds) The firm as a collaborative community: reconstructing trust in the knowledge economy. Oxford University Press, New York

McAfee RP, McMillan J (1986) Bidding for contracts: a principal-agent analysis. Rand J Econ 17 (3):326–338

McMillan J (1990) Managing suppliers: incentive systems in the Japanese and US industry. Calif Manage Rev 32:38–55

Nishiguchi T (1994) Strategic industrial sourcing: the Japanese advantage. Oxford University Press, Oxford

Okamuro H (2001) Risk sharing in the supplier relationship: new evidence from the Japanese automotive industry. J Econ Behav Org 45:361–382

Sako M (1996) Suppliers associations in the Japanese auto industry: collective action for technology diffusion. Camb J Econ 20(6):651–671

Sako M (2004) Supplier development at Honda, Nissan and Toyota: comparative case studies of organizational capability enhancement. Ind Corp Change 13(2):281–308

Smitka MJ (1989) Competitive ties: subcontracting in the Japanese automobile industry. Columbia University Press, New York

Stevens M (2008) Foreign influences on the Japanese automobile industry: the Nissan-Renault mutual learning alliance. Asia Pac Bus Rev 1(14):13–27

Tabeta N, Rahman S (1999) Risk sharing mechanism in Japan's auto industry: the keiretsu versus independent parts suppliers. Asia Pac J Manage 16:311–330

Womack JP, Jones DT, Roos D (1990) The machine that changed the world. Macmillan Press, New York

Yun M (1999) Subcontracting relations in the Korean automotive industry: risk sharing and technological capability. Int J Ind Org 17:81–108

Arnaldo Camuffo is Full Professor of Business Organization at Bocconi University, Milan, Italy, where he is also Director of ICRIOS—Invernizzi Center for Research on Innovation, Organization, Strategy and Entrepreneurship. Author of several books (the latest being *Lean Transformations for Small and Medium Enterprises*, New York, CRC-Productivity Press, 2016), his research has appeared in *Strategic Management Journal, Organization Science, Research Policy, Industrial and Corporate Change, MIT Sloan Management Review, IEEE Transactions on Engineering Management, Journal of Business Ethics, International Journal of Operations and Production Management, Human Resource Management, European Management Review, Industrial Relations, International Journal of Management Reviews, International Journal of Human Resource Management, Industry and Innovation, Industrial Marketing Management*, and *Entrepreneurship and Regional Development*. He is in the editorial board of *European Management Review, International Journal of Operations and Production Management, Human Resource Management*, and *International Journal of Innovation Management*.

Does Supply Chain Innovation Pay Off?

Jan Stentoft and Christopher Rajkumar

Abstract The purpose of this chapter is to investigate the relationship among supply chain innovation and performance in terms of market and operational performance. The chapter is built on empirically data subject to 187 useable responses from a questionnaire-survey among Danish manufacturers. A conceptual model was developed and subsequently two major hypotheses were formulated. Linear regression was performed using SPSS software 22.0 to tests the developed hypotheses. Supply chain innovation is unfolded through the components of business processes, networks structure and technology. Data reveals that supply chain innovation does pay off in terms of improved market and operational performance. The chapter also reveals that the strongest relationship is obtained with supply chain innovation and operational performance. Market performance may be influenced by a number of different factors beyond supply chain innovation. The chapter provides interesting findings of the network component with empirical evidence that it has a positive influence on both market and operation performance. The chapter concludes by suggesting new areas of research including also the relationship to financial performance.

1 Introduction

Supply chain innovation undoubtedly has become the most essential feature for any firm to survive in today's dynamic and competitive marketplace (Zimmermann et al. 2016). It has been widely acknowledged in both academia and practice that companies supply chains are vital sources for future competitiveness (Arlbjørn et al. 2011; Hazen et al. 2012; Narasimhan and Narayanan 2013). Innovation processes are important both from a single company perspective and from a network perspective with a focus on shared processes (Arlbjørn and Paulraj 2013; Ojha et al. 2016;

J. Stentoft (✉) · C. Rajkumar
Department of Entrepreneurship and Relationship Management, University of Southern Denmark, Kolding, Denmark
e-mail: stentoft@sam.sdu.dk; cra@sam.sdu.dk

A. C. Moreira et al. (eds.), *Innovation and Supply Chain Management*, Contributions to Management Science, https://doi.org/10.1007/978-3-319-74304-2_11

Wagner 2012). Supply chain innovation has received increased academic awareness (Arlbjørn et al. 2011; Lee et al. 2011; Vijayasarathy 2010; Yoon et al. 2016); however, with varied proposals for its content. Extant literature discusses supply chain innovation in relation with performance and has demanded this relationship further explored (Hazen et al. 2012; Panayides and Lun 2009). From a practical perspective, the supply chain area in general contains high cost impact in many companies and comprises much complexity about why a continued need to innovate in this area is important to remain competitive (DeTienne et al. 2015; Lee et al. 2011; Yoon et al. 2016). This chapter builds on the perception of supply chain innovation consisting of three components: (1) business processes, (2) network structure and (3) technology (Arlbjørn et al. 2011; Munksgaard et al. 2014). The major intention of supply chains is to build up their stability via continuous innovations as well as strategies to adapt to existing and new markets.

The role of supply chain innovation in developing the overall firm performance in terms of both market and operations seems still to be unexplored. Supply chain innovation helps the firms in sustaining their position in their market by providing original products, processes, and services. This in turn supports firms to also sustain their superior performance at an optimum level (Lee et al. 2011; Zimmermann et al. 2016). It is believed that supply chain innovation recommends firms to organize the three major components of business process innovation, network structure innovation, and technology innovation in order to achieve competitive edge and sustain superior performance by satisfying the needs of the customers and suppliers (Arlbjørn et al. 2011). This chapter induces supply chain innovation as an important capability that helps the firms in sustaining their overall performance in terms of market and operational performance.

Accordingly, the purpose of this chapter is to advance the understanding of supply chain innovation by testing how the overall supply chain innovation construct and its three individual components affect market and operational performance (Golicic and Smith 2013).

2 Theoretical Frame of Reference

This section describes the theoretical frame of reference which builds on supply chain innovation and market and operational performance. These two separate sections lead to the development of an overall theoretical model for the chapter presented in the third subsection.

2.1 Supply Chain Innovation

The phenomenon of supply chain innovation has been conceptualized by Arlbjørn et al. (2011) into three concurrent business components: (1) Business processes, (2) network structure and (3) technology. They define supply chain innovation as:

> a change (incremental or radical) within a supply chain network, supply chain technology, or supply chain process (or a combination of these) that can take place in a company function, within a company, in an industry or in a supply chain in order to enhance new value creation for the stakeholder (Arlbjørn et al. 2011, p. 8).

The framework has been used in various subject areas such as green supply chain innovation (Kronborg Jensen et al. 2013); humanitarian supply chain innovations (Heaslip et al. 2015), offshore wind energy sector supply chains (Stentoft et al. 2016a) and in relation to offshoring and backshoring of manufacturing (Stentoft et al. 2016b). In the following, the three components of the supply chain innovation framework are unfolded. We refer to Appendix for an operationalization of the different variables invested under each component.

2.1.1 Business Processes

The first component in the supply chain innovation framework is business processes. In the SCM literature, there is a strong agreement that business process thinking constitutes one of the backbones of supply chain management (Ellram and Cooper 2014; Lambert and Cooper 2000; Mentzer et al. 2001; Stock and Boyer 2009). This chapter uses the eight business processes developed by the Global Supply Chain Forum (Lambert and Cooper 2000) (see Appendix).

2.1.2 Network Structure

The second component in the SCI framework is the supply chain network structure. This component is about the how the focal company is positioned in the business network; the number of tiers across the supply chain (horizontal aspects), vertical aspects such as the number of dyads within tiers (Lambert and Cooper 2000) as well as internal alignment business different business functions. Furthermore, the component also includes aspects about the depth and width of relationships both upstream and downstream (Chen and Paulraj 2004) and different types members (e.g. customers, suppliers, competitors, universities and public agencies).

2.1.3 Technology

The third component of the supply chain innovation framework is supply chain technology. It is important to stress that SCI is not about the relevant technology itself [e.g. ERP, automation and additive manufacturing, and other disruptive technologies (Stentoft et al. 2017; Vyas 2016)] but it is in the novel use of technology in a supply chain context (Stentoft et al. 2016b). Technology may be applied in isolation or in combination with other technologies to create SCI (Munksgaard et al. 2014). Examples of technologies are enterprise resource planning (ERP) systems, identification systems (e.g. bar codes and radio frequency identification), analytical technologies, drone technology and Industry 4.0 technologies (e.g. robots, 3D printing and big data) (see Appendix).

2.2 Market and Operational Performance

Supply chain innovation supports the firms in effectively sustaining its competitive position and share in today's dynamic market and successively helps in sustaining their overall performance at an optimum level (Lee et al. 2011). Firm performance measurement describes the practice of evaluating firm's competence and effectiveness and it is crucial for effective firm management. Firm performance has in literature been used in several ways. In this chapter, we apply two of the three categories of firm performances as outlined by Golicic and Smith (2013). These two categories are market-based and operational-based performances. Market-based performance are concerned with indicators reflecting market goals such as meeting customer needs and includes market share, competitive advantage, customer loyalty, brand equity (see Appendix). Operational-based performance is concerned with operational efficiency with indicators such as process reliability, responsiveness, agility, costs and capacity utilization (see Appendix).

Market performance leads to superior customer value and profits (Flint et al. 2005; Min et al. 2007). Market performance measures increases the ability of the firms to assess the market condition and to accurately forecast the gains and performance (Cheng and Leung 2004). In addition, Ramaswami et al. (2009) states that firms should assess their market-based capabilities which include customer-driven development, cross functional integration, customer value, customer responsiveness, information sharing, and supply chain leadership. Market-based performance measures are not subjective to firm-specific traits (Ahmad and Jusoh 2014) instead they are more about external-oriented characteristics.

Operational performance relates to the activities that contribute towards consistency, responsiveness, productivity, costs and efficiency (Stank et al. 1999). Operational-oriented performance measures are more about internal-oriented traits and supports supply chain to continuously succeed in today's dynamic markets (Blome et al. 2013; Stank et al. 1999). In addition, Blome et al. (2013) describes operational performance measures as service-level accomplishments that lead to supply chain quality, supply chain efficiency, supply chain productivity, supply chain costs, and supply chain reliability. Operational performance also has positive impact on supply chain production planning and long-term firm perspectives (Brandon-Jones et al. 2014). Moreover, it is believed that high operational performance can be gained by networking with suppliers and customers (Patel et al. 2013; Rungtusanatham et al. 2003).

Above all, this chapter claims that supply chain innovation which includes business process, network structure, and technology leads to superior firm performance in terms of market-based as well as operational performance (Gunasekaran et al. 2004; Lee et al. 2011; Rungtusanatham et al. 2003).

2.3 Theoretical Model

The theoretical framework presented in Fig. 1 includes two major components, namely, supply chain innovation and performance. Supply chain innovation is a multidimensional construct which includes three dimensions business process (BP) innovation, network structure (NS) innovation, technology (TE) innovation. Likewise, the component performance includes market performance (PEMAR), and operational performance (PEOPR). The proposed model includes two major hypotheses (H1 and H2). In addition, this chapter will also investigate the relationship between the individual elements of supply chain innovation (business process, network structure, and technology) and performance (market and operational performance). The first dimension of supply chain innovation, business process includes customer relationship management (BPCRM), supplier relationship management (BPSRM), customer service management (BPCSM), demand management (BPDEM), order fulfilment (BPORF), manufacturing flow management (BPMFM), product development and commercialization (BPPDC), and returns management (BPREM). The second dimension of supply chain innovation, network structure includes internal functions (NSINT), customers (NSCUS), suppliers (NSSUP), third party provider logistics (NS3PL), competitors (NSCOM), consultants (NSCON), universities (NSUNI), and public authorities (NSPUB). The third dimension of supply chain innovation, technology includes planning and execution systems (TEPLA), identification systems (TEIDF), communication systems (TECOM), analytics technology (TEANA), electronic marketplaces (TEELM), advanced

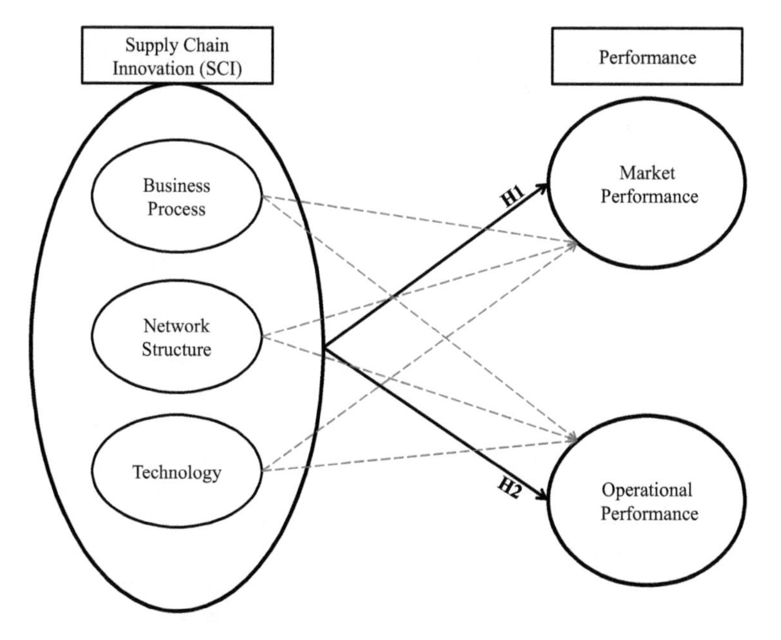

Fig. 1 Theoretical framework

manufacturing technologies (TEAMT), advanced materials (TEADM), big data (TEBIG), and drones (TEDRO). The various aspects of supply chain innovation were adopted from the extant literature (e.g., Arlbjørn et al. 2011). With respect to performance measures, the market performance measure comprises market share (PEMAR1), competitiveness (PEMAR2), customer loyalty (PEMAR3), and brand equity/value (PEMAR4) and the operational performance measure comprises reliability of supply chain processes (PEOPR1), supply chain responsiveness (PEOPR2), supply chain agility (PEOPR3), supply chain costs (PEOPR4), and effective capacity utilization (PEOPR5). The various aspects of performance measures were adopted from the existing literature (e.g., Golicic and Smith 2013).

A known fact, firms always aspire for innovation to achieve sustainable competitive advantage (Arlbjørn et al. 2011; Narasimhan and Narayanan 2013). Likewise, supply chain innovation help firms to achieve superior performance and new value creation (Arlbjørn et al. 2011; Arlbjørn and Paulraj 2013). Today, firms rely more on their supply chain partners to bring in greater innovation process (Bellamy et al. 2014) and therefore it is important to include all the three components of supply chain innovation (Arlbjørn et al. 2011) as well to concentrate equally on them to achieve higher firm performance. Most of the previous studies primarily concentrate on operational performance (e.g., Gligor and Holcomb 2012). On the contrary, this chapter planned to examine the firm performance in terms of both market and operational performance (Ahmad and Jusoh 2014; Patel et al. 2013; Swink et al. 2005). As discussed earlier, the primary objective of this chapter is to determine whether supply chain innovation could increase the firm performance in terms of market and operational performance. Accordingly, this chapter proposes the following two hypotheses:

H1 Supply chain innovation has a positive impact on market performance.

H2 Supply chain innovation has a positive impact in operational performance.

3 Method

This chapter is based on data gathered through a questionnaire-survey that was distributed among Danish manufacturing firms with at least 50 employees in the autumn 2016. The population of the companies was identified using the Danish company database "Names and numbers, business" (NN Markedsdata 2016). This chapter believes that medium and large enterprises work most systematically with supply chain innovation. The database allowed searching for these companies in a structured manner and the process resulted in a gross of 1580 companies. The selected companies were then telephoned and asked to be transferred to the person with the overall responsibility of supply chain management. This process provided us with a net population of 879 companies. Then email with a link to the electronic questionnaire (SurveyXact 2016) was sent to all the participating companies. Reminder e-mails were also sent to increase the response rate and allow comparison of early and late responses (before and after the initial deadline). This process finally resulted in 187 companies who provided valid responses with a response rate of approximately 21.3%.

The survey questionnaire was developed to test how different aspects of supply chain innovation affect different performance outcomes. The questionnaire included questions related to supply chain innovation in terms of business processes, network structure and technology (IT), and performance outcomes in terms of market based and operational based performance. The questions are grounded in the extant literature and validated by the industry representatives.

This chapter includes five constructs of which the first three are related to supply chain innovation and the other two are related to performance. The three constructs of supply chain innovation (the independent variables) was operationalized based on Arlbjørn et al. (2011) in terms of business process, technological and network innovations. The construct business process innovation was operationalized using the work of Lambert et al. (1998) and Lambert and Cooper (2000) in which the authors have defined eight supply chain processes. The respondents were asked to answer to what extent their company is pursuing innovation in relation to these eight supply chain business processes on a Likert-scale (from 1 very low degree to 5 very large degree). Technology usage in a supply chain management context is concerned with information technology (Arlbjørn et al. 2011; Vijayasarathy 2010) and this chapter believes that it is necessary to group the various technologies based on their purpose. First, identified measures within information management and operationalized it based on Vijayasarathy (2010) and Akkermans et al. (2003). Then, questions from general management literature were supplemented by including advanced manufacturing technologies and materials (Brennan et al. 2015; Vyas 2016) as well as analytics technologies and big data (Souza 2014; Wang et al. 2016). The respondents were specifically asked to answer to what extent their company is working with different technologies in their supply chain on a Likert-scale (from 1 very low degree to 5 very large degree). The construct network innovation was operationalized based on Pilav-Velić and Marjanovic (2016) as well as Fitjar and Rodríguez-Pose (2013) in which the authors have defined it as a company's external collaboration effort. The respondents were specifically asked to what extent their company innovates together with different supply chain actors on a Likert-scale (from 1 very low degree to 5 very large degree). Finally, the two constructs of performance (the dependent variables) were operationalized based on Golicic and Smith (2013) in which the authors identify market-based and operational-based performance as the two most frequently used dimensions of firm performance in business and supply chain management research (e.g. Gunasekaran and Kobu 2007; Hult et al. 2008a, b; Vachon and Klassen 2006). The respondents were specifically asked to indicate how they perceive their company's performance compared to their competitors on a Likert-scale (from 1 much worse to 5 much better).

This chapter uses the SPSS 22.0 software to evaluate the linear regression among the questions of interest. This analysis specifically identifies the relationship among the components of supply chain innovation and performance in terms of operational and market performance. As a first step of analysis, the relationship between the independent variable (supply chain innovation which includes business process, network and technology innovation) and the dependent variables market-based and operational-based performance was examined using linear regression. Then, the relationship between each individual components of supply chain innovation (business process, network and technology innovation) and performance (market-based and operational-based) was analyzed.

The complete list of indicators used to measure the various constructs are presented in Appendix. During the analysis, the indicator used to measure the construct business process (BP) innovation was supplier relationship management (BPSRM). The indicators used to measure the construct network structure (NS) innovation were public agencies (NSPUB), suppliers (NSSUP), third party providers (NS3PL), customers (NSCUS), competitors (NSCOM), universities (NSUNI), and consultants (NSCON). The indicators used to measure the construct technology (TE) innovation were identification systems (TEIDF), communication systems (TECOM), analytics technology (TEANA), electronic marketplaces (TEELM), advanced manufacturing technologies (TEAMT), and big data (TEBIG). The indicators used to measure the construct market performance (PEMAR) were market share (PEMAR1), competitiveness (PEMAR2), customer loyalty (PEMAR3), and brand equity/value (PEMAR4). The indicators used to measure the construct operational performance (PEOPR) were reliability of supply chain processes (PEOPR1), supply chain responsiveness (PEOPR2), supply chain agility (PEOPR3), supply chain costs (PEOPR4), and effective capacity utilization (PEOPR5).

4 Findings and Discussion

This section presents the results of the hypotheses tests (H1 and H2), including the standardized coefficient of each path in the proposed theoretical model. As a first step of analysis, reliability test was performed to observe the internal consistency and it is measured using the Cronbach's alpha value. This reliability designates the degree of correlation between the selected items. The reliability can be verified using Cronbach alpha value and the coefficient value of the construct should be 0.7 or higher (Hulland 1999; Nunnally and Bernstein 1994). In the reliability test, the Cronbach's alpha value for the construct market performance was 0.765 and operational performance was 0.796. The indicators for the constructs business process, network structure, and technology were directly used for the analysis except for market and operational performance.

Linear regression was performed to primarily examine the two proposed hypotheses. First, a simple linear regression was executed to predict the dependent/outcome variable (market performance) based on the independent/predictor variable (supply chain innovation). The result clearly indicates that supply chain innovation, without any doubt improves the market performance (see Table 1). In other words, the independent variable has a positive impact on the dependent variable and is statistically significant with an F-value of 1.866 ($p\text{-value} \leq 0.05$). Considering the

Table 1 Linear regression—supply chain innovation and market performance (H1)

	Dependent (market performance)
	Standardized coefficients Beta
BPSRM	−0.037
NSCUS	0.136
NSSUP	0.074
NS3PL	−0.171*
NSCOM	−0.059
NSCON	−0.195⁺
NSUNI	0.241*
NSPUB	0.093
TEIDF	0.021
TECOM	0.079
TEANA	0.098
TEELM	−0.176*
TEAMT	0.021
TEBIG	−0.070
Number of observations	186
F-value	1.866*
Adjusted R^2	0.061

*Significance at $p \leq 0.05$
⁺Significance at $p \leq 0.10$

components of the supply chain innovation, it is obvious that the emphasis is more on network with third-party logistics (NS3PL, p-value $\leq$ 0.05), network with consultants (NSCON, p-value $\leq$ 0.10), network with universities (NSUNI, p-value $\leq$ 0.05), and electronic marketplaces technology (TEELM, p-value $\leq$ 0.05) pertaining to market-based performance (see Table 1).

Then, a simple linear regression was executed to predict the dependent/outcome variable (operational performance) based on the independent/predictor variable (supply chain innovation). It is evident from the result that supply chain innovation helps in improving the operational performance (see Table 2). In particular, the independent variable has a positive impact on the dependent variable and is statistically significant with an F-value of 2.634 (p-value $\leq$ 0.01). Now, considering the components of the supply chain innovation, it is obvious that the emphasis is more on network with competitors (NSCOM, p-value $\leq$ 0.10), network with consultants (NSCON, p-value $\leq$ 0.01), network with universities (NSUNI, p-value $\leq$ 0.01), and electronic marketplaces technology (TEELM, p-value $\leq$ 0.10) concerning operational performance (see Table 2).

It is apparent from Tables 1 and 2 that supply chain innovation does payoff, however it can also be noticed that the more focus is on operational performance than that of market performance. Certainly, supply chain management is more of customer-driven, supply-driven, and market-driven, therefore, it is surprising that the results indicate there is less emphasis towards market performance. From a research

Table 2 Linear regression—supply chain innovation and operational performance (H2)

	Dependent (operational performance)
	Standardized coefficients Beta
BPSRM	0.020
NSCUS	0.109
NSSUP	0.028
NS3PL	0.012
NSCOM	0.148^{+}
NSCON	−0.293**
NSUNI	0.245**
NSPUB	0.014
TEIDF	0.096
TECOM	0.037
TEANA	0.128
TEELM	−0.146^{+}
TEAMT	0.036
TEBIG	0.063
Number of observations	186
F-value	2.634**
Adjusted R^2	0.110

**Significant at $p \leq 0.01$
$^{+}$Significance at $p \leq 0.10$

perspective, the extant literature also shows less importance on market-oriented measures while measuring the firm performance and demonstrates insignificant results while statistically examining the market performance (Ahmad and Jusoh 2014; Swink et al. 2005). Therefore, it is obvious that there is a potential gap in both research and practice. Now, this chapter claims that the research should focus on market-oriented performance measures and also firms should start concentrating equally on both operational and market performance. Furthermore, this chapter recommends that market oriented firms will experience increased customer focus which in turn helps in customer satisfaction, synchronized marketing to advance the competitiveness and market share as well as profit orientation (e.g., Min et al. 2007). As mentioned earlier, firms should also concentrate on market based performance as it increases their existing market oriented capabilities. Above all, market-oriented performance assists firms in modifying their firm and network capabilities on the basis of their opportunities of the future firm performance (e.g., Golicic and Smith 2013; Ramaswami et al. 2009). On the other hand, it is interesting to notice that firms are continuing their emphasis on operational performance however there is still potential for further improvement. The overall results of the major hypotheses are presented in Fig. 2.

In addition to the main hypotheses (H1 and H2), this chapter also made an attempt to examine the relationship between the individual components of supply chain

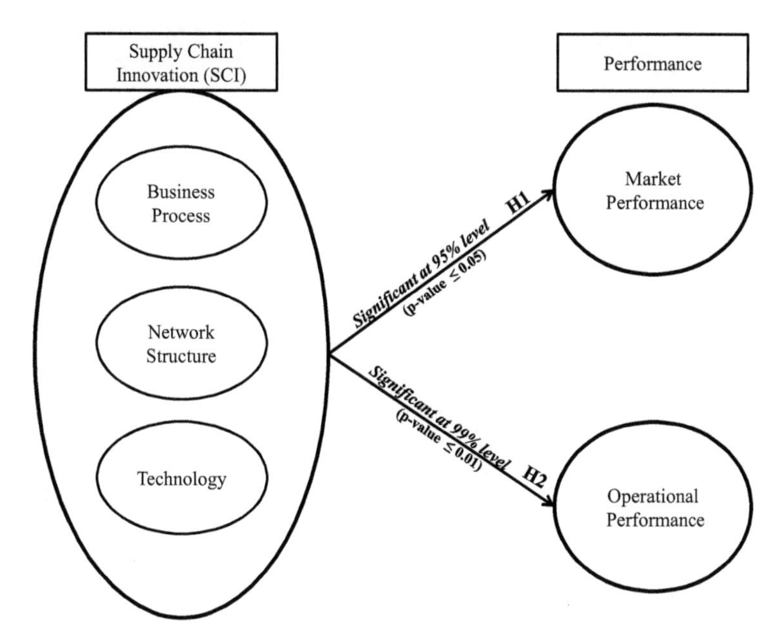

Fig. 2 Hypotheses—results

innovation such as business process, network structure as well as technology and performance in terms of market and operational performance. The results of the individual components of the supply chain innovation with respect to performance are presented in Fig. 3.

As an initial step, this study examined the relationship between business process and performance. In regard to the element business process, the only indicator considered for analysis was supplier relationship management (BPSRM). From Fig. 3, it is evident that business process has a positive relationship with only operational performance (F-value: 4.577, p-value $\leq$ 0.05) and not with market performance. Most of the earlier studies have concentrated more on operational performance measures than that of market performance and this could be the reason for this insignificant result with respect to business process and market performance. Another explanation for this could be that it might be easier to relate and isolate an innovation effort of a specific business processes to operational performance than to market performance. An improved market performance might be caused by other factors also than business process innovations. In contrast, operational performance improvements might a have stronger and direct relation to business process innovations. However, this chapter argues that firms should not consider only operational performance as long-term instead they should perceive both market performance and operational performance as long-term objectives.

As a next step, this study examined the relationship between network structure and performance. It is clear from Fig. 3 and Table 3 that network structure has a positive relationship with market performance (significant at 99% level). In addition,

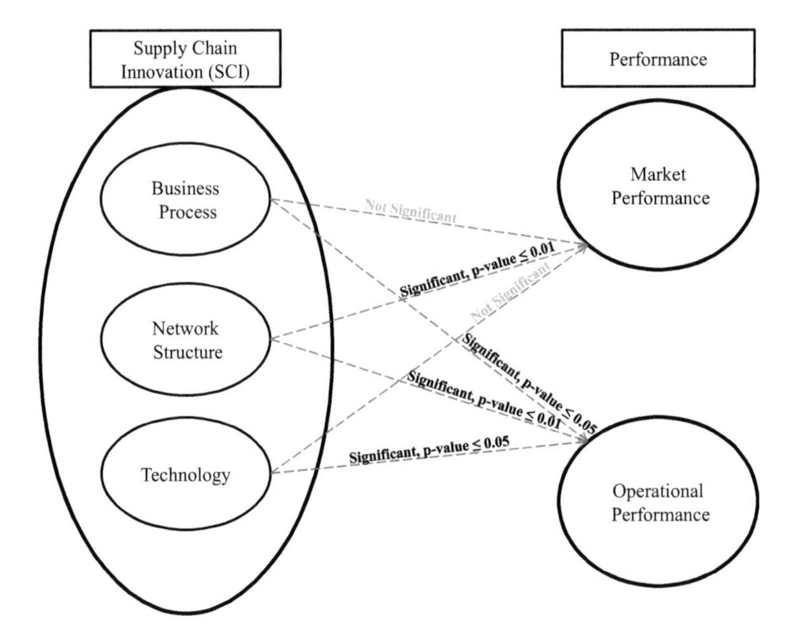

Fig. 3 Components of supply chain management

considering the indicators of network structure, it is obvious that the emphasis is more on third party provider logistics (NS3PL, p-value $\leq$ 0.05), consultants (NSCON, p-value $\leq$ 0.10), universities (NSUNI, p-value $\leq$ 0.05) pertaining to market performance.

From Fig. 3 and Table 4 it is evident that network structure has a positive relationship with operational performance (significant at 99% level).

In addition, considering the indicators of network structure, it is obvious that the emphasis is more on competitors (NCOM, p-value $\leq$ 0.10), consultants (NSCON, p-value $\leq$ 0.05), universities (NSUNI, p-value $\leq$ 0.01) concerning operational performance. Therefore, in general, the component network structure has a positive relationship with both market and operational performance. This is an interesting result since an earlier empirical study on this supply chain innovation framework found that the network structure component received the lowest mean value of 3.3 on a 5 point Likert scale on the respondents' perceptions of the components importance in creating supply chain innovations (Arlbjørn et al. 2013, p. 40). The technology component received an average of 3.5 and the business process component received an average of 3.8. The new findings of the survey reported in this chapter indicate that companies have become aware of the fact that they are dependent on their network actors' relationship in order to obtain both market and operational performance improvements.

Finally, this study examined the relationship between technology and performance. It is obvious from Fig. 3 and Table 5 that technology has no relationship with

Table 3 Linear regression—network structure and market performance

	Dependent (market performance)
	Standardized coefficients Beta
NSCUS	0.107
NSSUP	0.077
NS3PL	−0.195*
NSCOM	−0.082
NSCON	−0.180⁺
NSUNI	0.224*
NSPUB	0.116
Number of observations	186
F-value	2.885**
Adjusted R^2	0.066

**Significant at $p \leq 0.01$
*Significance at $p \leq 0.05$
⁺Significance at $p \leq 0.10$

Table 4 Linear regression—network structure and operational performance

	Dependent (operational performance)
	Standardized coefficients Beta
NSCUS	0.088
NSSUP	0.087
NS3PL	0.013
NSCOM	0.143⁺
NSCON	−0.241*
NSUNI	0.274**
NSPUB	0.032
Number of observations	186
F-value	3.911**
Adjusted R^2	0.099

**Significant at $p \leq 0.01$
*Significance at $p \leq 0.05$
⁺Significance at $p \leq 0.10$

market performance. This insignificant result could be because it might be difficult to relate a specific technology being the reason for improved market share and customer loyalty. Another explanation could be that companies still need to develop the strategic links between technology strategies and market performance.

On the other hand, it is evident from Fig. 3 and Table 6 that there is a positive relationship between technology and operational performance (significance at 95% level). Thus, it can be inferred that the respondents do perceive their technology innovations efforts and this efforts will certainly have an impact on their operational performances (e.g. more reliable processes, better cost performance and improved responsiveness).

Table 5 Linear regression—technology and market performance

	Dependent (market performance)
	Standardized coefficients Beta
TEIDF	0.031
TECOM	0.042
TEANA	0.122
TEELM	−0.177*
TEAMT	0.007
TEBIG	−0.057
Number of observations	186
F-value	0.934
Adjusted R^2	−0.002

*Significance at $p \leq 0.05$

Table 6 Linear regression—technology and operational performance

	Dependent (operational performance)
	Standardized coefficients Beta
TEIDF	0.133
TECOM	0.015
TEANA	0.136
TEELM	−0.068
TEAMT	0.036
TEBIG	0.079
Number of observations	186
F-value	2.230*
Adjusted R^2	0.038

*Significance at $p \leq 0.05$

Largely, the results of the analysis of the individual elements of supply chain innovation clearly indicate that there is more concentration on the element network structure than that of business process and technology. According to Arlbjørn et al. (2011), the supply chain innovation should include all the three elements and the firms should focus on all three elements equally to experience supply chain innovation. However, the firms at the moment are not focusing much on business process and technology pertaining to market performance. It is great that firms understand the importance of networking with their supply chain partners to innovate and to achieve greater performance in terms of market and operational performance. On the contrary, this chapter insists firms to realize the importance of business process and technology with reference to market performance. Firms need to start focusing equally on all the three elements of supply chain innovation to achieve higher performance in terms of both market and operational performance. Firms to achieve sustainable growth in terms of both market and operation should establish strong business process practice and employ robust technologies. Having said this, concerning network structure, it is evident that firms are not utilizing the entire available network. Firms again should recognize the value existing in the supply chain network to experience greater innovation and firm performance. Most importantly, taking all the three elements of supply chain innovation into account, supply chain innovation does pay off in terms of market and operation performance however the strongest relationship is for operational performance.

5 Conclusion

The objective of this chapter was to investigate the relationship between supply chain innovation and market and operational performance. In view of this objective, the data was analyzed and the results reveal that there is a positive relationship between supply chain innovation and market performance (significant at 95% level) and operation performance (significant at 99% level). Thus, the overall construct of supply chain innovation does pay off in terms of market and operation performance as measured in this chapter. The strongest relationship is for operational performance which indicates that the companies are aware of the fact that they need to innovate with their supply chains in order to lever their competitive parameters. It is also interesting to see that their innovations efforts also have a positive impact on market performance which indicates that the respondents have understood the importance of operating and developing market oriented supply chains (Green et al. 2006; Min et al. 2007; Min and Mentzer 2000).

When decomposing the overall supply chain innovation construct into its three constituting components: (a) the results reveal that business process innovation component seems to be more operational (significant at 95% level) focused than on market; (b) the results show that network structure innovation component has a positive relationship with both market and operation performance (both significant at 99% level); (c) the results show that technology innovation component seems to have a positive relationship with only operational performance (significant at 95% level) and not with market performance.

From a theoretical perspective, this chapter shows the positive relationship supply chain innovation and performance in terms of market and operational performance and in turn provides a road map for the researchers to continue their study focusing on other performance measures, for instance, financial performance measures. Supply chain innovation is an interesting and well established concept; therefore it is also opportunity to further develop this initial work grounding on various theories (e.g. dynamic capability, resource-based view, etc.). This will be a definite contribution to the prevalent literature to understanding the concept. The next phase of this study will be examining the individual indicators of business process, network structure and technology with respect to financial performance measures (including market and operational performance) (e.g., Shi and Yu 2013). In addition, several other hypotheses will be formulated and various other advanced statistical tests will be performance to evaluate the relationships between supply chain innovation (business process, network structure, and technology innovation) and performance (market, operational, and financial performance). Most importantly, it will be more stimulating to extend the current theoretical model with moderators such as firm size, technology intensity, industry clockspeed, innovation, ambidexterity, absorptive capacity etc. These moderating factors might have a significant impact and will provide more insights concerning the relationship between supply chain innovation and performance in terms of market, operation and finance. This study also tried to test the model with some of the above mentioned moderators and the results did show positive impact for some moderators and negative impact for some moderators concerning the explanation of the overall relationship between supply chain innovation and performance. However, this chapter did not include the results showing the impact of these moderators on the overall theoretical model. Furthermore, the extension of this current study will be certainly explored in the future research including extensive analysis of the moderators explaining the relationship between supply chain innovation and performance with reference to market, operation and finance. Besides theoretical contribution, this chapter recommends firms to observe supply chain innovation in connection with business process innovation, network structure innovation, and technology innovation to realize superior performance in terms of both market and operational performance. The results of this chapter inform firms that they are not focusing on market performance measures at the moment. Therefore, to achieve long-term objectives, firms should not just pursue supply chain innovation and measure their performance only in terms of operational; instead, they have to strategically integrate all the elements of supply chain innovation and measure their performance in terms of both market and operational.

Appendix

Business Processes

To what extent is your company pursuing innovations in the following supply chain management business processes?

- Customer Relationship Management (CRM)
- Supplier Relationship Management (SRM)
- Customer Service Management (CSM)
- Demand Management (DeM)
- Order Fulfilment (OrF)
- Manufacturing Flow Management (MFM)
- Product Development and Commercialization (PDC)
- Returns Management (ReM)

Network Structure

To what extent does your company innovate together with the following supply chain network actors?

- Internal functions
- Customers
- Suppliers
- Third party providers (e.g. logistics providers)
- Competitors
- Consultants
- Universities
- Public agencies

Technology

To what extent does your company work with the following technologies in your supply chain?

- Planning and execution systems (e.g. enterprise resource planning systems, advanced planning systems, material requirements systems)
- Identification systems (e.g. barcodes, radio frequency identification)
- Communication systems (e.g. electronic data interchange, web-based communication tools, mobile communication solutions, cloud technology)
- Analytics technology (e.g. business intelligence, statistics and analytics software, algorithms)
- Electronic marketplaces (e.g. e-portals, e-auctions, supplier collaboration tools)
- Advanced manufacturing technologies (e.g. advanced robotics, 3D-printing)
- Advanced materials (e.g. ultra-light or high-strength materials)
- Big data
- Drones

Market Performance

Indicate how you perceive your company's performance relative to your competitors?

- Market share
- Competitiveness
- Customer loyalty
- Brand equity

Operational Performance

Indicate how you perceive your company's performance relative to your competitors?

- Reliability of supply chain processes
- Supply chain responsiveness
- Supply chain agility
- Supply chain costs
- Effective capacity utilization

References

Ahmad AC, Jusoh MA (2014) Institutional ownership and market-based performance indicators: utilizing generalized least square estimation technique. Proc Soc Behav Sci 164:477–485

Akkermans HA, Bogerd P, Yücesan E, Van Wassenhove LN (2003) The impact of ERP on supply chain management: exploratory findings from a European Delphi study. Eur J Oper Res 146(2):284–301

Arlbjørn JS, Paulraj A (2013) Special topic forum on innovations in business networks from a supply chain perspective: current status and opportunities for future research. J Supply Chain Manag 49(4):3–11

Arlbjørn JS, de Haas H, Munksgaard KB (2011) Exploring supply chain innovation. Logist Res 3(1):3–18

Arlbjørn JS, Mikkelsen OS, Munksgaard KB, Schlichter J, Paulraj A (2013) Konkurrencekraft gennem supply chain innovation [Competitiveness through supply chain innovation]. Departmentof Entrepreneurship and Relationship Management, University of Southern Denmark

Bellamy MA, Ghosh S, Hora M (2014) The influence of supply network structure on firm innovation. J Oper Manag 32(6):357–373

Blome C, Schoenherr T, Rexhausen D (2013) Antecedents and enablers of supply chain agility and its effect on performance: a dynamic capabilities perspective. Int J Prod Res 51(4):1295–1318

Brandon-Jones E, Squire B, Autry CW, Petersen KJ (2014) A contingent resource-based perspective of supply chain resilience and robustness. J Supply Chain Manag 50(3):55–73

Brennan L, Ferdows K, Godsell J, Golini R, Keegan R, Kinkel S, Srai JS, Taylor M (2015) Manufacturing in the world: where next? Int J Oper Prod Manag 35(9):1253–1274

Chen IJ, Paulraj A (2004) Towards a theory of supply chain management: the constructs and measurements. J Oper Manag 22:119–150

Cheng LT, Leung TY (2004) A comparative analysis of the market-based and accounting-based performance of diversifying and non-diversifying acquisitions in Hong Kong. Int Bus Rev 13(6):763–789

DeTienne D, Golicic S, Swink ML (2015) Delivering successful supply chain innovations: lessons from CSCMP's supply chain innovation award winners. CSCMP Explores 12(Winter):1–14

Ellram LM, Cooper MC (2014) Supply chain management: it's all about the journey, not the destination. J Supply Chain Manag 50(1):820

Fitjar RD, Rodríguez-Pose A (2013) Firm collaboration and modes of innovation in Norway. Res Policy 42(1):128–138

Flint DJ, Larsson E, Gammelgaard B, Mentzer JT (2005) Logistics innovation: a customer value-oriented social process. J Bus Logist 26(1):113–147

Gligor DM, Holcomb MC (2012) Understanding the role of logistics capabilities in achieving supply chain agility: a systematic literature review. Supply Chain Manag Int J 17(4):438–453

Golicic SK, Smith CD (2013) A meta-analysis of environmentally sustainable supply chain management practices and firm performance. J Supply Chain Manag 49(2):78–95

Green KW, McGaughey R, Case KM (2006) Does supply chain management strategy mediate the association between market orientation and organizational performance? Supply Chain Manag Int J 11(5):407–414

Gunasekaran A, Kobu B (2007) Performance measures and metrics in logistics and supply chain management: a review of recent literature (1995–2004) for research and applications. Int J Prod Res 45(12):2819–2840

Gunasekaran A, Patel C, McGaughey RE (2004) A framework for supply chain performance measurement. Int J Prod Econ 87(3):333–347

Hazen BT, Overstreet RE, Cegielski CG (2012) Supply chain innovation diffusion: going beyond adoption. Int J Logist Manag 23(1):119–134

Heaslip G, Kovács G, Haavisto I (2015) Supply chain innovation: lessons from humanitarian supply chains. In: Stentoft J, Paulraj A, Vastag G (eds) Research in the decision sciences for global supply chain network innovation. New Jersey, Pearson, pp 9–26

Hulland J (1999) Use of partial least squares (PLS) in strategic management research: a review of four recent studies. Strateg Manag J 20(2):195–204

Hult GTM, Ketchen DJ, Griffith DA, Chabowski BR, Hamman MK, Dykes BJ, Cavusgil ST (2008a) An assessment of the measurement of performance in international business research. J Int Bus Stud 39(6):1064–1080

Hult GTM, Ketchen DJ, Griffith DA, Finnegan CA, Gonzalez-Padron T, Harmancioglu N, Cavusgil ST (2008b) Data equivalence in cross-cultural international business research: assessment and guidelines. J Int Bus Stud 39(6):1027–1044

Kronborg Jensen J, Balslev Munksgaard K, Stentoft Arlbjørn J (2013) Chasing value offerings through green supply chain innovation. Eur Bus Rev 25(2):124–146

Lambert DM, Cooper MC (2000) Issues in supply chain management. Ind Mark Manag 29 (1):65–83

Lambert DM, Cooper MC, Pagh JD (1998) Supply chain management: implementation issues and research opportunities. Int J Logist Manag 9(2):1–20

Lee SM, Lee D, Schniederjans MJ (2011) Supply chain innovation and organizational performance in the healthcare industry. Int J Oper Prod Manag 31(11):1193–1214

Mentzer JT, DeWitt W, Keebler JS, Min S, Nix NW, Smith CD, Zachariah ZG (2001) Defining supply chain management. J Bus Logist 22(2):1–25

Min S, Mentzer JT (2000) The role of marketing in supply chain management. Int J Phys Distrib Logist Manag 30(9):765–787

Min S, Mentzer JT, Ladd RT (2007) A market orientation in supply chain management. J Acad Mark Sci 35(4):507

Munksgaard KB, Stentoft J, Paulraj A (2014) Value-based supply chain innovation. Oper Manag Res 7(3–4):50–62

Narasimhan R, Narayanan S (2013) Perspectives on supply chain driven innovation. J Supply Chain Manag 49(4):27–42

NN Markedsdata (2016) Welcome to Navne and Numre® Erhverv. http://erhverv.nnmarkedsdata.dk/#. Accessed Aug 2016

Nunnally JC, Bernstein IH (1994) Psychological theory. MacGraw-Hill, New York, NY

Ojha D, Shockley J, Acharya C (2016) Supply chain organizational infrastructure for promoting entrepreneurial emphasis and innovativeness: the role of trust and learning. Int J Prod Econ 179:212–227

Panayides PM, Lun YHV (2009) The impact of trust on innovativeness and supply chain performance. Int J Prod Econ 122(1):35–46

Patel P, Azadegan A, Ellram LM (2013) The effects of strategic and structural supply chain orientation on operational and customer-focused performance. Decis Sci J 44(4):713–753

Pilav-Velić A, Marjanovic O (2016) Integrating open innovation and business process innovation: insights from a large-scale study on a transition economy. Inf Manag 53(3):398–408

Ramaswami SN, Srivastava RK, Bhargava M (2009) Market-based capabilities and financial performance of firms: insights into marketing's contribution to firm value. J Acad Mark Sci 37(2):97

Rungtusanatham M, Salvador F, Forza C, Choi TY (2003) Supply-chain linkages and operational performance: a resource-based-view perspective. Int J Oper Prod Manag 23(9):1084–1099
Shi M, Yu W (2013) Supply chain management and financial performance: literature review and future directions. Int J Oper Prod Manag 33(10):1283–1317
Souza GC (2014) Supply chain analytics. Bus Horiz 57(5):595–605
Stank TP, Goldsby TJ, Vickery SK (1999) Effect of service supplier performance on satisfaction and loyalty of store managers in the fast food industry. J Oper Manag 17(4):429–447
Stentoft J, Narasimhan R, Poulsen T (2016a) Reducing cost of energy in the offshore wind energy industry: the promise and potential of supply chain management. Int J Energy Sector Manag 10(2):151–171
Stentoft J, Mikkelsen OS, Jensen JK (2016b) Offshoring and backshoring manufacturing from a supply chain innovation perspective. Supply Chain Forum Int J 17(4):190–204
Stentoft J, Rajkumar C, Madsen ES (2017) Industry 4.0 in Danish Industry. Department of Entrepreneurship and Relationship Management, University of Southern Denmark
Stock JR, Boyer SL (2009) Developing a consensus definition of supply chain management: a qualitative study. Int J Phys Distrib Logist Manag 39(8):690–711
SurveyXact (2016) Main page. http://www.surveyxact.dk/. Accessed Aug 2016
Swink M, Narasimhan R, Kim SW (2005) Manufacturing practices and strategy integration: effects on cost efficiency, flexibility, and market-based performance. Decis Sci 36(3):427–457
Vachon S, Klassen RD (2006) Extending green practices across the supply chain: the impact of upstream and downstream integration. Int J Oper Prod Manag 26(7):795–821
Vijayasarathy LR (2010) An investigation of moderators of the link between technology use in the supply chain and supply chain performance. Inf Manag 47(7–8):364–371
Vyas N (2016) Disruptive technologies enabling supply chain evolution. Supply Chain Manag Rev 20(1):36–41
Wagner S (2012) Tapping supplier innovation. J Supply Chain Manag 48(2):37–52
Wang S, Wan J, Zhang D, Li D, Zhang C (2016) Towards smart factory for industry 4.0: a self-organized multi-agent system with big data based feedback and coordination. Comput Netw 101:158–168
Yoon SN, Lee D, Schniederjans M (2016) Effects of innovation leadership and supply chain innovation on supply chain efficiency: focusing on hospital size. Technol Forecast Soc Chang 113:412–421
Zimmermann R, Ferreira LMDF, Moreira AC (2016) The influence of supply chain on the innovation process: a systematic literature review. Supply Chain Manag Int J 21(3):289–304

Jan Stentoft is a Professor of Supply Chain Management at the Department of Entrepreneurship and Relationship Management at the University of Southern Denmark in Kolding. He holds a PhD in logistics and supply chain management (SCM). He is head of a strategic research program focusing on supply chain innovation in the offshore wind energy sector (ReCoE) (www.recoe.dk). His research and teaching interests are within SCM, supply chain innovation, and operations management. He primarily instructs MSc, PhD, and MBA students in SCM, operations management, and administrative information systems. He has practical industry experience from positions as Director (Programme Management Office) at LEGO Systems A/S, as ERP Project Manager at Gumlink A/S, and as management consultant in a wide number of industrial enterprises.

Christopher Rajkumar is a postdoctoral student at the Department of Entrepreneurship and Relationship Management, University of Southern Denmark, Kolding. His research is within the field of supply chain management. He is affiliated with the research program Reduced Cost of Energy (ReCoE). This multidisciplinary project aims to deliver solutions to reduce the Cost of Energy (CoE) in sustainable offshore wind power energy systems [including the wind turbine and the balance of plant (BoP)]. Christopher started his research in 2014 focusing on the strategic role of the sourcing function in creating innovative products, processes, and/or services (Sourcing Innovation) and obtained his PhD degree in 2017 from the Faculty of Business and Social Sciences at the University of Southern Denmark.

Part IV
Information and Technology

Technological Innovations: Impacts on Supply Chains

Cheryl Druehl, Janice Carrillo, and Juliana Hsuan

Abstract Supply chains have benefitted tremendously from digital and transportation technologies over the years. Advanced IT systems have enhanced inventory and demand visibility and facilitated communications with global partners and customers, while transportation technologies have improved the speed and efficiency necessary to transport goods globally. However, dramatic changes in both of these areas are on the horizon. The emergence of new technologies such as 3D printing, virtual reality, autonomous vehicles, drones, and the Internet of Things (IoT) will force the next big wave of changes in global supply chains. While some of these technologies have been adopted by individual firms, many questions remain concerning how these technologies will drive new supply chain policies, business models, and regulations in the future. To illustrate, while technologies such as autonomous vehicles and IoT facilitate supply chain efficiency and transparency, they also increase the risk of compromising data security. In this chapter, we offer a brief overview of each of these emerging technologies and summarize the impact on the supply chain. We intend for this chapter to spur interest and research into not only these technologies and their impact on supply chains, but also into envisioning the supply chains of the future.

C. Druehl (✉)
School of Business, George Mason University, Fairfax, VA, USA
e-mail: cdruehl@gmu.edu

J. Carrillo
Warrington College of Business, University of Florida, Gainesville, FL, USA
e-mail: jc@ufl.edu

J. Hsuan
Department of Operations Management, Copenhagen Business School, Frederiksberg, Denmark
e-mail: jh.om@cbs.dk

A. C. Moreira et al. (eds.), *Innovation and Supply Chain Management*, Contributions to Management Science, https://doi.org/10.1007/978-3-319-74304-2_12

1 Introduction

Much is being written in the popular press about emerging technologies and their impact on consumers. A self-driving car in testing might pass you on the road; your refrigerator might let you know on your smart phone that the milk is running low. While these technologies are coming closer to reality, much of their impact will be on the supply chain (SC) (World Economic Forum 2017). In this chapter, we focus on five—3D printing, virtual reality, autonomous vehicles, drones, and the Internet of Things (IoT)—of the more than 60 technologies identified by the World Economic Forum's (2017) Shaping the Future of Production Initiative. We chose these to allow a more in-depth view of each, their impact on the SC, and interesting future research questions.

Overall, we note that these technologies seem ubiquitous—they can be applied in many stages of the SC. They offer tremendous potential to improve SC transparency, reduce costs, and increase convenience for consumers. At the same time, they pose potential drawbacks such as privacy loss, as yet unspecified regulations, and job loss.

This chapter contributes a SC view of the uses and drawbacks of these technologies. Additionally, we reflect on these technologies and some commonalities. The chosen technologies are compelling and much experimentation with them is underway. However, most need to find appropriate business models to make them successful as parts of SC processes and solutions. Through this chapter, we intend to spur interest and research into not only the impacts of these technologies on the SC, but also into envisioning the SCs of the future.

2 Literature Overview

There has been significant literature on supply chains and technology. Here we identify some major reviews and thought pieces on the roles and intersections of supply chain management (SCM) and technology. Our intent is to include only those articles which are most relevant, as we do not provide an exhaustive review of the literature.

A SC involves flows of products, information and money and much of the literature focuses on the first two. Product flows are impacted mostly by autonomous vehicles, drones, and IoT, and the related literature is discussed in Sect. 3. Roth et al. (2016) suggest several promising opportunities in operations and SC research including the role of technologies including artificial intelligence and those underlying the sharing economy. In particular, information, information sharing, collaboration, and integration are important topics identified. Anderson and Parker (2013) review the literature on global knowledge networks, including distributed SCs and projects. Many topics they address including network design, information infrastructure design, and challenges of integration will apply to firms and industries as they realign SCs to adopt these technologies. Kamal and Irani (2014) review the SC integration literature and identify major themes. Of interest here is the theme of moving to IT-enabled SC to further integrate multiple firms. Swaminathan and Tayur (2003) review the emerging (at the time) literature on e-business and SC, noting that the issues can be divided into configuration and coordination (execution). An outcome of adoption, singly or in combinations, of the technologies discussed herein are likely to result in SC reconfigurations. For example, 3D printing may reduce the number of suppliers and drones offer a new mode of transportation, both issues of configuration. They also note that many existing SC issues are marginally changed by the rise of e-business, while new issues have also arisen, similar to the adoption of the technologies discussed here.

Services SCs will likely heavily utilize these emerging technologies as evidenced by Uber in transportation services and personal monitoring devices in IoT. Wang et al. (2015) offer a review of service SCs, delineating service-only SCs and product-service SCs. This chapter focuses on product-service SCs, but some, such as drones (monitoring), are applicable in both. Karmarkar (2015) offers a further perspective on services and information, advocating that information technologies improve productivity, and providing research opportunities.

SCs and suppliers play an increasing role in innovation. Carrillo et al. (2015) and Zimmermann et al. (2016) review the intersection of innovation and the SC, including both innovations in SCM (e.g., diffusion of ISO9000) and the role of the SC in new product/service development (e.g., early supplier involvement). The technologies herein will likely impact both of these.

Gaimon et al. (2017) introduce a special issue on Management of Technology (MOT), exploring "how firms develop and leverage internal and external knowledge-based resource capabilities to respond to the dynamic opportunities and threats created by innovations in technology" (p. 576). The external knowledge resources are those found in the SC, such as in alliances or suppliers. This chapter addresses the special issue themes of coordinating external knowledge sources, managing stakeholders, and how these technologies will impact productivity and platforms.

3 Technologies

There are many exciting technologies being developed and implemented to improve the SC. As mentioned, we attend to the five technologies of 3D printing, virtual reality, autonomous vehicles, drones, and IoT. We focus on the managerial, not technical, aspects of these technologies and bound this discussion to these aspects and literature.[1] Figure 1 presents a high level SC and highlights the stages in which these technologies are having or will have a large impact. Next we discuss each technology in turn, briefly introducing it, and identifying where and how it can be

[1]We acknowledge the substantial literature in technical arenas such as engineering and computer science journals and encourage those interested to seek out those articles.

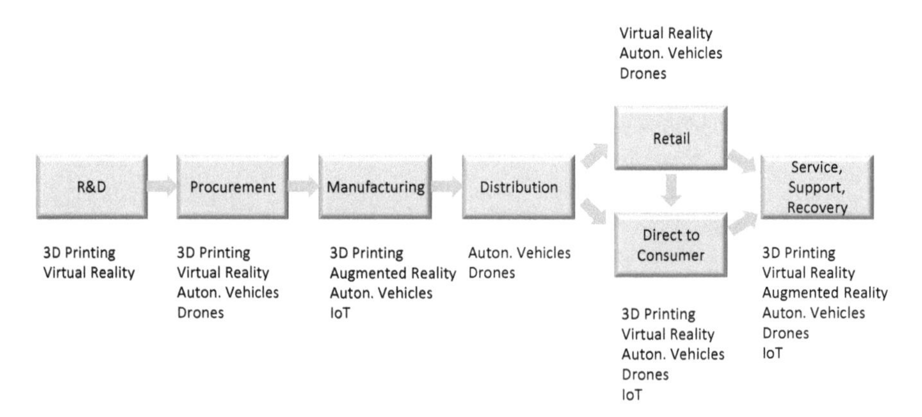

Fig. 1 Technology can be applied in many SC stages

used in the SC. Then we identify managerial, IS/IT and policy implications including benefits and risks, and mention any existing research, then lastly provide some potential future research areas.

3.1 3D Printing

3D printing (3DP), also known as additive manufacturing (AM) systems, is able to manufacture three-dimensional (3D) components and products directly from raw material and 3D design data (Baumers et al. 2016), usually layer upon layer. Because tools, molds, or cutting implements are not required, it allows a firm to make customized products without incurring any cost penalties in manufacturing (Weller et al. 2015). One can fabricate 3D components directly from a 3D (e.g., for consumer use) or AM (e.g., for industrial manufacturing application) printer, from raw materials specified by the 3D-CAD software. 3DP/AM can be considered as an evolution of rapid prototyping, as a tool to speed up the prototyping process (Gibson et al. 2010).

3.1.1 Impact on SC and Uses

3DP is at its infancy stage as it is not yet able to support high-volume production of end-use products (Baumers et al. 2016). But this trend is changing. Many industries, organizations (including educational institutions), and consumers are actively using 3DP. It is estimated that that by 2019, 10% of discrete manufacturers will apply 3DP/AM in their part-manufacturing operations (Gartner 2015). Furthermore, the market value of 3DP/AM is forecasted to grow from $4.1 billion in 2014 to $10.8 billion in 2021 (Wohlers 2015).

3DP/AM has been applied in various industries. In one aerospace application, it enabled a 50–80% reduction in component weight, achieving higher efficiency through less fuel consumption and lower buy-to-fly ratio (the weight ratio between the raw material used for a component and the component itself) (Haugom 2016). In the automotive industry, Ford was able to eliminate cost-intensive casting procedures in the development of a new engine, hence achieving considerable savings in tooling costs (Giffi et al. 2014). In the medical industry, AM is predicted to become an integral part of products offered to patients (Wohlers 2014). The hearing industry, for example, has been able to shorten lead times for patient-specific aids to one day (Ruffo and Hague 2007). In fashion and apparel, a variety of products, from shoes to dresses, have been produced from 3D-printed materials, primarily limited to high profile couture (Lewandrowski 2014), but the adoption rate seems slow. Recently, many running shoe companies started to unveil innovation programs with the application of 3D printing, such as Nike Zoom Superfly Flyknit, Futurecraft 3D by Adidas, and Zante Generate by New Balance.

3.1.2 Managerial, IS/IT and Policy Implications

AM potentially reduces the need for assembly work as it enables the production of complex and integrated functional designs in a single production step (Weller et al. 2015). Such flexibility enables mass customization of products (D'Aveni 2013). According to Pine (1993), the focus of mass customization is to support variety and customization through flexibility and quick responsiveness targeted at fragmented demand, heterogeneous niches, low cost/high quality of goods and services, short product development and life cycles. With CAD-based automated AM processes to construct parts, the production of customized goods can be moved upstream, enabling the manufacturing system to be more agile and capable of producing high product variety without high labor content (Tuck et al. 2008).

The increased adoption of 3DP/AM has tremendous implications on reconfiguration of SC networks. The individualized and customized production and design shifts the point of differentiation (the decoupling point) from make-to-stock to make-to-order, bypassing the distribution lead time. In mass customization applications, where customers are directly involved in the customization, ensuring IT integration and interoperability becomes critical. 3DP/AM fosters decentralization of production, and potentially requires more resources to mitigate risks and uncertainties due to increased coordination, communication, and monitoring (Manuj and Mentzer 2008). It also requires increased vendor support during the implementation (Mellor et al. 2014). In commercial application, the ownership of development and production processes is shifted from organizations to individuals. As such innovation activities become dispersed across SCs, quality control and monitoring become a challenge.

3.1.3 Existing Literature and Future Research Opportunities

Although there has been relatively fast adoption of 3DP/AM technology, literature on this topic has been dispersed in consultancy reports and engineering journals. Recently, Niaki and Nonino (2017) conducted a systematic literature review on the management of AM and identified eight factors: AM technology selection, SC management, product design and production cost models, environmental aspect assessment, strategic challenges, manufacturing system frameworks, open-source innovation and business and social impacts, and economics of AM. Holmström et al. (2010) investigate centralized and decentralized approaches for deploying rapid manufacturing in the aircraft spare parts SC, and propose that on-demand and centralized production is the most likely approach to succeed.

Future Research Opportunities

1. How does the development of rapid engineering and rapid manufacturing techniques impact the application of 3DP/AM for mass customization?
2. As consumers become the producers of their own designs, how will "democratized manufacturing" disrupt existing market structures (Weller et al. 2015)?
3. How do people and 3DP/AM interact in the organization?
4. How is 3DP/AM applied in the provision of services, such as preventive maintenance and inventory of spare parts?

3.2 Virtual Reality

Virtual reality (VR) and augmented reality (AR) are two types of computer-generated environments in the very early stages of their lifecycles. VR is an entirely different virtual, immersive, world such as those developed by Oculus Rift (now Facebook) and typified by headsets. AR layers digital content over the real world such as in Google Glass. Virtual reality has the potential to dramatically change many aspects of the SC and customer experiences and has been called the "Internet of Experiences" (Kelly 2016).

3.2.1 Impact on SC and Uses

In the SC, the stages likely to first use VR are R&D and retail (see Fig. 1). New product or service development (NPD/NSD) may utilize VR for prototyping (Kelly 2016) and a VR environment may allow for immersive customer feedback on designs without a physical prototype or the customer being physically present. Retailers are interested in virtual showrooms as Alibaba showed with a mobile application and cardboard headset for the Chinese celebration of Singles Day (Benton 2017). The VR store holds tremendous potential for data collection on consumer preferences (Benton 2017). VR showrooms could also further omni-channel retail. For instance, virtual showrooms will be able to efficiently allow customers to learn and experience products, traditionally a physical store function, and perform fulfillment using inventory pooling, traditionally an online store feature (Bell et al. 2015; Gao and Su 2017). VR is also being introduced into training for SC employees. For example, the material handling equipment manufacturer Raymond Corp. introduced a forklift operator VR training system (Anonymous 2017).

MIT researchers are building a Visual Analytics Lab in the Center for Transportation and Logistics to further develop AR for logistics and to encourage companies to adopt this technology (Eshkenazi 2016). One use case is in warehouse picking, where AR glasses can be used to provide the next pick information (Trebilcock 2017). Companies such as GE and Boeing have AR pilot programs. For instance, Boeing uses AR to provide diagrams and instructions to create complex wiring systems for airplanes (Castellanos 2016). VR/AR may potentially be used in after-sales service and support. VR could bring customer service representatives and customers together virtually, while AR could present directions or trouble-shooting advice to consumers.

3.2.2 Managerial, IS/IT and Policy Implications

Benefits of VR/AR implementation include the potential for improved inventory velocity or customization via integration with other SC systems that take the order and immediately generate supplier orders (MH&L 2015) or picking and shipping instructions. Additionally, VR showrooms will generate customer data on preferences, with the goal of a better customer experience as well as revenue growth (MH&L 2015).

Implementation of VR showrooms will require a fast way to scan objects instead of programming them and a seamless way to integrate with back-end applications to make the process efficient (ZDA LLC 2016). For fruitful retail applications, consumers need the appropriate technology. To date, the consumer acceptance of AR glasses has been low (Nicas 2016a), exemplified by Google no longer selling Glass. Additionally there are security and data privacy concerns, such as how actions in a VR session will be monitored, stored, and used and how privacy will be protected with facial recognition software in AR. Environmental issues are another concern as AR/VR devices will result in additional e-waste.

3.2.3 Existing Literature and Future Research Opportunities

There is some emerging SC literature on VR. Most articles that mention VR do so in passing as an example (e.g., Ba and Nault 2017; Keeney 1999; Roth et al. 2016). Others identify VR as a way to obtain customer data (e.g., Kim and Krishnan 2015), while others recognize VR as a means for customers to obtain product experience virtually, reducing uncertainty (Gao and Su 2017; Kim and Krishnan 2015;

Markopoulos and Hosanagar 2017). Anderson and Parker (2013) mention VR as a tool to be embedded in the SC information system to increase integration. Others have proposed VR as a means to improve collaboration and communication (e.g., Catalini 2017; Teodoridis 2017). VR is also seen as the future of simulation and prototyping (e.g., Jain et al. 2001; Tan and Vonderembse 2006; Zhang et al. 2016). Interestingly, there are no papers mentioning AR despite the fact that Forrester Research predicts that 14.4 million US workers will be using AR by 2025 (Eshkenazi 2016).

Future Research Opportunities

1. How can VR/AR be integrated with IoT to increase efficiency and effectiveness in operations?
2. What are appropriate quality/lean tools and techniques with VR/AR adoption? Do the measurements and/or do the tools change?
3. How can VR be integrated into the product/service design process for internal designers and to allow for consumer driven customization?
4. How can workforce management, training, incentives, and facility design be managed to maintain and increase productivity, morale, and safety during technology transition and new process adoption?
5. Will VR be a disruptive or sustaining innovation to retail and specifically to online or offline storefronts? Will it be a high-end encroachment (Schmidt and Druehl 2008)?

3.3 Autonomous Vehicles

Globally, over 44 major technology and auto companies have projects dedicated to autonomous vehicles (AV), also called driverless and self-driving vehicles (Supply Chain 24/7 2017). Autonomy is more than driverless, it means a vehicle "that is actually capable of thinking for itself, and making decisions based on the data it's received and the parameters it's been ordered to operate within" (p. 15, Futurenautics Ltd. 2016). A key technology is the vision system, such as light detection and ranging (lidar), cameras, and short range radar, where a dominant design is not yet established (Gates et al. 2017). Driver assistance systems and automated guided vehicles (e.g., in warehouses) fall short of an autonomous vehicle. In addition to cars, autonomous trucks and ships are being tested, all of which are early in the product lifecycle.

3.3.1 Impact on SC and Uses

AV will have the most impact on transportation (Fig. 1) with cars likely affecting end-consumer delivery most, trucks used in inbound and outbound logistics, and ships and railroads used for long-haul transport or internal operations. To illustrate,

Rio Tinto is using driverless trucks and a railroad in its "Mine of the Future" (Futurenautics Ltd. 2016).

Last mile delivery using AV has the potential to remove a significant portion of SC labor cost, similar to using drones. DHL has suggested using packstations, a self-driving repository for packages (DHL 2014). However, the question of how to move packages from the vehicle to the door remains (Schulz 2017).

Self-driving trucks for both long-haul driving and within-city trips have the biggest potential impact on SCs. Testing is ongoing; notably an automated truck delivered 50,000 cans of beer in Colorado in 2016 (Kane and Tomer 2017). Adoption may impact distribution network designs as most networks assume 500 miles per day (with one driver) to reach a customer. Autonomous trucks can drive 24 h per day, thereby increasing the single day range, and altering the number and locations of distribution centers (Potts 2016). This has the potential to increase inventory pooling with fewer locations, and increase savings in inventory, facility, and labor costs.

For unmanned ships, proponents see the first uses as close-to-shore applications such as ferries and tugs. For example, Kongsberg is working with Yara to develop an automated, electric ship which will sail between three Norwegian ports, replacing land-based transportation (Kongsberg n.d.). In the longer term, ocean-going autonomous ships are anticipated. These, along with data from smart ships, will improve transparency and enable more automated multimodal planning (Futurenautics Ltd. 2016).

3.3.2 Managerial, IS/IT and Policy Implications

The largest benefits from AV are seen as increased safety from reduced human errors and cost reduction from eliminating drivers (Kane and Tomer 2017). Currently, labor is estimated at approximately 35% of shipping cost (International Transport Forum 2017). AV are also likely to reduce congestion and emissions, and improve fuel efficiency (International Transport Forum 2017; Kane and Tomer 2017). Furthermore, autonomous trucks will help to alleviate the driver shortage, estimated at 48,000 in the US in 2015 and predicted to grow to over 1 million in US and Europe (International Transport Forum 2017). It is estimated that the same amount of freight could be delivered with approximately 40% fewer trucks due to increased truck utilization (Potts 2016). The reduction in operating costs should decrease shipping rates, further reducing SC costs. Due to these economic benefits, adoption in trucking is expected to be faster than in consumer vehicles (International Transport Forum 2017).

Potential benefits of autonomous ships include better real-time planning and realignment as demand or port availability changes. Another benefit is increased safety and shorter time away at sea for humans. Autonomous ships will also be designed without crew quarters, leaving more cargo space or a more fuel efficient, smaller ship. One estimate put the operational savings for unmanned ships at 40% compared to today's costs (Futurenautics Ltd. 2016).

The issues with adoption and implementation of AV include insurance, regulations, safety, and cybersecurity. Without drivers, who to insure or to hold responsible for accidents and damages is being debated. For example, a UK law holds insurers responsible to pay for damages and liability from accidents with AV, but leaves the question open as to whether automakers are responsible for reimbursing insurers (Out-law.com 2017). Current regulations in most countries allow testing of AV, but not actual operation without a human backup driver (International Transport Forum 2017). Transportation inherently crosses political borders, making consistent regulations difficult to achieve (Rolls Royce plc 2016). Highway and communication infrastructures may also need to change to manage large numbers of AV (International Transport Forum 2017; Markoff 2017). The potential for cybersecurity problems concerns the public and regulators, where safety could be compromised even if the AV was deemed safe otherwise (Montenegro 2015).

In trucking, the loss of jobs is one of the largest concerns. In the US, truck driver is one of the most common jobs (Bui 2015). As a result, labor unions are trying to slow adoption (Beene and Eidelson 2017) of AV. Overall, estimates for job loss in the US and Europe for truck drivers are 2 million by 2030 with additional jobs lost for taxi and bus drivers (International Transport Forum 2017).

3.3.3 Existing Literature and Future Research Opportunities

There is limited existing SC literature, yet interest is beginning to build. A few papers mention AV in passing as related to future technologies. Several articles discuss cities and the associated transportation and delivery systems with AV playing a major role (e.g. Mehmood et al. 2017; Savelsbergh and Van Woensel 2016). Other papers study transportation system impacts such as traffic operations and flow, such as highways for AV (e.g., Hall et al. 2001; Mahmassani 2016).

Future Research Opportunities

1. Where are changes required to current SCs to utilize AV and where can AV simply replace existing ones?
2. What will be the key reasons and barriers for adoption of AV in SCs? Are there new benefits to be uncovered?
3. What are the best ways to integrate data from AV into SC planning? How much planning should be done ahead versus real time based on current data and status?
4. What impacts will AV have on other transportation sectors such as railroads, airlines, and taxis?
5. How will AV impact third and fourth party logistics providers? Will firms develop their own closed transportation networks with AV or develop shared collaborative agreements?
6. What innovations and improvements will be needed to provide enough communication bandwidth and security to allow for truly AV?

3.4 Drones

Drones are essentially flying robots built on a hardware platform consisting of rotors, a battery, sensors and GPS devices. Otherwise known as "Unmanned Aerial Vehicles" (UAVs), these robots can be controlled autonomously or via remote control.

3.4.1 Impact on SC and Uses

Initially developed and utilized for military applications, drones have become more commonplace throughout the SC. Since drones can be employed in a similar manner to AV (Sect. 3.3), only the applications which are unique to drones are highlighted here.

The most high profile application to date is last mile delivery, providing many examples of drone use particularly from vehicle to door. Drones may be most effective for delivery in rural areas that are underserved due to economics of traditional transportation (Stern 2015; Wells and Stevens 2016). Amazon undertook a pilot delivery in rural England in 2016 to demonstrate the feasibility of a 30 min delivery time for a small package to a rural area (Wells and Stevens 2016). Also, Google's parent company Alphabet formed a research group to develop and test drones, partnering with Chipotle to deliver burritos at Virginia Tech (Nicas 2016b), while Dominos Pizza Enterprises joined with drone company Flirtey and completed a commercial delivery of pizzas in New Zealand (Lui 2016).

One distinct type of last mile delivery for which drones seem to be well suited concerns humanitarian requirements and/or natural disaster recovery. Drones have the potential to be an effective delivery mechanism where the terrain is difficult for humans and/or traditional equipment to access. UPS first tested drones for medical supplies delivery by delivering a medical inhaler to a rural children's summer camp (Reuters 2016). UPS also partnered with drone company Zipline and governmental organizations in Africa to coordinate emergency medical supplies delivery (such as blood) to Rwanda (UPS 2016). Additionally, the insurance industry has started utilizing drones to speed up claims for damage to homes after storms and improve safety for personnel after natural disasters (Marquand 2017).

Another role fulfilled by drones in the SC is to enhance transparency. Individual firms can utilize drones to simply measure and monitor their inventory and assets. Daimler recently bought drones from PINC Solutions to count and track finished vehicles (Banker 2016). Firms can also utilize drones to aid in the maintenance of expensive assets, such as the safety of bridges and railroads for BNSF Railway (Stern 2015). Another example is the use of drones for inspection, spare parts delivery, and to aid in maintenance for aircraft (Kenney 2015).

Utilizing cameras and other sensors onboard, drones can be used to physically monitor any point in the SC. Firms in the agriculture, oil, gas, and mining industries are utilizing drones to monitor upstream assets (Courtin 2015). One example in agricultural finance showed that the actual crops were much smaller than what the company claimed (Saraswathy 2016). Drones can also be utilized to gather data concerning potential customer needs during deliveries. Interestingly, Amazon has recently been granted a patent for analyzing data gathered by drones during home delivery to target future purchases (Price 2017).

3.4.2 Managerial, IS/IT and Policy Implications

As previously mentioned, drones have the potential to improve effective SC management in a variety of aspects, including improving transparency, monitoring, and delivery of goods. However, similar to AV, there are numerous drawbacks to this technology as well, including potential safety and nascent laws governing the operation of the drones.

Laws regulating commercial drone usage are evolving in response to new usage cases but are not yet well developed. In the US, there are age, security and training restrictions as well as rules on where drones can be flown (FAA 2017). In Europe, recent proposals utilize the concept of "geofencing," which delineates the airspace into zones which are legal and monitors them via GPS devices in order to establish virtual boundaries for the drones (EASA 2017). Regulations in China already allow for commercial delivery of agricultural goods from rural areas.

3.4.3 Existing Literature and Future Research Opportunities

Many review or conceptual articles make mention of the use of drones in SCs as a future topic of inquiry (Joglekar et al. 2016; Lee and Schmidt 2017; Lee and Tang 2017; Ransbotham et al. 2016; Savelsbergh 2015, 2016), although few articles are written specifically on the topic of drones within current SCM academic/theoretical literature. One exception is Chowdhury et al. (2017), who address the optimal locations for service depots for drones to deliver goods to areas where disasters have cut off other transportation methods. Kwon et al. (2017) introduce a semantic text mining technique to forecast the social impact of certain technologies and utilize drones as a sample case. This article collects data on drones from several popular journals and identifies certain clusters of policy/risk related scenarios (such as corrupted journalism, biomimetic drones, etc.).

Future Research Opportunities

1. Where in the SC can drones be most effective?
2. What kind of managerial policies are best for safe drone operation?
3. When in autonomous mode, what types of decisions should drones be allowed to make?
4. How can drones be best utilized to manage upstream transparency in the SC?
5. What is the best way to utilize drones to enhance the sales of products and services without compromising consumer privacy?

3.5 Internet of Things (IoT)

The Internet of Things (IoT) is a term used to describe the phenomenon of digitizing and customizing the delivery of goods and services by gathering, analyzing, and acting on data acquired via a network of interconnected semi-autonomous devices. These devices allow for semi-autonomous operation via analytics and algorithms which are then typically interconnected via the internet or other type of network such as the cloud (Reale 2017). Because the devices rely on artificial intelligence and can communicate directly with other devices to aid decision making, they are sometimes referred to as "smart" machines.

Technologies associated with IoT typically include the following: radio frequency identification (RFID), near field communication, wireless sensor networks, middleware, cloud computing, and enabling software (Lee and Lee 2015). In addition, note that other technologies discussed in this chapter can also be incorporated into IoT, thereby creating an amalgamation of devices and communication networks. Moreover, a key element of IoT concerns the effective management of the vast amounts of data captured and analyzed via IoT devices.

The term IoT in its most general form can refer to service, manufacturing, SC, R&D, and/or marketing applications. Two other closely related terms that refer more specifically to the manufacturing function include the Industrial Internet of Things (IIoT) and Industry 4.0, a European initiative.

A recent McKinsey report emphasizes that IoT essentially fuses "processes and devices," highlighting that materials will become "inextricably linked to their information" (Chui et al. 2017). IoT adoption is growing and companies are projected to spend $470 billion annually for the next 3 years on hardware, software, services and connectivity (Loten 2017). One UPS executive describes various phases in IoT adoption as descriptive (i.e., simply capturing the data), predictive (i.e., predicting data patterns), and prescriptive (i.e., optimizing the best action for the device) (SupplyChainBrain 2017a).

3.5.1 Impact on SC and Uses

There are numerous examples of successful applications of IoT which can be broadly classified into the following categories: smart factory, smart house, smart cities (including power grids), and smart agriculture. With regard to the SC, IoT is ubiquitous and can be adopted to enhance productivity at any stage, from marketing and product development to manufacturing, delivery and maintenance. In manufacturing, tasks such as inventory management, quality control, and routing can be automated, improving their efficiency/productivity.

Many elements of the "smart factory" portion of IoT have already been utilized by automobile manufacturers by automating the production process with CAD files and robots that monitor and control production via a series of sensors (McBeath 2015). Boeing and Airbus have adopted technologies to deploy IoT in aircraft maintenance to allow for the automatic identification of problems and to monitor for equipment failures (Cameron and Wall 2017).

The application of IoT technologies at the retail/customer phase can also indirectly impact NPD, simply by gathering the data concerning product utilization. Some foresee that the NPD process will become a more customized, automated interaction between the consumer and engineering processes, similar in nature to mass customization and servitization. The insurance industry increasingly relies on sensor data from vehicles or personal health devices to “more accurately model risks” (Norton 2017a). Disney World introduced “magic bands”, wearable devices for park guests that contain an RFID chip. These enable Disney to gather and analyze additional data concerning consumer usage patterns and to customize the consumers’ in-park experience (Barnes 2013; Lee and Lee 2015). Healthcare applications utilizing analytics from smart devices can also help to measure compliance of the patient and to predict potential outcomes with certain treatments (SupplyChainBrain 2017a).

More broadly, IoT is intrinsically integrated into the logistics involved with the ordering, delivering, and monitoring of goods throughout the SC, thereby enabling increased SC transparency in addition to efficiency. In agriculture, modern farming techniques utilize Global Positioning Systems (GPS) and weather systems to automatically dispense seed, fertilize, and water (McBeath 2015). While SC transparency is a touted benefit of IoT, interestingly, multi-firm applications of IoT are relatively rare, as the financial benefits are more difficult to quantify (Chui et al. 2017).

3.5.2 Managerial, IS/IT and Policy Implications

Security challenges are a crucial issue associated with IoT (Whitmore et al. 2015), with key elements including encryption and identity management. Because smaller devices do not have the capability to incorporate modern encryption algorithms, the devices can be easily manipulated via a third party (Lee and Lee 2015; Whitmore et al. 2015). In a recent interview, the Chief Digital Officer of elevator manufacturer Schindler Group states succinctly that the principal risk of IoT adoption is cybersecurity (Norton 2017b). Moreover, a recent survey of IT professionals showed that 96% of the respondents believed that increased security attacks will occur as a result of Industrial IoT, yet only 51% believed that they were prepared for such “malicious attacks” (Lapena 2017).

Another complication is to justify the return on investment for the adoption of IoT-based technologies (SupplyChainBrain 2017b), as many of the constituent technologies are not yet technologically mature (Gubbi et al. 2013; Loten 2016). For example, both General Electric and Siemens AB are developing competing cloud-based IoT systems (Norton 2017b).

Numerous other drawbacks concerning effective data management have been highlighted in the current press on IoT. Whitmore et al. (2015) highlight the privacy challenges associated with IoT, as increasingly personalized sensors and data have the potential to compromise personal information. Lee and Lee (2015) also point out the challenges associated with effectively mining and analyzing the vast amount of data associated with IoT. A recent survey from McKinsey reports that while managers recognized the potential value of IoT data, the majority of them responded that their companies utilize <10% of the information accumulated (Chui et al. 2017). This report also identifies numerous capability gaps (i.e., potential risks) associated with IoT, including integrating IoT into existing business processes. Finally, a recent survey identifies other obstacles for IoT adoption such as liability and regulatory issues (Loten 2016).

3.5.3 Existing Literature and Future Research Opportunities

Several review articles within the operations and SC literature identify IoT as an important future direction of investigation (Karmarkar 2015; Ransbotham et al. 2016; Roth et al. 2016). Lee and Lee (2015) offer an overview of the major types of technologies involved with IoT, and offer some insights into how firms can justify adoption based on options theory. Whitmore et al. (2015) provide an excellent introduction to IoT technologies and applications, and survey the current literature and trends in this area. Moreover, they highlight the fact that very few articles on this topic have addressed appropriate business models necessary to facilitate the effective use of IoT.

Future Research Opportunities

1. How will IoT influence ownership of tasks and liabilities in the SC? For example, will IoT enable further asset sharing in the SC?
2. How can IoT be financially justified among multiple SC partners?
3. How should IoT be effectively integrated into existing work processes? Which tasks can be automated and integrated?
4. How can IoT be utilized to create new business processes?
5. What are appropriate theories for effective IoT management (Whitmore et al. 2015)?

4 Discussion and Conclusions

These technologies offer tremendous benefits to the SC in terms of transparency, visibility, cost reduction, and convenience for consumers. They have a wide variety of applications in the SC, with new ideas likely to come. All have SC uses focused both at downstream/consumers and upstream/business-to-business as shown in Table 1, making them ubiquitous. However, there is still a great deal of uncertainty about them as well. The technology is still under development, the regulatory landscape is evolving, and dominant designs and platforms are not yet established.

All of the technologies will require changes to public or corporate infrastructure such as factories, SC networks, highways, or communication networks. Most require integration with existing SC information systems, as well as integration with

Table 1 Ubiquitous Technologies in the SC

Consumer focused	Business focused
3D printing	3D printing/AM
VR (retail)	AR
Autonomous (cars, delivery)	Autonomous (trucks/ships, transportation)
Drones (delivery)	Drones (monitoring and inspection)
IoT	IIoT

suppliers or customers' systems to gain full benefits. Effective implementation of all of these technologies requires changes to existing organizational routines as they are not straightforward replacements of an earlier technology. The dangers of lax cybersecurity come to the fore as well. As the technologies all combine hardware and software, they leave individuals and organizations vulnerable to data theft, fraud, or malicious intent. Another issue concerns the environmental impact of additional electronic devices and the waste generated.

Additionally, the technologies and their uses raise some fundamental questions about data safety and privacy. IS/IT such as analytics, internet, algorithms, machine learning and artificial intelligence enable these technologies and will provide some of the early test cases of widespread application. To date, consumers have been willing to share their data, relinquishing some privacy, for free services. However, the ability of drones to capture data while making a delivery or the ability of IoT smart devices to collect data without consent points to the increasing importance of defining privacy and establishing data ownership rights and laws. The technologies are developing rapidly, giving the environment a "Wild West" feel, and making it difficult for regulators to keep up.

While adoption is beginning for these technologies, how pervasive adoption will become and how long mass adoption will take is in question. Estimates vary in their optimism. The high cost of obtaining equipment, designing new products and services, and integration into processes and IS will slow adoption overall. Uncertainty in the technology and regulatory landscape is another major factor in adoption. Corporations in particular are hesitant to adopt without established standards to prevent an expensive investment that becomes obsolete quickly.

With clear technology standards, these technologies all benefit from positive network effects, likely resulting in two-sided platforms with potentially powerful intermediaries emerging that control them. Reconfigured supply chains may result in new networks and ecosystems of various stakeholders such as governments. For the artifacts of these technologies to interoperate with each other, and with other technological artifacts, a standard communication protocol must emerge, at least within geographic regions (like cell phone standards).

Currently these technologies are in an "era of ferment" as defined by Anderson and Tushman (1990). As standards and dominant designs emerge, there will likely be a period of consolidation in each industry and its supporting industries. More interesting is the question of how these technologies will combine, such as seen with cameras and phones. Combinations seem likely at some future date as these technologies address different needs in the SC, and where they can potentially interact, they seem to reinforce one another, making the other more useful. For example, as manufacturing becomes more disperse via 3DP/AM, drones could be used to monitor quality. Also, as AV deliver packages to the local depot, drones can complete the delivery to the door. As mentioned earlier, IoT can easily be imagined to subsume these technologies, truly integrating the SC functions.

Autonomy is a common theme across these technologies. A key question concerns the degree of autonomy to permit these devices. 3DP/AM allows consumers and companies to make parts and products easily without a supplier. The online availability of predesigned printing instructions for a large variety of items even allows consumers to be more independent of designers and engineers. Drones and AV enable machines to be autonomous, and IoT enables "things" to autonomously communicate and make decisions, freeing humans for other tasks.

This autonomy of consumers and technology make aspects of the SC harder to control. For example, 3DP/AM makes printing a counterfeit item relatively easy. The need for standards to allow interoperability also allow new entrants to more easily compete in the various stages of the SC. The number of players and their autonomy may make quality harder to control and regulations more difficult to enforce.

While not likely, this autonomy may make some parts of the SC disappear. For example, last mile delivery might rely on humans, or their AV or drones, to pick up from a depot. 3DP/AM could have a similar impact of eliminating, for example, manufacturers of simple plastic items, with consumers printing them at home. However, it is more likely that new and more efficient business models will emerge instead, such as shared transportation resources.

This chapter offers an overview of some of the many interesting technologies in the SC. We offer some areas for future research that we feel will be interesting and fruitful. Additionally, there are many general questions about technology development, adoption, and management that can be studied in the context of these particular ones. A key question for academics to address is what is different from earlier technology adoption? Therefore, what can we learn from past research that still holds true versus what new theories and studies are needed? Additionally, how will business models and business processes best support and utilize these technologies going forward? Finally, how will these reconfigured supply chains result in new ecosystems?

References

Anderson EG, Parker GG (2013) Integration of global knowledge networks. Prod Oper Manag 22(6):1446–1463

Anderson P, Tushman M (1990) Technological discontinuities and dominant designs: a cyclical model of technological change. Adm Sci Q 35(4):604–633

Anonymous (2017, May/June) A special supplement to Supply Chain Management Review. Supply Chain Manag Rev S76–S80

Banker S (2016, June 24) Drones and robots in the warehouse. Forbes. Retrieved from https://www.forbes.com/sites/stevebanker/2016/06/24/drones-and-robots-in-the-warehouse/
Ba S, Nault B (2017) Emergent themes in the interface between economics of information systems and management of technology. Prod Oper Manag 26(4): 652–666
Barnes B (2013, January 7) At Disney Parks, a bracelet meant to build loyalty (and sales). The New York Times. Retrieved from https://www.nytimes.com/2013/01/07/business/media/at-disney-parks-a-bracelet-meant-to-build-loyalty-and-sales.html
Baumers M, Dickens P, Tuck C, Hague R (2016) The cost of additive manufacturing: machine productivity, economies of scale and technology-push. Technol Forecast Soc Chang 102:193–201
Beene R, Eidelson J (2017, July 28) Unions urge slow down as self-driving car bills pick up speed. Retrieved from https://www.bloomberg.com/news/articles/2017-07-28/unions-urge-slow-down-as-self-driving-car-bills-pick-up-speed
Bell D, Gallino S, Moreno A (2015) Showrooms and information provision in omni-channel retail. Prod Oper Manag 24:360–362
Benton D (2017, February 21) Retail, virtual reality and the supply chain. Retrieved from http://www.supplychaindigital.com/scm/retail-virtual-reality-and-supply-chain
Bui Q (2015, February 5) Map: the most common* job in every state. Retrieved from http://www.npr.org/sections/money/2015/02/05/382664837/map-the-most-common-job-in-every-state
Cameron D, Wall R (2017, July 25) Boeing and airbus earnings: what to watch. Wall Street J. Available via https://www.wsj.com/articles/boeing-and-airbus-earnings-what-to-watch-1500987602
Carrillo JE, Druehl C, Hsuan J (2015) Introduction to innovation WITHIN and ACROSS borders: a review and future directions. Decis Sci 46:225–265
Castellanos S (2016, December 12) Augmented reality, hologram-like images enter the workplace. Wall Street J. Available via https://www.wsj.com/articles/augmented-reality-hologram-like-images-enter-the-workplace-1481551202
Catalini C (2017) Microgeography and the direction of inventive activity. Manag Sci (published online in Articles in Advance July 19, 2017)
Chowdhury S, Emelogu A, Marufuzzaman M, Nurre SG, Bian L (2017) Drones for disaster response and relief operations: a continuous approximation model. Int J Prod Econ 188:167–184
Chui M, Ganesan V, Patel M (2017, July) Taking the pulse of enterprise IoT. Retrieved from http://www.mckinsey.com/global-themes/internet-of-things/our-insights/taking-the-pulse-of-enterprise-iot
Courtin G (2015, February 24) Drones still useful in supply chains despite FAA regulations. Retrieved from http://www.zdnet.com/article/drones-still-useful-in-supply-chains-despite-faa-regulations/
D'Aveni R (2013) 3-D printing will change the world. Harv Bus Rev 91(5):22
DHL (2014) Self-driving vehicles in logistics. Retrieved from http://www.dhl.com/content/dam/downloads/g0/about_us/logistics_insights/dhl_self_driving_vehicles.pdf.
EASA (2017, May 5) EASA publishes a proposal to operate small drones in Europe. Retrieved from https://www.easa.europa.eu/newsroom-and-events/press-releases/easa-publishes-proposal-operate-small-drones-europe
Eshkenazi A (2016, December 16) Using augmented reality to improve supply chains. Retrieved from http://www.apics.org/sites/apics-blog/think-supply-chain-landing-page/thinking-supply-chain/2016/12/16/using-augmented-reality-to-improve-supply-chains
FAA (2017) Getting started. Retrieved from https://www.faa.gov/uas/getting_started/
Futurenautics Ltd. (2016) Autonomous ships. Retrieved from http://www.futurenautics.com/2016/11/white-paper-autonomous-ships/
Gaimon C, Hora M, Ramachandran K (2017) Towards building multidisciplinary knowledge on management of technology: an introduction to the special issue. Prod Oper Manag 26:567–578
Gao F, Su X (2017) Omnichannel retail operations with buy-online-and-pick-up-in-store. Manag Sci 63(8):2478–2492
Gartner (2015) Emerging technologies hype cycle for 2015. Available via http://na2.www.gartner.com/imagesrv/newsroom/images/emerging-tech-hc.png;wa0131df2b233dcd17

Gates G, Granville K, Markoff J, Russell K, Singhvi A (2017, June 16) The race for self-driving cars. Retrieved from https://www.nytimes.com/interactive/2016/12/14/technology/how-self-driving-cars-work.html
Gibson I, Rosen DW, Stucker B (2010) Additive manufacturing technologies, rapid prototyping to direct digital manufacturing. Springer Science & Business Media, Berlin, p 484
Giffi CA, Gangula B, Illinda P (2014) 3D opportunity for the automotive industry: additive manufacturing hits the road. Deloitte University Press
Gubbi J, Buyya R, Marusic S, Palaniswami M (2013) Internet of things (IoT): a vision, architectural elements, and future directions. Futur Gener Comput Syst 29(7):1645–1660
Hall RW, Nowroozi A, Tsao J (2001) Entrance capacity of an automated highway system. Transp Sci 35(1):19–36
Haugom MK (2016) Additive manufacturing: a new paradigm in manufacturing operations. Unpublished master thesis, Copenhagen Business School
Holmström J, Partanen J, Tuomi J, Walter M (2010) Rapid manufacturing in the spare parts supply chain: alternative approaches to capacity deployment. J Manuf Technol Manag 21(6):687–697
International Transport Forum (2017) Managing the transition to driverless road freight transport. Retrieved from https://www.itf-oecd.org/sites/default/files/docs/managing-transition-driverless-road-freight-transport.pdf
Jain S, Choong NF, Aye KM, Luo M (2001) Virtual factory: an integrated approach to manufacturing systems modeling. Int J Oper Prod Manag 21(5/6):594–608
Joglekar NR, Davies J, Anderson EG (2016) The role of industry studies and public policies in production and operations management. Prod Oper Manag 25:1977–2001
Kamal MM, Irani Z (2014) Analysing supply chain integration through a systematic literature review: a normative perspective. Supply Chain Manag Int J 19(5/6):523–557
Kane J, Tomer A (2017, April 06) Autonomous trucking overlooks skilled labor need. Retrieved from http://www.supplychain247.com/article/autonomous_trucking_overlooks_skilled_labor_need/Autonomous_Vehicles
Karmarkar U (2015) OM forum—the service and information economy: research opportunities. Manuf Serv Oper Manag 17(2):136–141
Keeney RL (1999) The value of internet commerce to the customer. Manag Sci 45(4):533–542
Kelly K (2016, April 18) The untold story of magic leap, the world's most secretive startup. Wired. Retrieved from https://www.wired.com/2016/04/magic-leap-vr/
Kenney R (2015, December 10) Top 4 uses for drones in aircraft maintenance. Aviation Week. Retrieved from http://aviationweek.com/advanced-machines-aerospace-manufacturing/top-4-uses-drones-aircraft-maintenance
Kim Y, Krishnan R (2015) On product-level uncertainty and online purchase behavior: an empirical analysis. Manag Sci 61(10):2449–2467
Kongsberg (n.d.) Autonomous ship project, key facts about YARA Birkeland – Kongsberg Maritime. Retrieved from https://www.km.kongsberg.com/ks/web/nokbg0240.nsf/AllWeb/4B8113B707A50A4FC125811D00407045?OpenDocument
Kwon H, Kim J, Park Y (2017) Applying LSA text mining technique in envisioning social impacts of emerging technologies: the case of drone technology. Technovation 60:5–28
Lapena R (2017, March 13) More than 90% of IT Pros expect more attacks, risk, and vulnerability with IIoT in 2017. Retrieved from https://www.tripwire.com/state-of-security/featured/90-pros-expect-attacks-risk-vulnerability-iiot-2017/
Lee I, Lee K (2015) The internet of things (IoT): applications, investments, and challenges for enterprises. Bus Horiz 58(4):431–440
Lee HL, Schmidt G (2017) Using value chains to enhance innovation. Prod Oper Manag 26:617–632
Lee HL, Tang CS (2017) Socially and environmentally responsible value chain innovations: new operations management research opportunities. Manag Sci (published online in Articles in Advance 31 Mar 2017)

Lewandrowski N (2014) 3D printing for fashion: additive manufacturing and its potential to transform the ethics and environmental impact of the garment industry. Available via http://ft.parsons.edu/skin/wp-content/uploads/2014/05/3Dprintingforfashion-final.pdf
Loten A (2016, July 27) Internet of things paying off for more tech providers: survey. Wall Street J. Retrieved from https://blogs.wsj.com/cio/2016/07/27/internet-of-things-paying-off-for-more-tech-providers-survey/
Loten A (2017, March 29) Industrial IoT puts CIOs 'center stage' in factories: report. Wall Street J. Retrieved from https://blogs.wsj.com/cio/2017/03/29/industrial-iot-puts-cios-center-stage-in-factories-report/
Lui K (2016, November 15) Watch the world's first commercial pizza delivery by drone. Fortune. Retrieved from http://fortune.com/2016/11/16/dominos-new-zealand-first-commercial-pizza-delivery-drone/
Mahmassani HS (2016) 50th anniversary invited article—autonomous vehicles and connected vehicle systems: flow and operations considerations. Transp Sci 50(4):1140–1162
Manuj I, Mentzer J (2008) Global supply chain risk management. J Bus Logist 29(1):133–155
Markoff J (2017, June 7) A guide to challenges facing self-driving car technologists. The New York Times. Retrieved from https://www.nytimes.com/2017/06/07/technology/autonomous-car-technology-challenges.html
Markopoulos PM, Hosanagar K (2017) A model of product design and information disclosure investments. Manag Sci (published online in Articles in Advance 08 Feb 2017)
Marquand B (2017, June 8) Meet your new insurance claims inspector: a drone. USA Today. Retrieved from https://www.usatoday.com/story/money/personalfinance/2017/06/08/meet-your-new-insurance-claims-inspector-drone/102560614/
McBeath B (2015, April 8) IoT distributed intelligence examples. Retrieved from http://www.clresearch.com/research/detail.cfm?guid=94E70C06-3048-79ED-99CC-0CF832390CBC
Mehmood R, Meriton R, Graham G, Hennelly P, Kumar M (2017) Exploring the influence of big data on city transport operations: a Markovian approach. Int J Oper Prod Manag 37(1):75–104
Mellor S, Hao L, Zhang D (2014) Additive manufacturing: a framework for implementation. Int J Prod Econ 149:194–201
MH&L (2015, November 12) Is virtual reality coming to the supply chain? Retrieved from http://www.mhlnews.com/technology-automation/virtual-reality-coming-supply-chain
Montenegro R (2015, June 7) Google's self-driving cars are ridiculously safe. Retrieved from http://bigthink.com/ideafeed/googles-self-driving-car-is-ridiculously-safe
Niaki MK, Nonino F (2017) Additive manufacturing management: a review and future research agenda. Int J Prod Res 55(5):1419–1439
Nicas J (2016a, January 6) WSJ.D technology – consumer electronics show: augmented reality draws respect from investors. Wall Street J, p B5
Nicas J (2016b, December 6) Silicon Valley stumbles in world beyond software. Wall Street J. Available via https://www.wsj.com/articles/silicon-valley-stumbles-in-world-beyond-software-1481042474
Norton S (2017a, April 5). Munich Re's big data effort paves way to new insurance products. Wall Street J. Available via https://blogs.wsj.com/cio/2017/04/05/munich-res-big-data-effort-paves-way-to-new-insurance-products/
Norton S (2017b, May 24) Cybersecurity is the biggest challenge to industrial IoT development: Schindler Digital Chief. Wall Street J. Available via https://blogs.wsj.com/cio/2017/05/24/cybersecurity-is-the-biggest-challenge-to-industrial-iot-development-schindler-digital-chief/
Out-law.com (2017, February 24). New UK laws address driverless cars insurance and liability. The Register. Retrieved from https://www.theregister.co.uk/2017/02/24/new_uk_law_driverless_cars_insurance_liability/
Pine BJ (1993) Mass customization. The new frontier in business competition. Harvard Business School Press, USA

Potts J (2016, June 23) How driverless trucks will change supply chain strategy. Retrieved from http://www.leanlogistics.com/blog/2016/06/23/driverless-trucks-will-change-supply-chain-strategy/

Price R (2017, July 26) Amazon's delivery drones could scan your house to sell you more products. Inc. Retrieved from https://www.inc.com/business-insider/amazon-drone-patent-deliveries-scan-your-house.html

Ransbotham S, Fichman RG, Gopal R, Gupta A (2016) Special section introduction—ubiquitous IT and digital vulnerabilities. Inf Syst Res 27(4):834–847

Reale A (2017, February 23) A guide to edge IoT analytics: internet of things blog. Retrieved from https://www.ibm.com/blogs/internet-of-things/edge-iot-analytics/

Reuters (2016, September 23) UPS tests drone delivery of emergency medical supplies. Fortune.com. Retrieved from http://fortune.com/2016/09/23/ups-tests-drone-delivery-of-emergency-medical-supplies/

Rolls Royce plc (2016) Remote and autonomous ships – the next steps. Retrieved from http://www.rolls-royce.com/~/media/Files/R/Rolls-Royce/documents/customers/marine/ship-intel/aawa-whitepaper-210616.pdf

Roth A, Singhal J, Singhal K, Tang CS (2016) Knowledge creation and dissemination in operations and supply chain management. Prod Oper Manag 25:1473–1488

Ruffo M, Hague R (2007) Cost estimation for rapid manufacturing – simultaneous production of mixed components using laser sintering. Proc IMech E Part B J Eng Manuf 221(11):1585–1591

Saraswathy M (2016, April 16) Insurers use drones for crop yields. Business Standard. Retrieved from http://www.business-standard.com/article/companies/insurers-test-drones-for-checking-crop-yields-116041500590_1.html

Savelsbergh M (2015) Editorial statement. Transp Sci 49(2):163–164

Savelsbergh M (2016) Editorial—a major milestone: transportation science turns fifty. Transp Sci 50(1):1–2

Savelsbergh M, Van Woensel T (2016) 50th anniversary invited article—city logistics: challenges and opportunities. Transp Sci 50(2):579–590

Schmidt GM, Druehl CT (2008) When is a disruptive innovation disruptive? J Prod Innov Manag 25(4):347–369

Schulz JD (2017, March 16) Truckers prepare for era of driverless trucks. Retrieved from http://www.supplychain247.com/article/truckers_prepare_for_era_of_driverless_trucks/Autonomous_Vehicles

Stern G (2015, October 5) Chris Caplice, Matthew Rose on what U.S. supply chains need. Wall Street J. Retrieved from https://www.wsj.com/articles/chris-caplice-matthew-rose-on-what-u-s-supply-chains-need-1444096893

Supply Chain 24/7 (2017, June 16) Cook: apple focusing on autonomous car systems. Retrieved from http://www.supplychain247.com/article/apple_focusing_on_autonomous_car_systems/Autonomous_Vehicles

SupplyChainBrain (2017a, April 19) The internet of things. Retrieved from http://www.supplychainbrain.com/content/index.php?id=7098&type=98&tx_ttnews[tt_news]=41641&cHash=da03e20e36

SupplyChainBrain (2017b, April 19) If a CNC machine calls for repair, but no one hears, does it make a sound? Retrieved from http://www.supplychainbrain.com/content/premier-sponsors/ups-inside-logistics/single-article-page/article/if-a-cnc-machine-calls-for-repair-but-no-one-hears-does-it-make-a-sound/

Swaminathan JM, Tayur SR (2003) Models for supply chains in E-business. Manag Sci 49(10):1387–1406

Tan CL, Vonderembse MA (2006) Mediating effects of computer-aided design usage: from concurrent engineering to product development performance. J Oper Manag 24(5):494–510

Teodoridis F (2017) Understanding team knowledge production: the interrelated roles of technology and expertise. Manag Sci (published online in Articles in Advance 28 Jul 2017)
Trebilcock B (2017, April 13) Materials handling clockspeed. Retrieved from http://www.mmh.com/article/materials_handling_clockspeed
Tuck CJ, Hague RJM, Ruffo M, Ransley M, Adams P (2008) Rapid manufacturing facilitated customization. Int J Comput Integr Manuf 21(3):245–258
UPS (2016, August) Launching lifesaving deliveries by drone in Rwanda. Retrieved from https://compass.ups.com/drone-medicine-delivery/
Wang Y, Wallace SW, Shen B, Choi T-M (2015) Service supply chain management: a review of operational models. Eur J Oper Res 247(3):685–698
Weller C, Kleer R, Piller FT (2015) Economic implications of 3D printing: market structure models in light of additive manufacturing revisited. Int J Prod Econ 164:43–56
Wells G, Stevens L (2016, December 14) Amazon conducts first commercial drone delivery. Wall Street J. Retrieved from https://www.wsj.com/articles/amazon-conducts-first-commercial-drone-delivery-1481725956
Whitmore A, Agarwal A, Da Xu L (2015) The internet of things—a survey of topics and trends. Inf Syst Front 17(2):261–274
Wohlers Associates (2014) Wohlers report: 3D printing and additive manufacturing state of the industry. Annual worldwide progress report. Fort Collins
Wohlers Associates (2015) Wohlers report: 3D printing and additive manufacturing state of the industry. Annual worldwide progress report. Fort Collins (executive summary only)
World Economic Forum (2017, March) Technology and innovation for the future of production: accelerating value creation. Retrieved from https://www.weforum.org/whitepapers/technology-and-innovation-for-the-future-of-production-accelerating-value-creation
ZDA, LLC (2016, January 6) What are virtual reality supply chains? Retrieved from http://www.zdaya.com/2016/01/06/virtual-reality-supply-chain-recruiters/
Zhang Y, Gregory M, Neely A (2016) Global engineering services: shedding light on network capabilities. J Oper Manag 42–43:80–94
Zimmermann R, Ferreira LMDF, Moreira AC (2016) The influence of supply chain on the innovation process: a systematic literature review. Supply Chain Manag Int J 21(3):289–304

Cheryl Druehl is currently an Associate Professor and Dean's Scholar at George Mason University's School of Business. She received her PhD in operations management from the Stanford Graduate School of Business, MBA from the University of Pittsburgh, and a BS in electrical engineering from UCLA. Prior to her PhD, she worked as a production engineer, IT consultant, and supply chain consultant at companies including Price Waterhouse, Watkins Johnson, and a technology start-up. Her research interests include product development, technology and innovation management, supply chain management, environmentally focused innovations, disruptive innovations, and innovation contests. Her work has appeared in journals such as *Production and Operations Management*, *Journal of Operations Management*, *Journal of Product Innovation Management*, and *Decision Sciences Journal.*

Janice E. Carrillo obtained her master's and doctorate degrees in operations management from the Georgia Institute of Technology. Her interests in technology management were fueled by her earlier work experience as an electrical engineer. Prior to her graduate studies, she worked at Clorox, Hughes Aircraft, Rockwell International, and McDonnell Douglas. Currently, Professor Carrillo is an Associate Professor and the PricewaterhouseCoopers Professor in the Warrington College of Business at the University of Florida. Her research has appeared in journals including *Management Science*, *IIE Transactions*, *Production and Operations Management*, *and Decision Sciences*. In the past, she served as President for the Technology Management Section (TMS) at the Institute for Operations Research and Management Sciences (INFORMS).

Juliana Hsuan is Professor of Operations and Innovation Management at Copenhagen Business School (CBS) and Guest Professor at Chalmers University of Technology. She worked as an automotive electrical design engineer with Motorola in the USA before joining academia. Her teaching and research interests include (service) operations management, servitization, supply chain management, innovation management, modularization strategies, mass customization, and portfolio management of R&D projects. Her research has been published in *International Journal of Operations and Production Management*, *Decision Sciences*, *Journal of Product Innovation Management*, *IEEE Transactions on Engineering Management*, *Production Planning and Control*, *Technovation*, *R&D Management*, *Journal of Cleaner Production*, among others. She has coauthored two textbooks: *Managing the Global Supply Chain* published by CBS Press and *Operations Management* published by McGraw-Hill.

The Role of Informational and Human Resource Capabilities for Enabling Diffusion of Big Data and Predictive Analytics and Ensuing Performance

Deepa Mishra, Zongwei Luo, and Benjamin T. Hazen

Abstract Big data and predictive analytics, or BDPA, has received great attention in terms of its role in making business decisions. However, current knowledge on BDPA regarding how it might link organizational capabilities and organizational performance remains unclear. Even more linted is knowledge regarding how human resources (HR) might also work to support this linkage. Drawing from the resource-based view, this chapter proposes a model to examine how information technology deployment (i.e., strategic information technology flexibility, business-BDPA partnership and business-BDPA alignment) and HR capabilities affect organizational performance through BDPA. Survey data from 159 Indian firms show that BDPA diffusion mediates the influence of IT deployment and HR capabilities on organizational performance. In addition, there is a direct effect of IT deployment and HR capabilities on BDPA diffusion, which also has a direct relationship with organizational performance. The findings suggest the important of HR capabilities, which are often overlooked in the quest for more and better technology situations. Informational capabilities are also shown to play an important role in diffusing BDPA, and driving subsequent performance.

D. Mishra (✉)
DeGroote School of Business, McMaster University, Hamilton, ON, Canada
e-mail: deepad@mcmaster.ca

Z. Luo
South University of Science and Technology of China, Shenzhen, Guangdong, China
e-mail: luozw@sustc.edu.cn

B. T. Hazen
Data Science Lab, Department of Operational Sciences, Air Force Institute of Technology, Dayton, OH, USA

A. C. Moreira et al. (eds.), *Innovation and Supply Chain Management*, Contributions to Management Science, https://doi.org/10.1007/978-3-319-74304-2_13

1 Big Data

Can we imagine a world where data cannot be stored? Can we imagine that data generated every second or less may get lost immediately after use? Impossible! Right? Because, in this scenario, it would not be feasible to extract valuable information and knowledge contained in the data. Thus, it is essential to store, harness and extract value from the growing deluge of data, known by the buzzword 'Big Data.' It is an evolving term that originated in mid-1990s during lunch-table conversations at Silicon Graphics Inc. (SGI) and is so popular that it was Google searched 252,000 times in November 2011, and then reached the impressive number of 270,000,000 hits in August 2017. Big data is ubiquitous and useful. It refers to the idea of analysing enormous volumes of information to make better business decisions, transform business processes, generate business insights, enhance performance and outperform competitors. According to McKinsey and Company,

> collecting, storing, and mining big data for insights can create significant value for the world economy, enhancing the productivity and competitiveness of companies and the public sector and creating a substantial economic surplus for consumers (Manyika et al. 2011: p. 1).

In recent years, the information deluge has created challenges for organizations interested in availing the benefits from analysing this huge amount of data. By realizing the potential and hidden values of big data, organizations can get a clear view of market trends, customer behaviour and many other decisions related to business. Undoubtedly, the best decisions are made when managers combine data and tools to obtain insights (Davenport 2006). Therefore, managers from across the globe are increasingly making decisions based on data rather than intuition (Lavalle et al. 2011).

1.1 Big Data Characteristics

Volume refers to the huge amount of data generated every second. The data is no more measured in terms of gigabyte or terabyte, but in petabyte, exabyte and zettabyte. If we consider all the data generated in the world starting from the beginning of time till 2008, the same amount of data is now generated every minute. In fact, every day 2.5 Quintillion bytes of data is created, and by 2020, the amount of data will be 50 times more than in 2011. The biggest contributor to this ever-expanding digital universe is the Internet of Things with sensors in all devices spread across the world. For instance, the sensors installed in airplane engines generate around 2.5 billion Terabyte of data every year, and self-driving cars generate 2 Petabyte of data every year.

Variety refers to the range of data types, domains and sources. It reflects that the data generated from various sources and formats will contain multidimensional data fields. The three types of data that are commonly used are structured, semi-structured, and unstructured. In the past, all the data was structured, i.e., it was stored in tables or relational databases, but nowadays, the data is unstructured, i.e., in the form of text, images, audio, and video. According to IBM's 2013 Annual Report, in 2012, 2.5 billion GB of data was generated every day, and out of which 80% was in unstructured format. Semi-structured data lies between these two data types and does not follow any particular standard. An example would be Extensible Mark-up Language that is used for exchanging data on internet.

Velocity refers to the rate at which data is generated and the speed at which it should be analysed and meaningful information should be extracted. Earlier, computers and servers required substantial time to process the data and update the databases, but in the big data era, data is created in real-time, and with the availability of Internet connected devices, data is passed the same moment it is created. Thus, technology helps us analyse the data while it is being generated without even storing it in databases. We can think of social media messages that get viral in seconds, or we can consider the retail company Wal-Mart which deals with more than one million transactions per hour.

Veracity refers to the unreliability present in some sources of data. It emphasizes the significance of data quality and the level of trust in a data source. We know that quality is an significant challenge for big data because the unpredictability inherent in the data cannot be removed even by adopting the best data cleansing methods.

Variability (and complexity) of big data were introduced by SAS. Variability refers to the variation in data flow rates when velocity of big data is inconsistent and undergoes continuous fluctuations. Complexity arises due to innumerable data sources. Therefore, it is important to connect, match, cleanse and transform data received from these sources (Gandomi and Haider 2015).

Value focuses on the economic benefits that can be availed from the data. Since big data contains great deal of information, it is necessary to capture meaningful information and use it for further analysis.

2 Big Data Analytics

Due to technological advancements, the amount of data generated is doubling each year and it is important for firms to make sense of it. Since the traditional database technologies cannot handle this huge amount of data, there is a need for advanced analytical techniques that can store, manage, analyse and visualize large and complex datasets. These techniques are referred to as Big Data Analytics (BDA). Owing to the potential benefits of BDA, it is also termed as the "next big thing in innovation," "the fourth paradigm of science," "the next frontier for innovation, competition and productivity" and "the management revolution."

Big data analytics has captured the interest of both industrial and academic professionals. Its decision-making capability motivates firms to adopt data-driven decision-making and advanced big data applications. In fact, huge amounts of structured and unstructured data, as well as powerful data mining tools, are available to managers and analysts. However, meaningful information cannot be extracted by just applying analytical tools to data. It requires intense collaboration between analysts and managers exploiting data and analytic tools to discover new knowledge. Moreover, it is known that data management and analytics are the necessary prerequisites for making sense out of data, where data management refers to processes and technologies which gather, store, prepare and retrieve data for analysis, and analytics refers to the techniques which analyse and acquire insights from big data. Scientists believe that great advancements can be made in the fields of medicine, commerce and national security by utilizing artificial intelligence tools to deal with, analyse and combine various data sets.

Amidst many research streams on the definition and function of BDA, one stream which has received considerable attention is strategy-driven analytics. This attention is driven by the fact that analytics that create sustainable value for business help in making better decisions. For instance, LaValle et al. (2011) remarked that the expectation for better decision making can be met by linking analytics-driven decisions to business strategy. The other stream of BDA research explains it through the perspective of identifying new opportunities with big data. This is the case, for example, in the paper by Davenport (2012) in which the author presents BDA as a key way to explore new products and value-added activities. Some other researchers defined BDA through behavioural elements, such as empathy, since it is considered to be crucial in improving the analytical ability of firms. From their point of view, BDA is a combination of business processes, technology optimization and emotional connections with the use of data.

The question of how BDA adds value can be examined via both transaction cost theory and resource-based theory. Viewing it through the lens of transaction cost theory, it can be argued that BDA assists online firms in improving market transaction cost efficiency (e.g., buyer-seller interactions online), managerial transaction cost efficiency (e.g., process efficiency-recommendation algorithms by Amazon) and time cost efficiency (e.g., searching, bargaining and after sale monitoring) (Devaraj et al. 2002). From the perspective of resource-based theory, BDA supports business needs: it identifies loyal and profitable customers, determines the optimal price, detects quality problems, and decides the lowest possible level of inventory of high-performance business processes (Akter and Wamba 2016).

2.1 *Big Data Analytics Techniques*

2.1.1 Text Analytics

Text analytics, also termed as text data mining, is the process of extracting valuable information from textual data. Organizations possess textual data in the form of social network feeds, emails, blogs, online forums, survey responses, corporate documents, news, and call centre logs (Gandomi and Haider 2015). Through text analytics, organizations can transform massive amounts of text generated by human beings into high quality information. For instance, Chung (2014) suggested that text analytics can be utilized for stock market prediction based on the information extracted from financial news. Information extraction, text summarization, question answering and sentiment analysis are some of the commonly used text analytics techniques.

2.1.2 Audio Analytics

Audio analytics, also termed as speech analytics, refers to techniques that are employed for analyzing and extracting information from unstructured audio data. Audio analytics are widely applied in call centres and healthcare. In call centres, audio analytics plays a key role in analyzing lengthy recorded calls which may even stretch up to millions of hours. This analysis benefits call centres in several ways, such as improving customer experience, evaluating agent performance, monitoring privacy and security policies and identifying issues related to products or service, just to name a few. Audio analytics also has the capability to identify the patient's communication patterns, such as depression, schizophrenia and cancer (Hirschberg et al. 2010). In addition, by using such analytical techniques one can analyze the physical and the emotional condition of an infant (Patil 2010). The transcript-based approach, also known as large vocabulary continuous speech recognition (LVCSR), and the phonetic-based approach are the commonly used speech analytics techniques.

2.1.3 Video Analytics

Video analytics, also termed as video content analysis, refers to techniques that are used to observe, analyze and extract valuable information from video data. Despite being in a nascent stage (Panigrahi et al. 2010), video analytics has received wide recognition in a sense that even real-time and pre-recorded videos can now be operated. The growth of the video analytics market can be attributed to the increasing use of closed-circuit television cameras and the trend of sharing videos online. Nonetheless, the main challenge that prevails is the size of video data. It is worthwhile to mention here that one second of a high-definition video is equivalent to over 2000 pages of text (Manyika et al. 2011). Now imagine the lengthy videos which are uploaded on YouTube every now and then. However, this challenge is transformed to opportunity, thanks to big data.

The applications of video analytics can be seen in automated security and surveillance systems. The automated security system, which is claimed to be less expensive, is more effective than a labour-based surveillance system (Hakeem et al. 2012). By applying video analytical techniques, we can perform surveillance functions effectively and efficiently. If any threat is found, the surveillance system may alert the security personnel using an alarm or turn on lights for safety. Server-based and edge-based are two commonly used video analytic techniques.

2.1.4 Social Media Analytics

Social media analytics refers to the techniques used to analyse the structured and unstructured data received from social networking platforms. This term encompasses a variety of online platforms which allows users to create and exchange content. Social media data can be obtained from a variety of sources: social networking websites, blogs, microblogs, social news, social bookmarking, wikis, question-and-answer sites, and many others (Barbier and Liu 2011; Gundecha and Liu 2012). In addition, mobile apps, such as WhatsApp and We Chat, act as social media channels since they offer huge platforms for interacting with friends, families and officials.

2.1.5 Predictive Analytics

Predictive analytics refers to the techniques which are employed for forecasting future outcomes based on past and current data. Predictive analytics has found applications in several fields ranging from predicting when jet engines might fail based on sensor data = to anticipating what and when the customers will buy. The primary aim of predictive analytics is to discover patterns and apprehend underlying relationships in data.

2.2 BDPA as a Supply Chain Innovation

Today's supply chain management (SCM) professionals are interested in finding ways to manage and leverage massive amounts of data using predictive analytics. By doing so, they can make better predictions and smarter decisions. In the context of SCM, BDPA can be defined as incorporating quantitative and qualitative methods in order to enhance supply chain design and competitiveness. BDPA plays an important role in improving visibility (Barratt and Oke 2007), resilience and robustness (Brandon-Jones et al. 2014), and organizational performance (Waller and Fawcett 2013). Thus, BDPA helps firms to achieve business value and firm performance, and the greater the degree to which BDPA is diffused throughout organizational processes, the greater the value it can have.

Supply chain innovation is defined as a change within a supply chain network, technology, or process that can take place anywhere within a firm or supply chain in an attempt to enhance new value creation for the stakeholder (Tan et al. 2015). Innovations like BDPA have the potential to improve customer response times, lower inventories, shorter time to market for new products, improve decision making process and enable a supply chain visibility. However, to realise these benefits of innovation, it is not sufficient to simply adopt the innovation. Instead, it must be accepted, routinized, and assimilated to some extent within the organization (Hazen et al. 2012). In the supply chain context, diffusion involves assimilation of shared technologies (to include the outcomes derived therefrom) across firms. When diffused properly, stakeholders are positioned to reap desired benefits from innovations such as BDPA.

After BDPA is assimilated into an organization or group of organizations, it can be used to discover new ideas about products, customers, and markets which are crucial to innovation. Manyika et al. (2011) note that firms can improve their supply chain operations and innovation with the assistance of BDPA. For instance, Amazon is a world-class leader in terms of assimilating and leveraging BDPA. Roughly 35% of Amazon's sales are generated from the personalized purchase recommendations suggested by Amazon via its expertise in BDPA. Another example is Netflix, an internet entertainment company, which currently has 103.95 million worldwide streaming customers. It has the advantage of knowing its customers well, and thus making better decisions and ultimately keeping users happier with their service (thereby retaining its customers). Walmart, with suppliers in about 70 countries, uses data that it receives from in-store and online sales-tracking and inventory management systems. Based on its BDPA capabilities, it predicts demand and consumer purchase behaviour, thereby minimizing product shortages and maximizing sales.

2.3 Applications of Big Data Predictive Analytics

BDPA is an emerging technique that has been successfully applied in marketing, supply chain management, manufacturing, logistics, human resources management and finance. Recognizing BDPA's potential, the World Economic Forum called it a "new class of economic asset" that is used as a tool by the top performing organizations to outperform their competitors. For instance, if BDPA is employed in the supply chain then the retailer can increase its operating margin by more than 60% (McAfee and Brynjolfsson 2012). Research conducted by BSA Software Alliance in USA reported that BDPA contributes to 10% or more of growth for 56% of firms (Columbus 2014). Moreover, the number of Fortune1000 companies investing in BDPA has increased by 85% and has reached the mark of 91% (Kiron et al. 2014a). A typical example of utilizing big data in improving business value and performance is SPEC, which is one of the leading eyeglass companies that generates new product ideas by analysing their social media (i.e., tweets, Google, Facebook, etc.) data.

Every company generates a large amount of data, which in the past was rarely put to use. Today, firms are realizing its importance and investing a huge amount of money and time in harvesting this big data, which is benefiting them in terms of long term competitive advantage. Its importance can also be recognized by the fact that the White House considered it as one of the national priority tasks in supporting healthcare and national security. By implementing analytics, firms can reduce costs, gain benefits and enhance their overall business value. Firms can also meet customer requirements, develop new products and services, expand into new markets and improve sales and revenue by employing analytics. As mentioned in the report published by Economist 2011, an organization can generate about 30% of its sales using analytics. In addition, Match.com reported more than 50% increase in revenue in the last 2 years, with more than 1.8 million paid subscribers in its core business (Kiron et al. 2014b). In a case study published in 2012 by IBM, it is stated that sharing large amounts of data and analytics helps in improving patient health. As can be seen from the aforementioned examples, electronic firms are increasingly adopting BDPA techniques to resolve their business issues. This excessive use of big data by electronic firms is probably "due [to] the social networking, the internet, mobile telephony and all kinds of new technologies that create and capture data" (Kauffman et al. 2012, p. 85). A study highlighted that US health care may get benefits of 300 dollars a year just by using big data effectively. Similarly, Amazon shared their success story of utilizing big data to generate sophisticated recommendation engines that deliver over 35% of all sales and automated customer service systems. This is an important component of customer satisfaction and dynamic pricing systems that reacts to changing pricing levels set by competitors by adjusting pricing on its own site.

3 Organizational Capabilities

Organizational capabilities can be categorized into two types: dynamic and operational. The dynamic capability reflects the ability of a firm to obtain new resource conditions in an uncertain market. Through dynamic capabilities, firms can achieve and sustain competitive advantage. It helps a firm in explaining why and how firms gain competitive advantage in an unpredictable environment. On the other hand, operational capability explains how firms operate to make a living in the present. It reflects the ability of a firm to perform and coordinate the tasks that are important to perform operational activities: distribution logistics and marketing campaigns. As the customers are very demanding in terms of time and cost-effective products, firms are building operational capabilities for superior firm performance.

3.1 Informational Technology Capabilities

IT capability is one of the most important capabilities that helps a firm in structuring its businesses as it handles activities like acquiring, deploying and leveraging of IT resources. It is a kind of structure that can be used to capitalize on a company's IT assets. IT capabilities are the high performing organizational processes that acquire, deploy and leverage IT assets, such as technical and human assets. Easily imitable IT assets cannot help a firm in improving its competitive advantage as these resources can be easily copied by competitors. Instead, advanced (and perhaps proprietary) capabilities and resultant business processes allow a firm to reconfigure the IT infrastructure by adding new IT components or existing information systems. IT resources can be obtained through outsourcing or through systems development. IT capabilities have been classified into different typologies. For instance, it can be categorized as, IT infrastructure capability, IT business spanning capability and IT proactive stance. It can also be explained through a sociomaterialistic perspective that considers it a function of IT management capability, IT personnel capability and IT infrastructure capability. Another way is to explain it through value, heterogeneity, and imperfect mobility where the first two reflect the necessary conditions for competitive advantage while the third one focuses on sustained advantage.

3.2 Human Resource Capabilities

Human resource (HR) is another important capability that can be used to improve firms' competitive advantage. It includes HR capabilities, resources, relationships and decisions that helps the firms in outperforming their competitors. In addition, HR roles should be placed at the centre of the activities with a view for the future. It is broadly accepted that the most important part of any firm is its employees as they are the ones driving innovation, which is not possible to achieve if employees are not well qualified or skilled. Therefore, it is very much important for firms to attract and retain their skilled employees so that they can help them in providing competitive edge. This is also supported by resource-based theory as it emphasizes the ability of the firm to appropriately manage employees' skills and knowledge. It focuses on attracting and retaining personnel; building and developing their expertise through development and learning systems and relationships; rewarding and sharing expertise; and learning. Thus, HR is seen as one of the major drivers for successful new product development.

3.3 Theoretical Framework and Hypotheses Development

Resource based theory highlighted that superior firm performance can be achieved through operational capabilities and dynamic capabilities. BDPA is considered a dynamic capability and this concept helps in understanding the implications of BDPA diffusion on organizational value creation. In addition, the concept of hierarchy of capabilities proposes that a higher-order capability develops from various lower order capabilities. Based on this argument, we develop a conceptual framework where lower-order capabilities (IT and HR) are leveraged to develop higher order capabilities (BDPA) that, in turn, directly affects organizational performance. Finally, we consider IT deployment capabilities in terms of three independent constructs: strategic IT flexibility, business-BDPA partnership and business-BDPA alignment (Fig. 1).

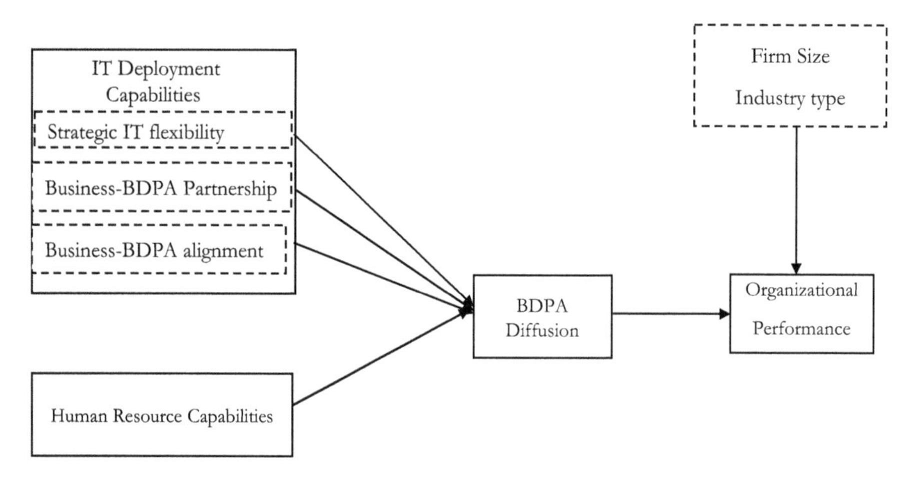

Fig. 1 Conceptual model

3.4 *Strategic IT Flexibility*

Strategic IT flexibility refers to the ability of an organization to manage various IT related activities easily and speedily and survive in the uncertain environment characterised by changing business and technologies. It is also known that new information technologies and services can be easily installed, applied and delivered if the firm has a flexible information infrastructure. Since deployment means an attempt to install and deliver IT in the adopter's organization, strategic IT flexibility is considered an IT deployment capability. Moreover, we propose BDPA diffusion as an organizational capability that aims to minimize uncertainties in demands, capacities, and supply availability. By implementing BDPA, organizations can develop information processing capabilities to understand and combine knowledge from different sources and can make better predictions about the future demands and requirements. We also know that BDPA plays a major role in making strategic and operational decisions in an uncertain environment and that dynamic capabilities are more likely to depend on the new knowledge created from a particular situation rather than on existing knowledge. Therefore, we hypothesize:

H1 *Strategic IT flexibility has a positive effect on BDPA diffusion.*

3.5 *Business-BDPA Partnership*

Business-BDPA partnership is an organizational capability that facilitates the smooth functioning of the complete IT deployment process. It is also known that IT resources can be successfully deployed if IT departments are interested in satisfying business needs and demands and maintaining a strong relationship with the business department. In addition, IT functions should be performed successfully so as to get an idea about the state-of-the-art information technologies that can be utilized to observe and seize emerging business opportunities. It is also reported that 33.9% of firms identified business and technology cooperation as the most crucial factor for enabling business adoption, and 23.2% identified strong business sponsorships as the next important factor. Thus, it can be clearly observed that partnership and cooperation with business leadership might be important keys to BDPA diffusion. Therefore, we hypothesize:

H2 *Business–BDPA partnership has a positive effect on BDPA diffusion.*

3.6 Business-BDPA Alignment

Business-BDPA alignment is an organizational capability that assigns resources and builds strategies to maintain alignment between BDPA and business. In fact, it is a bidirectional process since BDPA supports business, and business also capitalizes on strategic potentials of BDPA. Business-BDPA alignment can be defined at the process level, where it translates formal strategic plans into business activities and ways for IT to support the business, and at the organizational level: it is the degree to which the mission, objectives and plans of IT and business support each other. In addition, business and BDPA alignment poses a challenge to many organizations due to changing business needs, fluctuating market conditions, new technology, and the difficulty that the organization faces while dealing with these changes. Further, it is believed that enterprises should emphasize on bridging the gap between analytics and operational needs. Hence, maximum insights from BDPA can be obtained if business, IT and data professionals work together. Therefore, we hypothesize:

H3 *Business–BDPA alignment has a positive effect on BDPA diffusion.*

3.7 Human Resource (HR) Capabilities

HR has been identified as a key source of competitive advantage. It is considered as the reason for the success of an organization and as an indicator of enhanced organizational effectiveness. In addition, HR capabilities play an important role in building dynamic capabilities, particularly in redeveloping the resource base. A wrong decision taken by a manager may lead to a wrong dynamic capability which could be very dangerous for a firm. Moreover, availability of skilled employees along with the ability to perform business analytics reflects upon the willingness of organizations to diffuse BDPA. For this reason, top management commitment and support is required, and this is possible only if the organizations have enough resources and capabilities. Therefore, we hypothesize:

H4 *HR capabilities have a positive effect on BDPA diffusion.*

3.8 Impact of Capabilities on Organizational Performance

Resource based theory focuses on the fact that lower order capabilities help a firm in developing higher order capabilities that drive performance. In other words, IT capabilities are considered as antecedents of higher order business capabilities, including knowledge management and agility capabilities. In addition, capabilities mediate the relation between resources and performance. Therefore, we suggest that lower-order capabilities (IT deployment and HR) improve organizational performance when mediated by BDPA diffusion (higher-order capability). We hypothesize:

H5a *Strategic IT flexibility under the mediation effect of BDPA diffusion is positively related to organizational performance.*

H5b *Business-BDPA partnership under the mediation effect of BDPA diffusion is positively related to organizational performance.*

H5c *Business-BDPA alignment under the mediation effect of BDPA diffusion is positively related to organizational performance.*

H5d *HR capabilities under the mediation effect of BDPA diffusion is positively related to organizational performance.*

3.9 BDPA Diffusion and Organizational Performance

Organizations can outperform their competitors when they effectively utilize their resources and capabilities which are unique and difficult to imitate. These resources and capabilities are the major determinants of business value and organizational performance. In addition, resources and capabilities are also considered as the main drivers for improving efficiency and effectiveness, which further leads to the organizational performance. We claim that BDPA diffusion, being a higher order capability, helps a firm in improving its performance if utilized effectively. Therefore, we hypothesize:

H6 *BDPA diffusion has a positive impact on organizational performance.*

Table 1 Constructs

Constructs	Definitions	References
Strategic IT flexibility	It is defined as the organizational capability which aids in modifying information systems according to environmental changes by either combining new IT components with the existing IT infrastructure or, by re-building the existing information systems	Tian et al. (2010)
Business-IT partnership	It is considered as an organizational social capital which helps in smoothing the complete IT deployment process	Tian et al. (2010)
Business-IT alignment	It is the capability to assign resources and build strategies to attain and maintain alignment between IT and business so that it can support and shape business	Tian et al. (2010)
Human resource capability	It describes the role of executives in building dynamic capabilities, specifically in redeveloping the resource base	Wiklund and Shepherd (2003), Teece (2007), Augier and Teece (2009)
BDPA diffusion	BDPA are the analytical techniques that are used to gain insights from large and complex data sets to make better decisions	Hazen et al. (2012), Chen et al. (2012), Watson (2014)
Organizational performance	It is measured by the average market share, sales volume and sales growth	Whitten et al. (2012)

4 Research Methodology

For conducting this study, we divided the survey instrument into three broad sections. In the first section, we covered the constructs related to organizational capabilities (IT deployment and HR), and in the second and third sections, we included the constructs BDPA diffusion and organizational performance, respectively. The items were assessed by participants based on Likert scales ranging from "strongly disagree" to "strongly agree". Beside these constructs, we considered two control variables, namely, the industry type and firm size. The final items used in the survey instrument and their sources are listed in Table 1.

Our survey mainly targeted the manufacturing, consulting, e-commerce and technology companies. The correspondence address for senior managers working in 383 unique Indian firms was obtained from an online database. We received 159 usable responses resulting in a 41.5% response rate, which is generally typical of survey-based studies. Table 2 shows the detailed demographics of the respondent firms.

Table 2 Firm breakdown

Title	Number	Percentage
Annual sales revenue		
Under USD10 Million	7	3.41
USD10–USD25 Million	12	5.85
USD26–USD50 Million	16	7.80
USD76–USD100 Million	42	20.49
USD101–USD250 Million	18	8.78
USD251–USD500 Million	12	5.85
Over 251 Million	53	25.85
	160	
Number of employees		
0–50	3	1.46
51–100	6	2.93
101–200	13	6.34
201–500	11	5.37
501–1000	74	36.10
1001+	53	25.85
	160	
Industry		
Manufacturing	60	29.27
Consulting	30	14.63
E-commerce	13	6.34
Technology company	57	27.80
	160	

5 Results

Before testing hypotheses, we have checked for psychometric properties and discriminant validity between the constructs (Tables 3 and 4). In Table 4, the bold values are the square root of average variance extracted (AVE) and it is calculated to check the discriminant validity. As the values in the leading diagonal are greater than all the values in the same column and row, we can say that discriminant validity is possessed by the constructs. We tested our research hypotheses using multiple regression analysis with mediation tests (Table 5). H1 suggests that strategic IT flexibility is positively associated with BDPA diffusion. Since the value of path coefficient is 0.446 ($p < 0.001$), we find support for H1. H2 suggests that business-BDPA partnership has a positive association with BDPA diffusion. As the value of path coefficient is 0.672 ($p < 0.001$), we find support to H2. H3 argues that business-BDPA alignment is positively associated with BDPA diffusion. As the value of path coefficient is 0.809 ($p < 0.001$), we find support to H3. H4 suggests that HR capabilities have a positive association with BDPA diffusion. As the value of path coefficient is 0.935 ($p < 0.001$), we find support to H4.

For H5a, we first performed regression analysis with strategic IT flexibility as independent variable and organizational performance (OP) as dependent variable

Table 3 Scales and item performance

Scale	Items To what extent do you agree that your corporation's information systems can easily and quickly perform the following business actions?	λi	SCR	AVE
Strategic IT flexibility Cronbach's alpha = 0.96	Expand into new regional or international markets	0.94	0.96	0.87
	Change (i.e. expand or reduce) the variety of available applications	0.94		
	Adopt new technologies to produce better, faster and cheaper information services	0.94		
	Switch to new suppliers to enjoy lower costs, better quality or improved delivery times	0.91		
Business-IT partnership Cronbach's alpha = 0.94	Our BDPA department and business units understand the working environments of each other very well	0.82	0.96	0.87
	There is high degree of trust between our BDPA department and business units	0.97		
	The goals and plans for BDPA projects are jointly developed by both the analytics department and other functional departments	0.98		
	Conflicts between analytics department and other functional departments are always resolved through dialogue and mutual adjustment	0.94		
Business-IT alignment Cronbach's alpha = 0.94	BDPA plans reflect the business plan goals	0.51	0.96	0.81
	BDPA plans support the business strategies	0.96		
	BDPA plans recognize external business environment forces	0.97		
	Business plan refer to BDPA plans	0.96		
	Business plans refer to specific technologies	0.94		
	Business plans have reasonable expectations of BDPA	0.96		
Human resource capability Cronbach's alpha = 0.93	In your firm employees have excellent business knowledge; they have a deep understanding of business priorities and goals	0.95	0.95	0.83
	Your firm has highly productive employees	0.95		
	In your firm, employees are willing to contribute with ideas for new products and services	0.86		
	In your firm, employees have a positive commitment to the company's development	0.90		
BDPA diffusion Cronbach's alpha = 0.94	Supporting accounting management	0.84	0.95	0.81
	Supporting product and service delivery management	0.96		
	Supporting warehousing and inventory management	0.96		
	Supporting production and operations management	0.95		
	Facilitating purchase ordering and fulfilment management among supply chain partners	0.76		

(continued)

Table 3 (continued)

Scale	Items To what extent do you agree that your corporation's information systems can easily and quickly perform the following business actions?	λi	SCR	AVE
Organizational performance Cronbach's alpha = 0.63	Your organization has experienced high average market share growth due to use of big data and predictive analytics (BDPA)	0.70	0.72	0.50
	Your organization has experienced high average sale volume growth due to use of big data and predictive analytics (BDPA)	0.71		
	Your organization has experienced high average sales growth due to use of big data and predictive analytics (BDPA)	0.61		

Table 4 Discriminant validity matrix

Constructs	1	2	3	4	5	6
Strategic IT flexibility	**0.93**					
Business-IT partnership	0.28	**0.93**				
Business-IT alignment	−0.12	−0.07	**0.90**			
HR capability	0.46	0. 34	0.00	**0.91**		
BDPA diffusion	−0.12	−0.11	0.39	0.08	**0.90**	
Organizational performance	0.17	0.06	−0.11	−0.13	−0.07	**0.68**

Table 5 Results of mediation test

Path	R	R^2	β	P	VIF
Strategic IT flexibility → OP	0.467	0.218	0.467	0.000	1.000
Strategic IT flexibility → BDPA	0.446	0.199	0.446	0.000	1.000
BDPA → OP	0.281	0.079	0.281	0.000	1.000
(Strategic IT flexibility + BDPA) → OP	0.474	0.225	0.090	0.255	1.249
Business-BDPA partnership → OP	0.394	0.156	0.394	0.000	1.000
Business-BDPA partnership → BDPA	0.672	0.452	0.672	0.000	1.000
BDPA → OP	0.281	0.079	0.281	0.000	1.000
(Business-BDPA partnership + BDPA) → OP	0.395	0.156	0.028	0.778	1.825
Business-BDPA alignment → OP	0.167	0.028	0.167	0.034	1.000
Business-BDPA alignment → BDPA	0.809	0.655	0.809	0.000	1.000
BDPA → OP	0.281	0.079	0.281	0.000	1.000
(Business-BDPA alignment + BDPA) → OP	0.299	0.089	0.422	0.001	2.892
HR capability → OP	0.224	0.050	0.224	0.005	1.000
HR capability → BDPA	0.935	0.875	0.935	0.000	1.000
BDPA → OP	0.281	0.079	0.281	0.000	1.000
(HR capability + BDPA) → OP	0.301	0.091	0.569	0.009	7.974

(path C) and found that strategic IT flexibility has significant influence on OP (a coefficient of 0.467; $p < 0.001$). The next step was to test the effect of strategic IT flexibility on BDPA diffusion (path A), which showed significant influence with a coefficient of 0.446 ($p < 0.001$). In the third step, influence of BDPA diffusion on OP (path B) was tested and the result was found to be significant with a coefficient of 0.281 ($p < 0.001$). Finally, the last step tested the effect of strategic IT flexibility on OP by controlling BDPA (path C′), which resulted in a coefficient of 0.090 ($p > 0.1$). It can be observed that the relations in the first three steps are significant while the relation in the last step is not significant. Thus, we find support to H5a since BDPA diffusion acts as a full mediator between strategic IT flexibility and OP (see Table 5).

For H5b, we follow the steps of H5a and find that in the first three steps, the relations are significant while it is not significant in the last step 0.028 ($p > 0.1$). Hence, we find support to H5b since BDPA diffusion acts as a full mediator between business-BDPA partnership and OP (see Table 5). For H5c, we follow the steps of H5a and find that in all the four steps, the relations are significant, hence indicating partial mediation. Hence, we find support to H5c (see Table 5).

For H5d, we follow the steps of H5a and find that in all the four steps, the relations are significant. Hence, we find support to H5d since BDPA diffusion acts as a partial mediator between HR capabilities and OP (see Table 5). Finally, H6 argues that BDPA diffusion has a positive impact on OP. Since the value of path coefficient is 0.281 ($p < 0.001$), we find support to H6.

6 Practical Implications

From a practical point of view, this chapter provides guidance to managers involved in the BDPA implementation process. The mediating role of BDPA diffusion clearly highlights how it can be leveraged as a source of organizational performance. The finding that the four IT deployment capabilities, strategic IT-flexibility, business-BDPA partnership, business BDPA alignment and HR capabilities, strongly influence BDPA diffusion and improve the performance of the firm indicates that managers need to concentrate on organizational capabilities. Moreover, our finding that IT deployment and HR capabilities positively affect BDPA diffusion is beneficial for firms that invest heavily in IT and HR to achieve superior organizational performance as these investments may not be fruitful if organizations do not leverage their IT and HR capabilities to achieve superior dynamic capabilities and to derive performance. Thus, it is crucial for managers to leverage lower order capabilities (IT and HR) to build higher order organizational capabilities (BDPA diffusion) and improve organizational performance. Also, this research provides insights to managers on how BDPA diffusion can directly influence organizational performance. Although potential benefits of BDPA diffusion are well recognized in industries, there are some who are reluctant to use it due to their insufficient knowledge about the way to proceed ahead and implement BDPA. Our finding provides the necessary guidance and assurance that BDPA usage can benefit the organization.

7 Conclusions

The main objective of this chapter is to identify the indirect relation between IT and HR capabilities on organizational performance as this relation is mediated through BDPA diffusion. From this chapter, we can conclude that strategic IT flexibility, business–BDPA partnership, business–BDPA alignment and HR capabilities have direct impact on BDPA diffusion, whereas, these constructs have an indirect impact on organizational performance. Through our results, we can also see the strong support for the full mediation effect of BDPA diffusion on the relation between strategic IT flexibility, business–BDPA partnership and organizational performance while it partially mediates the relation between business-BDPA alignment, HR capabilities and organizational performance.

References

Akter S, Wamba FS (2016) Big data analytics in E-commerce: a systematic review and agenda for future research. Electron Mark J 26(2):173–194
Augier M, Teece DJ (2009) Dynamic capabilities and the role of managers in business strategy and economic performance. Organ Sci 20(2):410–421
Barbier G, Liu H (2011) Data mining in social media. In: Aggarwal CC (ed) Social network data analytics. Springer, USA, pp 327–352
Barratt M, Oke A (2007) Antecedents of supply chain visibility in retail supply chains: a resource-based theory perspective. J Oper Manag 25(6):1217–1233
Brandon-Jones E, Squire B, Autry CW, Petersen KJ (2014) A contingent resource based perspective of supply chain resilience and robustness. J Supply Chain Manag 50(3):55–73
Chen H, Chiang RH, Storey VC (2012) Business intelligence and analytics: from big data to big impact. MIS Q 36(4):1165–1188
Chung W (2014) BizPro: Extracting and categorizing business intelligence factors from textual news articles. Int J Inf Manag 34(2):272–284
Columbus L (2014) Making analytics accountable: 56% of executives expect analytics to contribute to 10% or more growth in 2014. Forbes. Available via http://www.forbes.com/sites/louiscolumbus/2014/12/10/making-analytics-accountable-56-of-executivesexpect-analytics-to-contribute-to-10-or-more-growth-in-2014/#761c65a95b56. Accessed 2 Feb 2016
Davenport TH (2006) Competing on analytics. Harv Bus Rev 84:98–107
Davenport TH (2012) The human side of big data and high-performance analytics. International Institute for Analytics, pp 1–13
Devaraj S, Fan M, Kohli R (2002) Antecedents of B2C channel satisfaction and preference: validating e-commerce metrics. Inf Syst Res 13:316–333
Gandomi A, Haider M (2015) Beyond the hype: big data concepts, methods, and analytics. Int J Inf Manag 35(2):137–144
Gundecha P, Liu H (2012) Mining social media: a brief introduction. Tutor Oper Res 1(4)
Hakeem A, Gupta H, Kanaujia A, Choe TE, Gunda K, Scanlon A et al (2012) Video analytics for business intelligence. In: Shan C, Porikli F, Xiang T, Gong S (eds) Video analytics for business intelligence. Springer, Berlin, Heidelberg, pp 309–354
Hazen BT, Overstreet RE, Cegielski CG (2012) Supply chain innovation diffusion: going beyond adoption. Int J Logist Manag 23(1):119–134

Hirschberg J, Hjalmarsson A, Elhadad N (2010) You're as sick as you sound: using computational approaches for modeling speaker state to gauge illness and recovery. In: Neustein A (ed) Advances in speech recognition. Springer, USA, pp 305–322
Kauffman RJ, Srivastava J, Vayghan J (2012) Business and data analytics: new innovations for the management of E-commerce. Electron Commer Res Appl 11:85–88
Kiron D, Prentice PK, Ferguson RB (2014a) The analytics mandate. MIT Sloan Manag Rev 55:1–25
Kiron D, Prentice PK, Ferguson RB (2014b) Raising the bar with analytics. MIT Sloan Manag Rev 55:29–33
LaValle S, Lesser E, Shockley R, Hopkins MS, Kruschwitz N (2011) Big data, analytics and the path from insights to value. MIT Sloan Manag Rev 52:21–32
Manyika J, Chui M, Brown B, Bughin J, Dobbs R, Roxburgh C, Byers AH (2011) Big data: the next frontier for innovation, competition and productivity. McKinsey Global Institute
McAfee A, Brynjolfsson E (2012) Big data. The management revolution. Harv Bus Rev 90 (10):61–67
Panigrahi BK, Abraham A, Das S (2010) Computational intelligence in power engineering. Springer, Berlin
Patil HA (2010) Cry baby: using spectrographic analysis to assess neonatal health status from an infant's cry. In: Neustein A (ed) Advances in speech recognition. Springer, USA, pp 323–348
Tan KH, Zhan Y, Ji G, Ye F, Chang C (2015) Harvesting big data to enhance supply chain innovation capabilities: an analytic infrastructure based on deduction graph. Int J Prod Econ 165:223–233
Teece D (2007) Explicating dynamic capabilities: the nature and micro foundations of (sustainable) enterprise performance. Strateg Manag J 28:1319–1350
Tian J, Wang K, Chen Y, Johansson B (2010) From IT deployment capabilities to competitive advantage: an exploratory study in China. Inf Syst Front 12(3):239–255
Waller MA, Fawcett SE (2013) Click here for a data scientist: big data, predictive analytics, and theory development in the era of a maker movement supply chain. J Bus Logist 34(4):249–252
Watson HJ (2014) Tutorial: big data analytics: concepts, technologies, and applications. Commun Assoc Inf Syst 34:65
Whitten DG, Green KW Jr, Zelbst PJ (2012) Triple-A supply chain performance. Int J Oper Prod Manag 32(1):28–48
Wiklund J, Shepherd D (2003) Knowledge-based resources, entrepreneurial orientation, and the performance of small and medium-sized businesses. Strateg Manag J 24:1307–1314

Deepa Mishra is a postdoctoral fellow at De Groote School of Business, McMaster University, Canada. She completed her PhD in operations management from Indian Institute of Technology Kanpur, India. She has interest in supply chain management, risk management, big data, and Internet of Things and has done empirical as well as review papers. Deepa has published around 12 peer-reviewed articles in journals like *International Journal of Production Economics*, *Annals of Operations Research*, *Industrial and Data Management Systems*, *European Journal of Operation Research*, *Total Quality Management: Business Excellence*, and *Journal of Organizational Change Management*.

Zongwei Luo is a researcher and an associate professor in Southern University of Science and Technology (SUSTC) located in Shenzhen, China, with over 15 years R&D and project management experience. Dr. Luo obtained his PhD from the University of Georgia, USA, focusing on enterprise computing. After that, Dr. Luo was with IBM TJ Watson Research Center in Yorktown Heights, NY, USA, before coming to E-Business Technologies Institute, the University of Hong Kong. Dr. Luo has been actively working on big data, cognitive informatics, service innovation, and mechanism design and their applications. Dr. Luo has over 100 publications with one best paper award in an IEEE conference. Dr. Luo is the founding Editor-in-Chief of *International Journal of Applied Logistics*.

Benjamin T. Hazen is an Associate Professor of Logistics and Supply Chain Management at the Air Force Institute of Technology (USA). He enjoys doing research in sustainability, data science, closed-loop supply chains, supply chain information systems, and innovation. His current research centers on how firms can derive business value from data science innovations. Ben has published more than 60 peer-reviewed articles in top journals across a variety of disciplines. He serves as a senior associate editor of *International Journal of Physical Distribution and Logistics Management* and an associate editor of *Global Journal of Flexible Systems Management*, and he is a past Editor-in-Chief of *International Journal of Logistics Management*. Ben now serves as Editor-in-Chief of the new *Journal of Defense Analytics and Logistics*.

Adoption of Industry 4.0 Technologies in Supply Chains

Gustavo Dalmarco and Ana Cristina Barros

Abstract The widespread use of internet is changing the way supply chain echelons interact with each other in order to respond to increasing customer requests of personalized products and services. Companies acquainted with the concept of industry 4.0 (i4.0) embrace the use of internet to improve their internal and external processes, delivering the dynamic and flexible response customers want. This chapter aims to discuss how supply chains may benefit from the adoption of i4.0 technologies by their partners and highlights some of its implementation challenges. Eight technologies cover most of i4.0 applications: additive manufacturing; big data & analytics; cloud computing; cyber-physical systems; cyber security; internet of things; collaborative robotics; and visual computing. At individual level, technologies such as additive manufacturing, collaborative robots, visual computing and cyber-physical systems establish the connectivity of a certain company. However, the integration of the whole supply chain, based on the principles of i4.0, demands that information provided by each company (Big Data) is shared through a collaborative system based on Cloud Computing and Internet of Things technologies. To safely share useful information, Cyber Security techniques must be implemented in individual systems and cloud solutions. Summing up, even though the adoption of i4.0 demands an individual initiative, it will only raise the supply chain's competitive advantage if all companies adapt their manufacturing and supply chain processes. The main advantage foreseen here is based on an improved communication system of the whole supply chain, bringing consumers closer to the production process.

G. Dalmarco
INESC TEC, Porto, Portugal

PUCRS, Business School, Porto Alegre, RS, Brazil
e-mail: gustavo.dalmarco@inesctec.pt

A. C. Barros (✉)
INESC TEC, Porto, Portugal
e-mail: ana.c.barros@inesctec.pt

A. C. Moreira et al. (eds.), *Innovation and Supply Chain Management*, Contributions to Management Science, https://doi.org/10.1007/978-3-319-74304-2_14

1 Introduction

The consumer behaviour of the Internet-Based Society is changing the way companies interact with customers (Koufaris 2002). Following this path, the interaction with clients, suppliers and competitors is also improving by both governance changes (Langlois 2013) and the application of new interactive technologies (Daugherty et al. 2014). Several manufacturing improvements by the application of concepts such as Cyber-Physical Systems (CPS), Additive Manufacturing (AM), Internet of Things (IoT) and Cloud Computing, among others, are being used by companies who want to keep themselves competitive in such dynamic scenarios (Lu 2017; Suri et al. 2017).

However, to be up to date with current trends, companies need to develop products and processes that comply with new standards in productivity, sustainability and competitivity. The integration of such technologies into an autonomous, knowledge—and sensor-based—self-regulating production system is currently known as Industry 4.0 (i4.0) (Lasi et al. 2014; Hofmann and Rüsch 2017). I4.0 is a concept used to characterize the new strategic positioning of the German industry, based on a flexible internet-based production system (Kagermann et al. 2013). In the US, although the integration of manufacturing equipment into a flexible production system connected to the internet is called Industrial Internet of Things (IIoT), it encompasses the same principles of i4.0 (Daugherty et al. 2014). Consequently, both concepts i4.0 and IIoT (adding to Smart Factory, Manufacturing 4.0, and SMART Manufacturing, among others) are used to define companies that are applying more flexible and autonomous production systems (Sniderman et al. 2016). It is mainly based on communication improvements that allow a more decentralized production process, integrating sensors and actuators through internet connection (Hermann et al. 2016). The features of i4.0 also include horizontal integration (to facilitate the collaboration among companies), vertical integration of hierarchical subsystems (to improve flexibility of the production line), and end-to-end engineering integration, supporting product customization through the entire supply chain (Brettel et al. 2014; Wang et al. 2016).

Consequently, to be competitive in an i4.0 business environment, industries cannot be stand-alone institutions. Suppliers and customers are already an active part of the production process through customization and collaboration in the product development process, and need to be part of the integration of new production technologies as well (Autry et al. 2010; Patterson et al. 2003). The application of i4.0 will foster changes in technology and customer experience, enabling a more dynamic connection of different partners within the supply chain, with cooperation and coordination beyond organizational boundaries (Glas and Kleemann 2016; Stevens and Johnson 2016).

In order to analyse the adoption of i4.0 in supply chains, this chapter identifies by means of literature review the main technologies related to i4.0 that may be applied to the integration of companies in a supply chain. Posada et al. (2015) and Hermann et al. (2016) had performed recently such a review, identifying the main technologies being described by both scientific and technical documents available online. After analysing these two articles we complemented the authors' findings with reports from consulting companies and governmental organizations that discussed the potential advantages with the adoption of i4.0 and its implications for industry (Daugherty et al. 2014—Accenture; Davies 2015—European Union; Rüßmann et al. 2015—Boston Consulting; Sniderman et al. 2016—Deloitte), organizing then a list of technologies related to i4.0 and supply chain. Table 1 presents the main technologies evidenced by this review, with the respective description and main references.

Although the use of these technologies by different factories is already described in the literature, our objective in this chapter is to discuss their adoption in the supply chain. It can be observed that at individual level, technologies such as additive manufacturing, collaborative robots, visual computing and cyber-physical systems establish the connectivity of a certain company. However, the integration of the whole supply chain based on the principles of i4.0 demands that information provided by each company (Big Data) feed a collaborative system that allows technologies such as Cloud Computing, Internet of Things and Cyber Security to safely share and optimize information which may improve the whole production system.

To better explain these technologies and their integration into a supply chain 4.0, the next section will presents their advantages, limitations and examples of application by different supply chains.

2 Technologies for Supply Chain 4.0

The management of technology in companies, and consequently in supply chains, is a strategic process that requires the involvement of different hierarchical levels inside the companies of a supply chain. From a small group of decision makers to the operators at the production line, the adoption of new technologies into the production process and expanded to the supply chain includes planning, directing, control and coordination of the company's technological development capabilities to shape and accomplish the strategic and operational objectives of each organization of a supply chain (Liao 2005). The adoption of new technologies includes three main stages: decision, implementation, and assimilation (Greenhalgh et al. 2004; Rogers 2003). It is initiated by a decision to adopt a certain technology, frequently made by a restricted group of decision makers (Gallivan 2001). Then, during the implementation stage, efforts are initiated to include the technology in the routine operations of the adopter and to align the adopter and the technology to better fit the operations and the expected outcomes (Gallivan 2001; Greenhalgh et al. 2004; Rogers 2003). At this stage the technology is being gradually adopted by the users, with the assistance of training sessions and other efforts to promote the acceptance of the technology (Gallivan 2001). The implementation and assimilation stages are intermingled. In the assimilation stage, efforts to routinize and incorporate the technology continue, but

Table 1 Description of the technologies related to industry 4.0 and supply chain

Technology	Description	References
Additive Manufacturing	Is a technology that allows 3D CAD models to be printed, layer by layer, into one solid piece. It can be printed in different materials such as metal, wax, plastics, and ceramics. It also allows to produce mechanical parts that couldn't be fabricated by regular processes	Berman (2012), Harris and Director (2011), Oettmeier and Hofmann (2017), Scott and Harrison (2015)
Big Data & Analytics	Is the information available by the acquisition of different sensors, gathered into a historical and real time dataset. Analytics is the use of tools and statistical method to use information in managerial decisions	Hazen et al. (2014), Lee et al. (2013, 2014), Megahed and Jones-Farmer (2013), Posada et al. (2015), Shrouf et al. (2014), Witkowski (2017)
Cloud Computing	Is the use of data and software available through networks instead of being installed physically on a local computer	Bughin et al. (2010), Marston et al. (2011), Rüßmann et al. (2015)
Cyber-Physical Systems	Is an automated system that orchestrates the communication among several devices and equipment through a computing infrastructure. It includes smart machines, storage systems, and production facilities that can exchange information with autonomy and intelligence, are able to decide and trigger actions, and can control each other independently	Baheti and Gill (2011), Hermann et al. (2016), Lee (2008), MacDougall (2014), Posada et al. (2015)
Cybersecurity	Is the protection of the information available by the devices connected to a computer network, adding to the security of the user connected to the network and their assets	Von Solms and Van Niekerk (2013), Wang et al. (2016), Waslo et al. (2017), Witkowski (2017)
Internet of Things	Is characterized by the interconnection of equipment and devices (things) through the Internet. Equipment provides information (such as their status, environment, production processes and maintenance schedule, among others) to the network by embedded electronics (RFID tags, sensors, etc.), being also able to perform actions based on the information of other devices	Hermann et al. (2016), Hofmann and Rüsch (2017), Shrouf et al. (2014), Rüßmann et al. (2015)
Collaborative Robotics	It can be defined as robots who can interact with human operators and other robots in an intuitive self-learning behaviour	Awais and Henrich (2013), Rüßmann et al. (2015)
Visual Computing	It can be defined as the entire field of acquiring, analysing, and synthesizing visual data by means of computers that provide relevant-to-the-field tools	Paelke (2014), Posada et al. (2015), Shellshear et al. (2015)

the technology is already fully working within the adopter's operations and begins to lose its external identity by becoming an ongoing element of those operations (Rogers 2003).

Following these three stages—decision, implementation and assimilation, the adoption of i4.0 technologies by companies in a supply chain is an ongoing process in both emergent and developed countries (Daugherty et al. 2014; Davies 2015; Saldanha et al. 2015). To assist companies and researchers interested in i4.0 for supply chains, this chapter summarizes the main technologies applied to i4.0 and examples of their adoption by different industries.

2.1 *Additive Manufacturing*

Additive manufacturing (AM) is a technique that consists of the reproduction of CAD drawings into solid 3D parts by the fusion of several layers of a specific material, either in the form of plastic or metal powder (Scott and Harrison 2015). It is also known as "rapid manufacturing", "rapid prototyping", "direct manufacturing" and "digital manufacturing" (Holmström et al. 2010; Hopkinson and Dickens 2001; Oettmeier and Hofmann 2017). The 3D parts can be used as parts of an assembly, final products, or in the manufacturing of parts for maintenance, repair and overall (MRO) operations (Hopkinson and Dickens 2001; Khajavi et al. 2014). This method differs from other manufacturing processes since it adds material to form a new piece, instead of removing material from a raw piece, reducing waste in the production line.

Besides this difference in the manufacturing process, AM presents some further advantages when compared to other methods (Holmström et al. 2010; Laureijs et al. 2017; Oettmeier and Hofmann 2017). First of all, as the parts are printed on its final form, tolling is not significantly necessary, reducing production time and expenses. It is also economically feasible to produce small batches (especially one-of-a-kind), as it is to perform quick changes in design and customization. It allows lightweight design with complex geometries that without AM would be impossible to be produced. Analysing the supply chain, AM has potential to reduce the number of suppliers, lead time, transportation services and inventories (Holmström et al. 2010; Laureijs et al. 2017; Oettmeier and Hofmann 2017).

Researchers analysing AM have observed as main restrains its technical limitations and process costs (Bonnín-Roca et al. 2017; Scott and Harrison 2015). For example, AM still has limitations regarding the availability of material for printing, time to print one piece and finishing quality (Scott and Harrison 2015). Even though multiple material can be used for printing, the number of material suited for 3D printing is still limited. This also influences the time for printing, which is slow when compared to other methods, especially when producing big parts. The quality of parts produced still needs improvements as well, since the layering process may display a "stair effect", affecting the top finishing. Complementing these issues, Bonnín-Roca et al. (2017) describe the limitations of the three main aspects of AM:

printing complex pieces, development of local printing services, and mass customization. Printing complex pieces presents the limitations in reliability and staff capabilities for producing such parts. The development of local printing services is still incipient due to complex post production steps and the costs of local services. Mass customization is also on initial stages since companies are still able to develop flexible products and efficient supply-chain management, limiting the demand for mass customization.

Even though more research is needed to explore the full possibilities and applications of AM, some implications have already been pointed out. The first analyzes the costs to invest in this technology. Laureijs et al. (2017), describing the use of metal AM against forging, conclude that even with simple designs AM parts are lighter and, on the application of lighter parts in the aerospace industry, parts are cheaper than similar forged parts, even in small series. Analyzing the supply chain, Holmström et al. (2010) argue that AM may create a market for local service providers who could manufacture different types of parts, becoming a supplier to different industries. They also reinforce that companies could reduce the number of suppliers by the use of AM technology, being more flexible to attend customer demands.

2.2 *Big Data & Analytics*

The fast development of the internet and, consequently, the number of devices and instruments connected to the network, has increased the amount of data produced and collected every day (Witkowski 2017). This large, diverse, complex and longitudinal amount of information is called Big Data (Megahed and Jones-Farmer 2013). The volume, variety and velocity of new data generation brings the attention to the development of new processing techniques and tools in order to add value to the available information (Witkowski 2017). Bringing to the supply chain level the use of big data analytics raises opportunities to improve the production system of each individual supply chain echelon, leading the whole supply chain into higher levels of performance and also to the identification of new patterns for new products and services (Shrouf et al. 2014).

In fact, the use of big data & analytics is very much aligned with the new pattern of production and connectivity of i4.0. Machines should, in the near future, all be connected through internet protocols, improving production, flexibility and efficiency (Lee et al. 2014). Data from past and future behaviour of the storage and flow of inventory in logistics may optimize the integration of business processes through the supply chain, reducing associated costs and improving service levels. Customer information regarding decision and purchasing behaviour, items browsed and bought, frequency, dollar value and timing are some of the data that can be used to improve performance and efficiency (Waller and Fawcett 2013).

However, to take advantage of the big amount of data generated by these machines, processes and people, the use of prediction tools is a natural requirement.

Data is only good if it can be acknowledged, measured, monitored, and controlled, improving its reliability (Hazen et al. 2014). Only by the use of new tools data can be translated into information, mitigating process uncertainties by the identification of failure patterns (Lee et al. 2014). The information that flows between the production line and business management levels spills over to suppliers and clients, making supply chain management more transparent, organized and efficient.

Some of the benefits of applying big data & analytics to the supply chain have been confirmed in the literature. For example, Waller and Fawcett (2013) present logistic improvements that could be implemented by companies in the supply chain through the use of information such as traffic, weather and characteristics of the driver. Witkowski (2017) confirms the improvement of the delivery information sent to customers by presenting a tool that analyses manufacturing information from each company in the supply chain. In another analysis, Shrouf et al. (2014) uses big data to improve energy management by storing and analysing power consumption information from each machine in the production process.

2.3 *Cloud Computing*

Applications based on cloud computing allow multiple access to data, being available to the whole supply chain in real time (Rüßmann et al. 2015). Cloud computing aims to deliver all functionalities of existing information and services, increasing reliability of data and reducing operational costs (Marston et al. 2011).

The cloud computing environment is divided in two functions (Zhang et al. 2010): (1) Infrastructure provider, who invests and manages the hardware available for the cloud platform; and (2) Service provider, who uses the resources of different providers to offer their service to the end user. Such division is important when working with cloud computing, since it has a direct impact on the available infrastructure and resources invested. In an initial stage, companies may choose to hire a service provider, thus reducing costs. On a second stage, to improve functionalities and security, companies may prefer to integrate the whole cloud infrastructure in the supply chain, becoming both provider and user.

Regarding cloud computing key advantages, Marston et al. (2011) lists some aspects that motivate companies to adopt such technology: (1) it reduces the initial costs of start-ups who rely on compute-intensive services, also reducing barriers to innovation; (2) it provides instant access to hardware resources; (3) it makes it easier for companies to scale-up their services or upgrade their access to data; and (4) it opens a wide range of possibilities for new services. Cloud computing is an easy access and low cost technology, which in big networks such as supply chains may help to improve the possibilities of i4.0, upgrading technologies and services provided.

Considering that internet is accessible in different parts of the world, both local and international companies are investing in cloud solutions to improve relations with partners and subsidiaries. For example, Rüßmann et al. (2015) describes the use of an online platform for collaborative design and manufacturing, managing the exchange of products and production among multiple partners. Marston et al. (2011) describes other services that are available through cloud application, from office tools such as e-mail, calendar organizer, word processing and web site creation to solutions that integrate enterprises data centre with cloud storage.

2.4 Cyber-Physical Systems

Cyber-Physical Systems (CPS) allow the connection between the physical world and the virtual (internet based) one. The current dynamics of the production process raises the need for integration of the supply chain, which can be facilitated by the use of CPS (Wang et al. 2015). CPS uses computational systems to monitor and control physical processes, involving smart machines and embedded sensors with feedback loops where one affects the other and vice versa (Davies 2015; Hermann et al. 2016).

CPS is an updated view of industrial automation systems (IAS), since this model started to present several limitations to system integration and internet connectivity, two major technological trends in actual manufacturing systems (Leitão et al. 2016). CPS can leverage the interconnectivity of different kinds of machines, turning them into intelligent, resilient and self-adapting units (Lee et al. 2015). This system can analyse its own degradation, sharing this information with other peers for smart maintenance decisions to avoid potential issues (Lee et al. 2014). Consequently, an interconnected system may improve both maintenance and manufacturing decisions by sharing information among partners.

Still, there are several challenges for CPS implementations, such as assessing data from the system, making sense of data, and lack of off-the-shelf solutions (Barros et al. 2017). Security is a permanent concern due to data generated by CPS and shared among companies within a supply chain (Monostori 2014). The full exploitation of CPS technologies is mainly restrained by the companies themselves, who do not make the necessary allowances for the uncertainties inherent to the implementation of new technologies (Wang et al. 2015).

Regarding the application of CPS, Barros et al. (2017) analyses CPS implementations for production lead-time reduction in two manufacturing contexts, namely footwear and natural cork stoppers. In the footwear industry a system is implemented to collect data from sensors and actuators available at the manufacturing line for production monitoring and predictive maintenance. In the natural cork stopper company, RFID tags and environmental sensors are used to provide data about cork piles (e.g. temperature and humidity). Data is used by a management software and two other equipment that adjust their settings according to the characteristics of the cork.

2.5 Cyber Security

Communication and information exchange are the main characteristics of i4.0. When analysing its application in supply chains, where information should flow outside the company's network, security issues arise. The flow of information includes not only data about consumers, processes, business partners and commercial strategies, but also data about a new set of devices, machines and other physical objects which will be connected to an external network (Wang et al. 2016). To ensure the safety of data, cyber security encompasses both internal IT infrastructure and the digital network of the whole supply chain, including internet (Sharma 2012). As mentions Witkowski (2017), there is a big concern about privacy and security information on internet-based solutions.

In this scenario, the focus of cyber security is the protection of general information, not just systems. To embrace the growing amount of devices connected to the network, security will be granular, approaching companies and users, machines and computers, to protect the information stored (Sharma 2012). The use of data encryption and system authentication is useful, but new tools and techniques specifically developed for i4.0 will be necessary (Wang et al. 2016).

To prevent the uncertainty of cyber-security in i4.0 applications, some conservative procedures can be implemented. For example, companies should: avoid keeping devices connected to the network more than necessary; regularly update industrial production network; and keep critical information stored in the company's private network (Wang et al. 2016; Waslo et al. 2017). Adding to this, device manufacturers and users should share security responsibility in order to prevent system breaches and failures (Waslo et al. 2017).

By analysing the implication of cybersecurity for supply networks, smart factories and connected devices, Waslo et al. (2017) address some issues that executives must be aware of when adopting i4.0. On the supply network, the authors reinforce the need to assure the integrity of private data when sharing common information among partners. They also emphasized aspects of the smart factory such as safety for employees' data, company's reputation (in case of cyber attack) and guarantees for continuous production and recovery of critical systems. Regarding connected devices, Waslo et al. (2017) identify the need of the security software development life cycle (to keep software updated on new threats) and procedures to quickly restore operations and security after an incident.

2.6 Internet of Things

The concept of Internet of Things (IoT) encompasses the integration of physical items with embedded electronics (Radio Frequency Identification—RFID—tags and sensors, among others), all connected to the internet (Shrouf et al. 2014). The exchange of information among "things" improves their functionality, decentralizing analytics and enhancing the information available to the user (Hermann et al. 2016; Rüßmann et al. 2015).

There are three distinguishing features of IoT: context, omnipresence and optimization (Witkowski 2017). The first refers to the interaction sensor-environment that provides real-time monitoring (location, physical or atmospheric conditions) and, consequently, the possibility of instant response. Omnipresence describes the wide range of possibilities for objects connected to the network. Nowadays they are mostly information providers for the user, but the interaction among objects will grow substantially in a near future. The last one—optimization—expresses the wide range of functionality each object possesses. These features, among other characteristics of IoT, may open up several business opportunities, being considered one of the most promising technologies of i4.0 with huge innovative potential (Hofmann and Rüsch 2017).

Among companies of a supply chain, IoT may help manufacturers with more accurate information, even in real time. The interconnection of machines and users can help companies to better understand the flow of materials and process timetable (Shrouf et al. 2014). Through IoT companies can improve the production plan and integrate process planning.

Discussing the application of IoT in supply chain, Hofmann and Rüsch (2017) have presented an integrated solution for a Just in Time production line, when RFID tags warn the system that a specific station is empty (based on Kanban). This warning signal is forwarded to the supplier, who will then deliver new parts to the station. Another example is given by Rüßmann et al. (2015), who have described an autonomous production line of Bosch Rexroth. According to the authors, the parts on the production line are identified with RFID tags, enabling workstations to recognize what product is being produced and adapting themselves to it.

2.7 Collaborative Robots

Robots are vastly used by manufacturers in different industries to perform complex activities. However, the concept of a fixed robot in a production cell, repeating the same task continuously, is changing (Pedersen et al. 2016). Robots are evolving to a greater utility, becoming more autonomous, flexible and cooperative (Rüßmann et al. 2015). In i4.0, robots have a collaborative role, interacting with humans and other robots. Improvements in Human-Machine interaction will promote new ways of operation of factories aiming for the adoption of i4.0 (Posada et al. 2015).

Collaborative robots are not considered as a simple replacement for human workers, since humans have an important role as creative problem-solvers (Grote et al. 2014). For instance robots may be used in hazardous or labour-intensive activities, leaving humans as supervisors and role models for autonomous units. In order to perform smooth movements, robots may be trained by the observation of human movement, even being able to predict future movements and actions (Awais and Henrich 2013). This may keep employees from performing dangerous and repetitive activities, improving the factory's safety and efficiency.

Discussing the adaptation of robots to a more flexible behaviour, Pedersen et al. (2016) argues that besides a set of sensors robots should also be upgraded with certain skills. The authors state that upgraded robots should have an improved perception of their current state, be self-sustained and flexible enough to perform any task. In this sense, Pedersen et al. (2016) lists a set of skills necessary for new autonomous robots to perform any task: evaluate if the task can be performed according to the input parameter and environmental conditions; perform the task regardless of variations in the input parameter; evaluate if the task was implemented in accordance with input parameters. These skills are important when adapting robots to i4.0 standards.

By the implementation of a set of skills (such as movement inside an adjustable area, or moving different kinds of objects and shapes), Pedersen et al. (2016) have programmed and tested an auxiliary robotic arm that was used as a complementary manipulator for a flexible production line. A similar solution was presented by Rüßmann et al. (2015), who described the functionalities of autonomous interconnected robots that can adjust their settings according to the kind of part being produced. In both cases, cameras were used for calibration and security interaction with humans working on the production line. Collaborative robots may also have an important role in distribution centres, where they can work on unloading, sorting, organizing and loading trucks on a 24 h shift, interacting with humans who will supervise and make key operational decisions (Bonkenburg 2016).

2.8 Visual Computing

The adoption of i4.0 impacts different parts of the manufacturing process. From robots and 3D printers to its integration to cyber-physical systems and internet, there are some steps that should be carefully implemented for a smooth change. To mitigate the impact of these changes, tools such as visual computing are used. Posada et al. (2015) argues that visual computing is an important technology for the adoption of i4.0, since it can be used as a unifying element of different applications.

The use of visual tools allows the creation of a virtual representation of the whole production system and its interactions with supply chain partners. One can virtually test the modification and adoption of new technologies without disturbing ongoing production (Shellshear et al. 2015). Posada et al. (2015) summarizes the tools and techniques that involve visual computing, such as: Visual analytics; Human-machine interfaces; Virtual engineering; Virtual reality and virtual environments; Augmented reality; 3D Reconstruction; Cognitive vision; 3D geometric modelling; Simulation/visualization; IoT in 3D/Web3D; GIS/visualization; and Multimedia.

Regarding the use of visual computing, the creation of 3D models and their application into a reliable simulation tool are important to explore the full possibilities of this technique. For this reason, 3D CAD research has provided many industries with competitive advantage (Posada et al. 2015), while augmented reality allows the visualization of these 3D models directly within the spatial context of the factory (Paelke 2014). Adding to this, 3D scanning of the production line improves process analysis, since it not only simulates the process as a whole but also analyses physical limitations, preventing collisions (Shellshear et al. 2015). Summing up, the application of visual techniques of products and processes is a key aspect for the product configuration and manufacturing flexibility required for the adoption of i4.0 (Posada et al. 2015).

Describing the applications of visual computing, Rüßmann et al. (2015) present the use of augmented reality glasses by workers in the production line. According to the authors, this technology enables the operator to visualize the location of each part in the assembly line. It also displays information about logistics and manufacturing in their field of vision, reducing failures and enhancing quality control. Another example is presented by Posada et al. (2015), who describes a game-like 3D environment that can be used for training operators and maintenance staff in different situations, even before the development of the equipment. The author also presents the use of cameras to capture the movement of robotic manipulators, using this material for programming new robots. Visual computing tools can also be used to improve supply chain visibility, since activities such as commercial demands, planning activities and product availability, among others, can be displayed in real-time to managers and operators, mitigating manufacturing problems (Messina et al. 2016).

3 Concluding Remarks

The concept of Industry 4.0 originated in Germany has evolved from other terms such as Industrial IoT, Smart Factories, Manufacturing 4.0 or SMART Manufacturing and encompasses communication, flexibility and productivity. This new technological paradigm comprises the exploitation of the internet to ultimately increase the competitiveness of industrial companies.

However, as discussed in this chapter, companies are not unique islands, but are part of archipelagos involving suppliers, manufacturers, distributors and retailers. For this reason, the discussion about the adoption of i4.0 technologies has to move from only one specific company, as has been the focus of many articles, to the supply chain, improving the competitive advantage of supply chains as a whole.

According to our review, there are eight technologies that cover most i4.0 applications: additive manufacturing; big data & analytics; cloud computing; cyber-physical systems; cybersecurity; internet of things; collaborative robotics; and visual computing. These technologies have different loci of adoption and application in supply chains, as Fig. 1 shows.

The use of additive manufacturing, collaborative robots, visual computing and cyber-physical systems are part of the manufacturing upgrade each company should

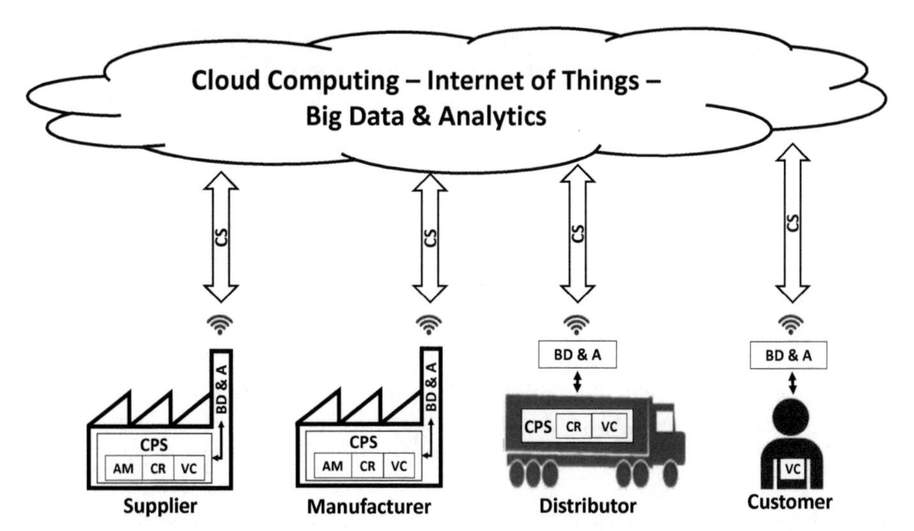

Legend: AM: Additive Manufacturing; CR: Collaborative Robotics;VC: Visual Computing; CPS: Cyber-Physical Systems; BD&A: Big Data & Analytics; CS: Cyber Security

Fig. 1 Application of Industry 4.0 technologies in the supply chain. Legend: *AM* additive manufacturing; *CR* collaborative robotics; *VC* visual computing; *CPS* cyber-physical systems; *BD&A* big data & analytics; *CS* cyber security

implement to become more flexible and able to communicate with external partners. Visual computing tools give inputs for the decision and implementation phases of the adoption of new technologies, while also supporting the manufacturing process. Additive manufacturing and collaborative robots contribute to process optimization as well. In turn, cyber-physical systems allow the connection and communication of manufacturing equipment and sensors with the internet, improving the integration of the production line. The adoption of CPS contributes to the adoption of technologies such as big data, analytics, cloud computing and the internet of things, since CPS is the main responsible one for making manufacturing information available to these online platforms. Consequently, CPS is the main path to connect each company to a network which encompasses all partners of the supply chain.

While online, the supply chain can improve its manufacturing performance by the use of big data & analytics, internet of things and cloud computing. Big data collects available information from the supply chain and stores it online, while analytics provides knowledge about patterns and bottlenecks of the full supply chain process. The internet connection of all partners' devices and sensors (things) increases the reliability of analytics, which is then used to improve the whole process. The internet of things adds to the analytics process by supplying information about the end customer, since it is the main channel to reach clients. Adding to the online technologies, cloud computing allows the supply chain to process all the big data information online before downloading the results. Cloud computing can also improve access to systems and software, which can be shared among companies of the supply chain. However, to prevent data loss or security breaches, cyber security measures should be applied by all companies individually, and also to the internet access of the supply chain. The adoption of cyber security technologies makes it possible for companies to combine and share information online, ensuring the use of the concepts of i4.0 by the whole supply chain.

Besides improving the productivity of the supply chain, the adoption of i4.0 technologies adds possibilities of new business models (Thoben et al. 2017). The integration and expansion of the supply chain, combination of products and services available to other companies or to the final customer, among others, are some of the possibilities available. The development of innovative projects among supply chain companies is also easier when partners are already digitally integrated. In the end, the use of internet to share and absorb data is the new trend of the internet-based society, and the adoption of technologies related to i4.0 is the first step supply chains should take to stay competitive.

Acknowledgements This work is financed by the FCT—Fundação para a Ciência e a Tecnologia (Portuguese Foundation for Science and Technology) within project CMUP-ERI/TPE/0011/2013 of the CMU Portugal Program and by the Project "TEC4Growth—Pervasive Intelligence, Enhancers and Proofs of Concept with Industrial Impact/NORTE-01-0145-FEDER-000020" financed by the North Portugal Regional Operational Programme (NORTE 2020), under the PORTUGAL 2020 Partnership Agreement, and through the European Regional Development Fund (ERDF).

References

Autry CW, Grawe SJ, Daugherty PJ, Richey RG (2010) The effects of technological turbulence and breadth on supply chain technology acceptance and adoption. J Oper Manag 28(6):522–536

Awais M, Henrich D (2013) Human-robot interaction in an unknown human intention scenario. In: 11th international conference on frontiers of information technology (FIT). IEEE, pp 89–94. Retrieved from https://doi.org/10.1109/FIT.2013.24

Baheti R, Gill H (2011) Cyber-physical systems. In: The impact of control technology, vol 12. pp 161–166. Retrieved from http://www.ieeecss.org/sites/ieeecss.org/files/documents/IoCT-Part3-02CyberphysicalSystems.pdf

Barros AC, Simões AC, Toscano C, Marques A, Rodrigues JC, Azevedo A (2017) Implementing cyber-physical systems in manufacturing. In: Paper presented at the 47th international conference on computers and industrial engineering, Lisbon, Portugal. Retrieved from http://www.cie47.com/

Berman B (2012) 3-D printing: the new industrial revolution. Bus Horiz 55(2):155–162

Bonkenburg T (2016) Robotics in logistics: a DPDHL perspective on implications and use cases for the logistics industry. DHL Customer Solutions & Innovation. Retrieved from http://www.dhl.com/content/dam/downloads/g0/about_us/logistics_insights/dhl_trendreport_robotics.pdf

Bonnín-Roca J, Vaishnav P, Mendonça J, Morgan G (2017) Getting past the hype about 3-D printing. MIT Sloan Manag Rev 58(3):57

Brettel M, Friederichsen N, Keller M, Rosenberg M (2014) How virtualization, decentralization and network building change the manufacturing landscape: an industry 4.0 perspective. Int J Mech Ind Sci Eng 8(1):37–44

Bughin J, Chui M, Manyika J (2010) Clouds, big data, and smart assets: ten tech-enabled business trends to watch. McKinsey Q 56(1):75–86

Daugherty P, Banerjee P, Negm W, Alter AE (2014) Driving unconventional growth through the industrial internet of things. Accenture Technology. Retrieved from http://www.mcrockcapital.com/uploads/1/0/9/6/10961847/accenture-driving-unconventional-growth-through-iiot.pdf.

Davies R (2015) Industry 4.0. Digitalisation for productivity and growth. Briefing from EPRS. European Parliamentary Research Service. Retrieved from http://www.europarl.europa.eu/RegData/etudes/BRIE/2015/568337/EPRS_BRI(2015)568337_EN.pdf

Gallivan MJ (2001) Organizational adoption and assimilation of complex technological innovations: development and application of a new framework. ACM Sigmis Database 32(3):51–85

Glas AH, Kleemann FC (2016) The impact of industry 4.0 on procurement and supply management: a conceptual and qualitative analysis. Int J Bus Manag Invention 5(6):55–66

Greenhalgh T, Robert G, Macfarlane F, Bate P, Kyriakidou O (2004) Diffusion of innovations in service organizations: systematic review and recommendations. Milbank Q 82(4):581–629

Grote G, Weyer J, Stanton NA (2014) Beyond human-centred automation–concepts for human–machine interaction in multi-layered networks. Ergonomics 57(3):289–294

Harris ID, Director AMC (2011) Development and implementation of metals additive manufacturing. DOT International, New Orleans. Retrieved from https://ewi.org/eto/wp-content/uploads/2013/06/Additive-Manufacturing-DOT-Paper-2011.pdf

Hazen BT, Boone CA, Ezell JD, Jones-Farmer LA (2014) Data quality for data science, predictive analytics, and big data in supply chain management: an introduction to the problem and suggestions for research and applications. Int J Prod Econ 154:72–80

Hermann M, Pentek T, Otto B (2016) Design principles for industrie 4.0 scenarios. In: System sciences (HICSS), 2016 49th Hawaii international conference on system sciences. IEEE, pp 3928–3937. Retrieved from https://doi.org/10.1109/HICSS.2016.488

Hofmann E, Rüsch M (2017) Industry 4.0 and the current status as well as future prospects on logistics. Comput Ind 89:23–34

Holmström J, Partanen J, Tuomi J, Walter M (2010) Rapid manufacturing in the spare parts supply chain: alternative approaches to capacity deployment. J Manuf Technol Manag 21(6):687–697

Hopkinson N, Dickens P (2001) Rapid prototyping for direct manufacture. Rapid Prototyp J 7 (4):197–202

Kagermann H, Wahlster W, Helbig J (2013) Recommendations for implementing the strategic initiative Industrie 4.0: final report of the Industrie 4.0 Working Group. Retrieved from http://www.acatech.de/de/publikationen/stellungnahmen/kooperationen/detail/artikel/recommendations-for-implementing-the-strategic-initiative-industrie-40-final-report-of-the-industr.html

Khajavi SH, Partanen J, Holmström J (2014) Additive manufacturing in the spare parts supply chain. Comput Ind 65(1):50–63

Koufaris M (2002) Applying the technology acceptance model and flow theory to online consumer behavior. Inf Syst Res 13(2):205–223

Langlois RN (2013) Business groups and the natural state. J Econ Behav Organ 88:14–26

Lasi H, Fettke P, Kemper HG, Feld T, Hoffmann M (2014) Industry 4.0. Bus Inf Syst Eng 6 (4):239–242

Laureijs RE, Roca JB, Narra SP, Montgomery C, Beuth JL, Fuchs ER (2017) Metal additive manufacturing: cost competitive beyond low volumes. J Manuf Sci Eng 139(8):081010–081011

Lee EA (2008) Cyber physical systems: design challenges. In: Proceedings of the 11th IEEE international symposium on object/component/service-oriented real-time distributed computing, pp 363–369. Retrieved from https://doi.org/10.1109/ISORC.2008.25

Lee J, Lapira E, Bagheri B, Kao HA (2013) Recent advances and trends in predictive manufacturing systems in big data environment. Manuf Lett 1(1):38–41

Lee J, Kao HA, Yang S (2014) Service innovation and smart analytics for industry 4.0 and big data environment. Proc CIRP 16:3–8

Lee J, Bagheri B, Kao HA (2015) A cyber-physical systems architecture for industry 4.0-based manufacturing systems. Manuf Lett 3:18–23

Leitão P, Colombo AW, Karnouskos S (2016) Industrial automation based on cyber-physical systems technologies: prototype implementations and challenges. Comput Ind 81:11–25

Liao SH (2005) Technology management methodologies and applications: a literature review from 1995 to 2003. Technovation 25(4):381–393

Lu Y (2017) Industry 4.0: a survey on technologies, applications and open research issues. J Ind Inf Integr 1(6):1–10

MacDougall W (2014) Industrie 4.0: smart manufacturing for the future. Germany Trade & Invest. Retrieved from https://www.gtai.de/GTAI/Content/EN/Invest/_SharedDocs/Downloads/GTAI/Brochures/Industries/industrie4.0-smart-manufacturing-for-the-future-en.pdf

Marston S, Li Z, Bandyopadhyay S, Zhang J, Ghalsasi A (2011) Cloud computing: the business perspective. Decis Support Syst 51(1):176–189

Megahed FM, Jones-Farmer LA (2013) A statistical process monitoring perspective on "big data". In: Knoth S, Schmid W (eds) Frontiers in statistical quality control. Springer, New York

Messina D, Santos C, Soares AL, Barros AC (2016) Risk and visibility in supply chains: an information management perspective. In: Jamil GL, Soares AL, Pessoa CRM (eds) Handbook of research on information management for effective logistics and supply chains. IGI Global, Hershey

Monostori L (2014) Cyber-physical production systems: roots, expectations and R&D challenges. Proc CIRP 17:9–13

Oettmeier K, Hofmann E (2017) Additive manufacturing technology adoption: an empirical analysis of general and supply chain-related determinants. J Bus Econ 87(1):97–124

Paelke V (2014) Augmented reality in the smart factory: supporting workers in an industry 4.0. environment. Emerging technology and factory automation (ETFA). IEEE, pp 1–4. Retrieved from https://doi.org/10.1109/ETFA.2014.7005252

Patterson KA, Grimm CM, Corsi TM (2003) Adopting new technologies for supply chain management. Transp Res Part E Logist Transp Rev 39(2):95–121

Pedersen MR, Nalpantidis L, Andersen RS, Schou C, Bøgh S, Krüger V, Madsen O (2016) Robot skills for manufacturing: from concept to industrial deployment. Robot Comput Integr Manuf 37:282–291

Posada J, Toro C, Barandiaran I, Oyarzun D, Stricker D, de Amicis R, Pinto EB, Eisert P, Döllner J, Vallarino I (2015) Visual computing as a key enabling technology for industrie 4.0 and industrial internet. IEEE Comput Graph Appl 35(2):26–40

Rogers EM (2003) Diffusion of innovations, 5th edn. Free Press, New York

Rüßmann M, Lorenz M, Gerbert P, Waldner M, Justus J, Engel P, Harnisch M (2015) Industry 4.0: the future of productivity and growth in manufacturing industries. Boston Consulting Group. Retrieved from https://www.bcgperspectives.com/content/articles/engineered_products_project_business_industry_40_future_productivity_growth_manufacturing_industries/

Saldanha JP, Mello JE, Knemeyer AM, Vijayaraghavan TAS (2015) Implementing supply chain technologies in emerging markets: an institutional theory perspective. J Supply Chain Manag 51 (1):5–26

Scott A, Harrison TP (2015) Additive manufacturing in an end-to-end supply chain setting. 3D Print Addit Manuf 2(2):65–77

Sharma R (2012) Study of latest emerging trends on cyber security and its challenges to society. Int J Sci Eng Res 3(6):1–4

Shellshear E, Berlin R, Carlson JS (2015) Maximizing smart factory systems by incrementally updating point clouds. IEEE Comput Graph Appl 35(2):62–69

Shrouf F, Ordieres J, Miragliotta G (2014) Smart factories in Industry 4.0: a review of the concept and of energy management approached in production based on the internet of things paradigm. In: Industrial engineering and engineering management (IEEM), 2014 I.E. international conference on industrial engineering and engineering management. IEEE, pp 697–701. Retrieved from https://doi.org/10.1109/IEEM.2014.7058728

Sniderman B, Mahto M, Cotteleer MJ (2016) Industry 4.0 and manufacturing ecosystems: exploring the world of connected enterprises. Deloitte Consulting. Retrieved from https://dupress.deloitte.com/content/dam/dup-us-en/articles/manufacturing-ecosystems-exploring-world-connected-enterprises/DUP_2898_Industry4.0ManufacturingEcosystems.pdf

Stevens GC, Johnson M (2016) Integrating the supply chain... 25 years on. Int J Phys Distrib Logist Manag 46(1):19–42

Suri K, Cuccuru A, Cadavid J, Gerard S, Gaaloul W, Tata S (2017) Model-based development of modular complex systems for accomplishing system integration for industry 4.0. In: Proceedings of the 5th international conference on model-driven engineering and software development, vol 1. pp 487–495

Thoben KD, Wiesner S, Wuest T (2017) "Industrie 4.0" and smart manufacturing – a review of research issues and application examples. Int J Autom Technol 11(1)

Von Solms R, Van Niekerk J (2013) From information security to cyber security. Comput Secur 38:97–102

Waller MA, Fawcett SE (2013) Data science, predictive analytics, and big data: a revolution that will transform supply chain design and management. J Bus Logist 34(2):77–84

Wang L, Törngren M, Onori M (2015) Current status and advancement of cyber-physical systems in manufacturing. J Manuf Syst 37(Part 2):517–527

Wang S, Wan J, Li D, Zhang C (2016) Implementing smart factory of industrie 4.0: an outlook. Int J Distrib Sens Netw 12(1):1–10

Waslo R, Lewis T, Carton R (2017) Industry 4.0 and cybersecurity: managing risk in an age of connected production. Deloitte University Press. Retrieved from https://dupress.deloitte.com/dup-us-en/focus/industry-4-0/cybersecurity-managing-risk-in-age-of-connected-production.html

Witkowski K (2017) Internet of things, big data, industry 4.0: innovative solutions in logistics and supply chains management. Proc Eng 182:763–769

Zhang Q, Cheng L, Boutaba R (2010) Cloud computing: state-of-the-art and research challenges. J Internet Serv Appl 1(1):7–18

Gustavo Dalmarco is a postdoctoral researcher at INESCTEC—Portugal, associate professor at the Business School of PUCRS–Brazil, and consultant on technology transfer and innovation projects at Innovaspace—England. His main research interests are in academic entrepreneurship, innovation, technology transfer, and university–industry relations topics. He has a PhD in business at UFRGS—Brazil with focus in university–industry relations. He has experience in organizing entrepreneurship tournaments and business plan mentoring for start-ups; worked as business developer for PUCRS (2013–2017), establishing research partnerships; and was a visiting researcher at the Technology Transfer Program Office of the European Space Agency—Netherlands.

Ana Cristina Barros is a senior researcher at INESC TEC. Her research focuses on supply chain strategy, innovation networks, and technology management. She has been visiting researcher at Carnegie Mellon University (2014/2015; 2012), Massachusetts Institute of Technology (2008–2011), and Cornell University (2002). She teaches technology management in the Porto Business School and supervises master's and PhD students of the University of Porto. Ana Barros obtained a PhD in engineering and management (Technical University of Lisbon, 2011), an MBA in logistics and entrepreneurship (Technical University of Munich, 2004), and an M.S. and B.S. in chemical engineering (University of Porto, 2000). She has several years of work experience in the procurement and production planning departments of German and Portuguese companies.

Advanced Supply Chains: Visibility, Blockchain and Human Behaviour

Alexander Kharlamov and Glenn Parry

Any sufficiently advanced technology is indistinguishable from magic.

Arthur C. Clarke

Abstract Technological advances over the last decade saw the rise of ICT and IoT, paving the way for the Supply Chain of Things. Blockchain technology was one of the most recent and potentially most significant developments. Blockchain technology are secure by design and can enable decentralization and visibility, with application in cryptocurrency transactions, historical records, identity management, traceability, authentication, and many others. However, successful adoption of such technology requires that the people, process and technology are ready. We propose a conceptual framework where the concept and technology can balance between positive and negative manifestations depending on human behavior, therefore determining the success of Blockchain technology application in supply chains. While both the concept and technology are relatively ready, human behavior is a challenge as it is known that people suffer from habits and perform poorly when exposed to large volumes of data. Therefore, the development of advanced supply chains with much greater visibility enabled by Blockchain technology must take into consideration people in order to succeed.

A. Kharlamov (✉) · G. Parry
Business Management, Bristol Business School, Faculty of Business and Law, University of the West of England, Bristol, UK
e-mail: alex.kharlamov@uwe.ac.uk; glenn.parry@uwe.ac.uk

A. C. Moreira et al. (eds.), *Innovation and Supply Chain Management*, Contributions to Management Science, https://doi.org/10.1007/978-3-319-74304-2_15

1 Introduction

Recent advances in digital technology and Internet of Things (IoT) have opened new opportunities for industry, manufacturing, and service provision, reshaping entire supply chains and the modus operandi. Through a multiplicity of technologies, connected devices, otherwise called "smart" devices or "things", allowed providers and users of goods, services, and technology to significantly increase efficiency in many tasks and operations that took much effort and time in the pre-IoT days.

Back in 1990s hardly anyone could imagine having a personal computer. Today, the overwhelming majority of the human population in the developed world carries a computer in their pocket. By doing so, we all deal with information and communications technology (ICT) on a daily basis. ICT usually refers to a system of unified communications that include telecommunications; computers; software;, storage; and audio-visual solutions. This system enables users to access, store, transmit, and manipulate information. Technology helps us to *socialize*, *learn*, *organize*, *administer*, *shop*, and *entertain* ourselves.

We use technology to *socialize* by e-mailing and sharing with others on social media. We *learn* through browsing, keeping up-to-date with the news and pursue our individual interests and hobbies through reading and writing blogs. Technology helps us to *organize* our lives through digital calendars, journey planners and map services. We also perform *administrative* tasks by using online banking as well as paying bills and submitting tax returns using digital technology. Online *shopping* becomes more and more popular. We also use technology to entertain ourselves by making use of online video portals, listening to the music and playing games. Are these behaviors affected by ICT? In order to answer this question, we need to separate the effect of ICT through assisting us with various activities and its impact through data that we generate.

If we look at ICT and activities—the answer is yes! Only a few years ago, in order to schedule an event, you would need to pick up the phone and make several rounds of phone calls to invite your colleagues to a meeting. Today, ICT allows you to do this quickly and efficiently by sharing your electronic calendar with others or by using a dedicated online scheduling system. Your fitness and diet app helps you make healthier choices in the grocery store by ditching cakes and ice cream for fruit and vegetables. However, ICT may also influence your behavior in a negative way. For example, many people develop dangerous addictions to gaming, social media or taking selfies.

However, if we look at data, the answer is not so clear. Many of us have distinct data habits, traces of which could be found in our smartphones or in social media. However, do we actually know how much data we generate? In addition, how often do we look back at the data we have generated to give these data a second thought and use them to change our behavior? It seems that our data is primarily used by various companies, but very rarely by us. These companies collect and use our data to develop new products and services, suggest new products and services to us, and

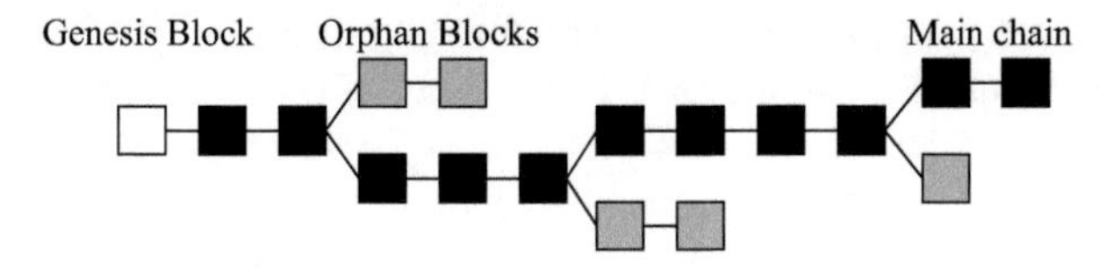

Fig. 1 Illustration of a Blockchain

even trade our data with third parties. Going further, personal data hubs[1] and markets are being developed that are likely to be the next big innovation affecting supply chains.

Several market players have gained significant advantage by adopting ICT innovations early. For example, in the Business-to-Consumer model, Amazon.com acts as a digital market platform for buyers and sellers to meet and carry out transactions. Through the adoption of digital technology, Amazon not only manages to offer a large range of products, but also uses this technology to provide customized and personalized experience to every user. In the Business-to-Business model, IBM and Cisco champion the creation of smart communication systems, through which businesses connect and manage daily operations more optimally.

With the emergence of players such as Amazon.com, IBM, and Cisco, amongst others, the rise of digital platforms reached a completely different level, signaling the need for the emergence of new ways of dealing with transactions.

2 Blockchain in the New Digital Economy

2.1 What Is Blockchain?

One of the recent developments of the technological revolution is Blockchain technology. In essence, a Blockchain is a list of records, usually financial transactions. More precisely, a Blockchain is a distributed ledger which captures an accumulating a list of records stored in a particular (usually chronological) order. These records are compiled into small groups called "blocks", that are similar to pages in a ledger. The blocks are listed in linear time order in what is called a "chain" (see Fig. 1). Using cryptographic technology each block is analysed and a code of fixed length generated, named a hash. The hash from the previous block is placed in the following block, locking them together in the chain. By locking the records together a continuous record or "roadmap" of a process is created (examples of such a process include monetary transactions, delivery transactions, etc.). It is difficult to change a block as it requires the unpicking of the chain and its reassembly at a rate faster than that of new block creation.

---

[1] See https://hubofallthings.com/

Blockchain grew out of research on cryptography from the 1990s (e.g., Bayer et al. 1993; Haber and Stornetta 1990). The concept of Blockchain was coined and the system described by Satoshi Nakamoto in 2008 (Nakamoto 2008). Nakamoto proposed the use of a peer-to-peer dynamic record of timestamped blocks which are stored on a distributed server and cannot be deleted as a way to simplify multi-actor transactions and create cryptocurrencies that do not require a central trusted entity like a bank.

Nakamoto's work was applied with the Blockchain cryptocurrency named Bitcoin. The original purpose of Bitcoin was to address a problem of double spending, often faced by digital artefacts. Digital tokens can be easily copied and in the case of money, spent multiple times. This is fraudulent, and causes inflation by creating amounts of money that did not previously exist. While traditional ways of dealing with these issues were blind signatures and secret splitting (solutions often provided by intermediaries such as large financial institutions or banks and their affiliated security companies), Bitcoin offered a new way of solving the problem of double spending without using a trusted intermediary (e.g., secure financial institution). By creating a distributed ledger of transaction records, Bitcoin allowed a network to trace monetary transactions "on the fly" through an open ledger of records which made an intermediary redundant. Bitcoin became an example of a very successful Blockchain experiment growing from 22,247 megabytes in October 2014 to 130,624 megabytes in July 2017.[2] It is estimated that the market capitalization of Bitcoin on August 31st 2017 reached $77,259,270,173.[3]

2.2 *Why Blockchain?*

Why is Bitcoin so popular and why does Blockchain seems to be such an attractive concept? The main benefit of the Blockchain idea is that it allows significant simplification of interaction and reduces the noise in communication between agents, thus allowing them to communicate directly without the need for trusted intermediary.

It is easy to illustrate Blockchain benefits through the use of the so called Two Generals Problem (Lamport et al. 1982). Imagine that there are two generals each in charge of an army planning to jointly attack a city (Fig. 2). They know that the attack will only succeed if carried out by both armies simultaneously. The target city is located in a valley surrounded by steep hills and the two armies are also located in valleys surrounded by hills. The two generals need to coordinate with each other in order to attack the city simultaneously. The problem is that while they have agreed to attack the city together, they failed to agree on the time of the attack and need to communicate with each other to start the attack at the same time. In order to send

[2]See https://blockchain.info/charts/blocks-size?timespan=3years

[3]See https://coinmarketcap.com/ for more recent figures.

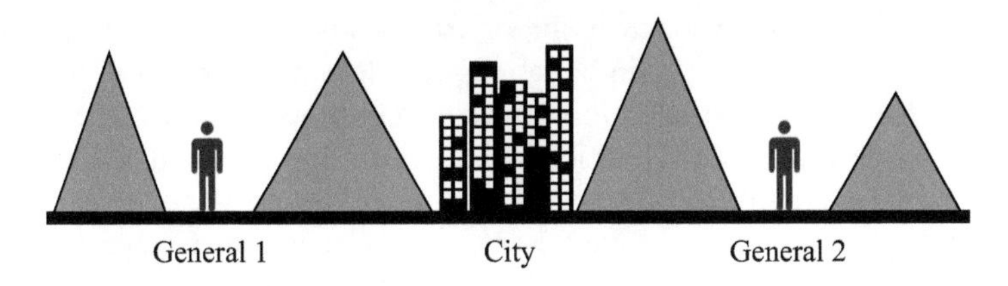

Fig. 2 Two general problem: setup

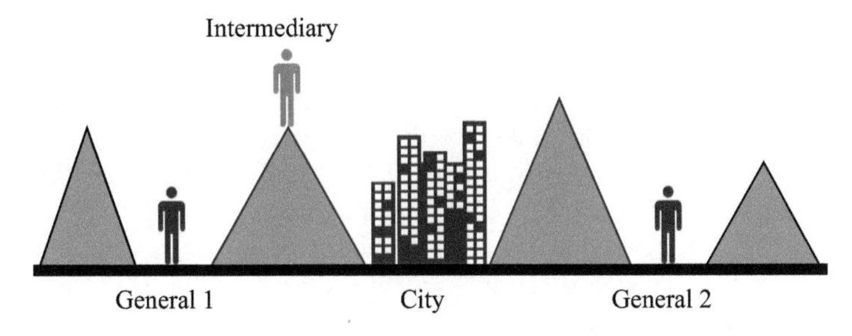

Fig. 3 Two general problem: intermediary

messages to each other, they have to send messengers and these messengers have to pass the city to reach the ally army. Yet, there is a danger that the messenger could be captured and that the message could be altered or falsified. Additionally, even if one general successfully send the message to the other, it is not enough to ensure joint action. The other general needs to acknowledge the receipt of the message by sending a reply confirming the time of the attack. Each time of the attack has to be not only communicated, but also confirmed, and each time a messenger passes through the city the message could be compromised. In theory, the generals will never attack a city, as they will never be sure that their ally will be coordinated with them.

Now, imagine that there is an intermediary who guarantees to both generals that she can safely deliver messages across the city (Fig. 3). This intermediary will charge a fee for her services, but will guarantee the trustworthiness of the service. Would this solve the problem? At first sight the answer is "yes"—the messenger will securely deliver the message about the time of the attack as well as the reply and the attack can go ahead. Yet, on second thought, the answer depends on the price that the intermediary is planning to charge for her service. In the extreme, the price could be so high that it would nullify the whole purpose of the attack e.g., an intermediary could demand the city in exchange for her services.

Under these circumstances, the technology of Blockchain might be very useful. Imagine now that the two generals can use a new technology to communicate directly by keeping an electronic ledger and this ledger cannot be altered. This would allow them not only to agree on the time of the attack but also do so (1) safely—the messages cannot be tampered with; (2) quickly—using just one step of communication as once one of the generals sends a message to the other, the information about this message will be stored in the ledger so both generals will see it; (3) cost-effectively—generals will not have to engage with a costly intermediary.

Similarly to the Two Generals' Problem, businesses need to coordinate to engage in transactions, be it delivery of physical goods, provision of services, or money transfers. At each step, actions of businesses need to be authorized and verified for the transaction to take place. For example, imagine that a large retailer SuperMartCo wants to purchase trolleys for several of its stores. It then contacts Trolleyzl who supply trolleys. They sign a contract and SuperMartCo orders 1000 trolleys for five stores each worth £30. Trolleyzl receives the order and asks SuperMartCo for an advance pre-payment e.g., 25% of the order or £7500. SuperMartCo then authorizes a payment of advance, transfers money to Trolleyzl's bank account, and informs Trolleyzl of the transfer. Trolleyzl then verifies the receipt of the advance payment, produces trolleys and informs SuperMartCo that the trolleys were produced. Trolleyzl also asks SuperMartCo to confirm which five stores should the order be delivered to. SuperMartCo communicates the addresses of the stores to Trolleyzl. Once the trolleys are produced, Trolleyzl delivers them to the five SuperMartCo stores and informs SuperMartCo that the deliver was made. Each of the five SuperMartCo stores then confirm to SuperMartCo that the delivery was made and that all delivered trolley stock is produced in accordance with SuperMartCo requirements, after which point SuperMartCo authorizes the release of the remaining 75% payment balance to Trolleyzl.

The SuperMartCo example illustrates that even delivering such a simple order involves multiple messaging steps. At each step the communication can go wrong and delay either the delivery of trolleys e.g., Trolleyzl fails to confirm the product requirements with SuperMartCo, or payment for the fulfilled order e.g., SuperMartCo stores fail to confirm delivery of trolleys and, as a result, Trolleyzl will have to chase SuperMartCo for the final payment. In other words, the communication between companies can be incredibly noisy and relies of mutual trust, trust that at each point of the transaction the process will work as it should. To alleviate this noise, companies often engage an intermediary e.g., the bank, whose job is to decrease noise by communicating with both parties, to chase both parties for updates and to automatize some of the transaction steps. For example, both SuperMartCo and Trolleyzl may engage a trusted intermediary (bank) who will automatically pay Trolleyzl for the order in two installments upon receipt of the relevant authorizations from SuperMartCo.

The main advantage of Blockchain is that, in theory, it allows decentralization of control in the supply chain system and by-passing of intermediaries, allowing parties to conduct transactions directly (Fig. 4). Consider a multisided market shown below, without the Blockchain technology, an intermediary (e.g., the bank) will keep the ledger between various actors on the market and coordinate transactions receiving payment for their services. With Blockchain technology, actors are able to communicate directly: they all will access a secure electronic ledger where each transaction is verified automatically multiple times and these verifications cannot be altered or deleted.

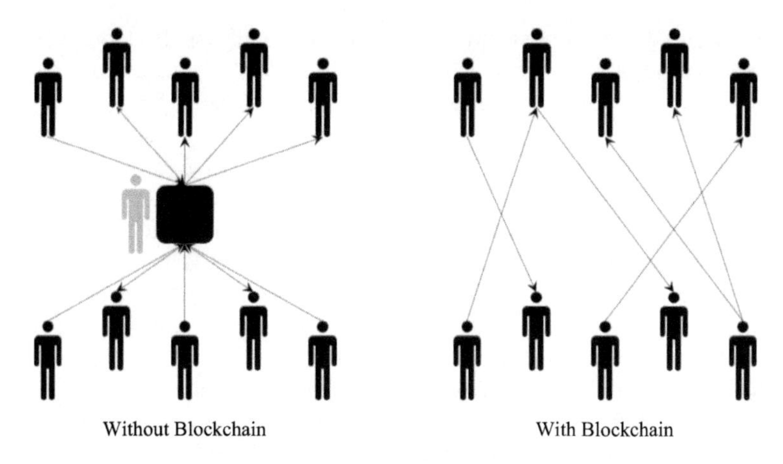

Fig. 4 Blockchain technology versus a third party

Being able to conduct business transactions without an intermediary is beneficial, and not only because it allows businesses to eliminate transaction costs. While, in theory, intermediaries should reduce noise, in practice, they may increase noise. For example, since an intermediary needs to communicate with agents at both ends of a transaction, errors, miscommunication, and delays may occur not only between these agents but also between agents and the intermediary. The Blockchain technology allows businesses to avoid these problems and can be applied to other purposes, for example, traceability and visibility.

2.3 *What Is Blockchain Technology Used for?*

Blockchain technology can have many applications with significant implications on the supply chains and supply chain management. The best known application of Blockchain technology is for Bitcoin. The Bitcoin "pilot" paved the way for new Blockchain (or Blockchain 2.0) which now are now the new "trending" market technology. The concept of Blockchain 2.0 incorporates new innovations such as *Smart Contracts*. The development of the Blockchain allowed the creation of *Smart Contracts* Platform—*Ethereum*. This platform allowed transacting agents to incorporate computer programs directly into Blockchain. These programs, in turn, allowed agents to build in various financial instruments (e.g., bonds, loans, etc.) directly into the digital ledger further simplifying transactions. *Ethereum* market capitalization as of September 2017 was over $28 billion[4] and it is likely to continue growing.

[4]See https://coinmarketcap.com/currencies/ethereum/ for the latest figures.

Bitcoin is not the only cryptocurrency enabled by Blockchain technology; in fact, there are over one thousand cryptocurrencies.[5] In theory, anyone can create a new cryptocurrency with different characteristics or purposes. This opens up a completely new set of opportunities and challenges for supply chains. For example, Zcoin[6] implemented the Zerocoin[7] protocol which is a protocol extension fixing the current issue of Bitcoin, the lack of privacy, therefore allowing true anonymity. There are also industry-specific cryptocurrencies. For example, Potcoin[8] was created for the Cannabis enthusiasts market. Sexcoin[9] is another industry-specific cryptocurrency created for the adult entertainment industry. Going beyond industries, a large supply chain can create its own cryptocurrency and enforce its use end to end, from consumers to raw material suppliers.

While we see a rapid development of Blockchain on financial markets with major ICT players like IBM building Blockchain for large banks and financial institutions,[10] there is a growing realization that Blockchain could potentially be useful virtually to any company in any industry.

For example, in order to reduce fraud, risk, as well as to assist insurers and open markets, Blockchain technology has found application in the gems industry (specifically, in controlling the origins and supply of diamonds). In 2015, a start-up *Everledger*[11] was created to build a global digital ledger for tracking and protecting diamonds across the whole supply chain. Everledger proposes that it could make diamond transactions simpler and more secure, and significantly increase the transparency of the entire diamond trade. Specifically, if it gains scale it could make it much harder to exchange blood diamonds due to the visibility of gems' origins that is built into the system.

The Art market, similarly to the diamond industry, often deals with high-value assets and is exposed to threats, e.g., forgery. In fact, 2016 has been named "The Year of the Fake".[12] One particularly relevant situation involved the artist Lee Ufan whose work was counterfeited and when required to identify the fakes, the artist claimed several of the fakes as his own original work.[13] The case started a momentum with South Korean government introducing a new law in August 2017[14] that makes it compulsory to have licenses to sell art and to maintain proper records. To

[5]See https://cryptopedia.wiki/Main_Page

[6]https://zcoin.io

[7]http://zerocoin.org/

[8]http://www.potcoin.com/

[9]https://www.sexcoin.info/

[10]See for example https://www.cnbc.com/2017/06/26/ibm-building-blockchain-for-seven-major-banks-trade-finance.html

[11]See https://www.everledger.io/

[12]See https://news.artnet.com/art-world/biggest-art-forgeries-2016-783464

[13]See https://news.artnet.com/art-world/lee-ufan-verifies-denies-forgeries-540461

[14]See http://koreabizwire.com/new-law-to-root-out-counterfeit-artwork-in-korea/67555

solve this kind of challenges companies such as ArteQueswta,[15] Verisart[16] and Ascribe[17] apply Blockchain technology that allows works to be verified online and in real-time, providing provenance to art works that underpins their value. Fully implemented, Blockchain technology can potentially surpass other systems and human experts.

Governments are also investigating the use of blockchain with different applications, such as land ownership registration, personal identity and elector registration. For example, the UK Government is considering the application of Blockchain technology to fight government corruption.[18] The Republic of Georgia is one of the first countries to implement online land registry on blockchain.[19] Identity is a key area of development for blockchain.[20] Online voting enabled by Blockchain technology has been gaining ground[21] in an effort to reduce election's fraud. There are numerous solutions linked to finance such as Banqu.com, Veridu.com, and Civic.com, whilst others have developed purely identification systems such as Bitnation's online global citizenship passport.[22]

3 Blockchain and Supply Chain of Things

In the modern digitized economy where the majority of "things" are connected and our use of ICT is extensive, we can talk about Supply Chain of Things rather than just Supply Chain (Geerts and O'Leary 2014). The technology of Blockchain fits perfectly into Supply Chain helping to transform it into the Supply Chain of Things. Specifically, we see more and more evidence of Supply Chain components being "smart" (equipped with sensor technology) allowing businesses to track transactions in real time. For example, "smart" containers ship the latest clothes collections from China to Paris, "smart" beer barrels allow bars to make decisions about when to replenish their lager supplies; there is "smart" rubbish that makes it easier to inform waste disposal and recycling services. This creates the closed loop for consumption, linking forward and reverse supply chains via data (Parry et al. 2016).

The Blockchain technology through its distributed mechanisms offers many opportunities for the further development of Supply Chain of Things. Yet, as any "change" concept, Blockchain is prone to all threats that such a change may bring.

[15]See http://www.artequesta.com/

[16]See https://www.verisart.com/

[17]See https://www.ascribe.io/

[18]See http://uk.businessinsider.com

[19]See https://www.forbes.com/sites/laurashin/2017/02/07/the-first-government-to-secure-land-titles-on-the-bitcoin-blockchain-expands-project/#296654334dcd

[20]See https://hbr.org/2017/03/blockchain-will-help-us-prove-our-identities-in-a-digital-world

[21]See https://followmyvote.com/

[22]See https://bitnation.co/world-citizenship-id/

Fig. 5 People process technology

Specifically, the change literature tells us that in order for any organizational change to succeed (Cooper 1964), one needs to make sure that people, process, and technology within the organization are prepared such that change is a part of organizational culture (Fig. 5).

We propose a Blockchain success framework for supply chain (see Fig. 6) based on the People-Process-Technology model from organization research literature (originally proposed in Cooper 1964; and extended in Prodan et al. 2015). We identify three groups of Blockchain success determinants, where each determinant can have either a positive or a negative manifestation. There are determinants relating to Concept [C], Technology [T], and Behavior [B]. Human behavior that in many ways identifies organizational behavior and determines actions of businesses across the entire supply chain is a key component capable of either ensuring the Blockchain success or setting it up for failure. Concept [C] and Technology [T] determinants can be affected by behavioral components [B] as discussed below. Assuming that behavioral determinant [B] are favorable, Blockchain will succeed if the majority of Concept [C] and Technology [T] determinants have positive manifestations.

3.1 Concept Determinants

Concept determinants refer to factors that, in theory, should contribute to the Blockchain success, yet, in practice; they may have a reverse effect hindering processes in supply chains.

Blockchain should lead to a *Simplification* of supply chains because they should offer transparency and dynamic verification of transactions allowing companies to skip many stages of communication. Yet, in practice, human behavior may lead Blockchain to further *Complication* resulting in delay-prone, noisy systems. Consider, for example, that despite having a general ledger, agents in the supply chain continue to keep individual ledgers and, as a result, keep crosschecking their

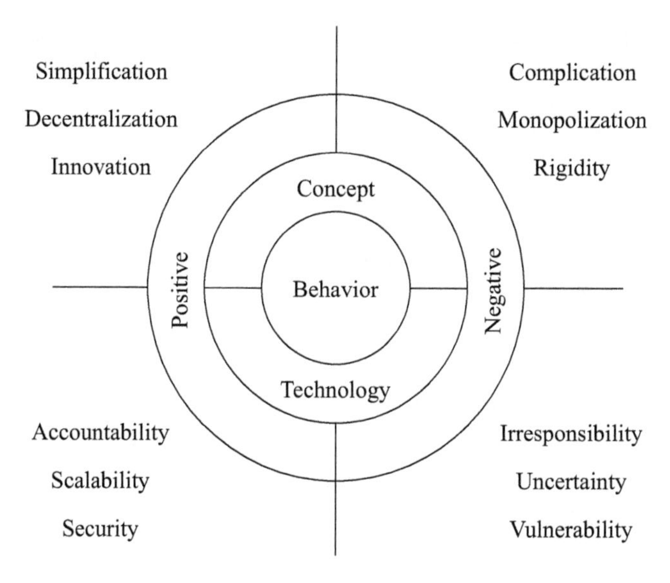

Fig. 6 Conceptual framework of Blockchain technology success

individual ledgers with the general ledger. Under these circumstances, errors and delays are likely to creep into the processes making supply chains less efficient.

Blockchain should lead to *Decentralization* of control in the supply chain, yet, human behavior may instead result in *Monopolization*. Consider the Internet system that emerged as a completely free and decentralized environment and, in a matter of a decade turned into a realm of large players (e.g., Google, IBM, Microsoft.) Signs of such monopolization are already observable. For example, the current Blockchain are guaranteed by the so-called "Proof-of-Work" systems. These systems entitle the group of agents with the largest computing power (often called "miners") to make the majority of important decisions. "Miners" emerged as Blockchain superpowers because they are able to act as secure data centers in exchange for cryptocurrency payments. As a rule, "miners" are large financial institutions. For example, People's Bank of China is exploring digital currencies and the possibility of having its own cryptocurrency enabled by Blockchain technology.[23] Central Bank group is managing cryptocurrency conversion in India.[24] It is expected that "Proof-of-Work" systems will be eventually replaced by "Proof-of-Stake" systems that will make such monopolies redundant replacing them with complex cryptocurrency instruments. Yet, the conversion towards "Proof-of-Stake" systems is slow and has yet to prove its worth.

[23]See https://cointelegraph.com/news/ripple-talks-with-peoples-bank-of-china-key-to-chinese-blockchain-market

[24]See https://www.coindesk.com/central-bank-backed-group-plans-blockchain-platform-launch-india/

While Blockchain intend to promote inherent *Innovation*, human behavior may create *Rigidity* in the system. For example, the creation of Blockchain initially led to optimism about the potential of additive manufacturing opening doors for any individual to become an entrepreneur. Yet, with cryptocurrency being a subject to speculative market swings, it became next to impossible for new entrepreneurs to engage in cryptocurrency exchange. For example, in September 2017, Bitcoin reached a record high trading $5000 to 1 Bitcoin. Clearly, small entrepreneurs can hardly afford many bitcoins, not to mention engaging in any exchange transactions with cryptocurrencies.

3.2 Technology Determinants

There are also three technological determinants important for the Blockchain success in the Supply Chains of the future:

It is said that Blockchain should bring better *Security* to the entire system. Yet, as practice shows, human nature invents more sophisticated tools for bridging or leveraging on the system enriching mechanisms of cybercrime.[25] Therefore, instead of better *Security*, Blockchain may lead to more *Vulnerability* of the supply chain system. There is no doubt that cybercriminals of the future will become more and more sophisticated and inventive in exploring the vulnerabilities of Blockchain.

Scalability is another advantage of Blockchain. Currently, every computer engaged in the Blockchain network is engaged in processing every transaction that is going through this network. Obviously, this makes the processing slow. It is expected that in the future new distributed technology will be in place that will automatically calculate the optimal computer power necessary to verify each transaction and engage exactly the right number of computers at each stage. Making such an optimization will open Blockchain to new types of *Uncertainty* because it would be difficult to match the speed of VISA, MASTERCARD, or SWIFT systems without compromising security.

4 Blockchain and Human Behavior

As discussed previously, human behavior that affects organizational behavior is a key component determining Blockchain success in supply chains. Blockchain technology comes with a promise of greater efficiency in large supply chains.[26]

[25]See for example https://ftalphaville.ft.com/2017/06/01/2189634/its-not-just-a-ponzi-its-a-smart-ponzi/

[26]See: https://hbr.org/2017/01/the-truth-about-blockchain

Table 1 Persistence biases

Decision bias	Description	References
Habit	Tendency to conserve the same choices out of habit	Hogarth (1987); Sage (1981); Slovic (1975)
Persistence	Tendency to conserve preference towards recent choices	Boldrin et al. (2001); Kaplan and Schoar (2005)
Status quo	Tendency to retain the status quo and perceive the disadvantages of change higher over advantages	Fernandez and Rodrik (1991); Kahneman et al. (1991); Samuelson and Zeckhauser (1988)

However, the advantages of Blockchain can be its disadvantages when we take into consideration the human being(s) whose job is managing operations.

The main danger is that humans are prone to habits that are very rigid and make take years to change. Consider, for example, how economic progress in Eastern Germany was hindered for many years after unification of Germany, mainly because the human psychology of Eastern Germans was very different from that of Western Germans.[27] Another example is the introduction of the ERP system SAP. Managers often adopted the new technology by using it in the same way as the old one. In essence, SAP is often used as a repository of data and that is downloaded, processed in a separate spreadsheet and (sometimes) uploaded back to update the system. Such use completely defeats the original purpose of the SAP and potentially creates a bigger problem.

4.1 Habits

Habits are one of the greatest determinants of human behavior. In a more general form, habits are part of the persistence bias group. Although we expect that humans will easily adopt decentralized systems, we see more and more evidence of them artificially creating new intermediaries, tending to make systems more centralized.

Persistence biases listed in Table 1 are those linked with the natural tendency to make judgements and choose conserving past choices, relying on habits and aversion to change. Namely, habit bias (Hogarth 1987; Sage 1981; Slovic 1975) and persistence bias (Boldrin et al. 2001; Kaplan and Schoar 2005) regard the conservation of choices, but are different because the first has a longer memory than the second. Finally, status quo bias (Fernandez and Rodrik 1991; Kahneman et al. 1991; Samuelson and Zeckhauser 1988) regards to the overall tendency to conserve what is already in place while judging the disadvantages of change higher than the

[27] See: https://www.theguardian.com/world/2015/oct/02/german-reunification-25-years-on-how-different-are-east-and-west-really

advantages of change for equivalent situations, therefore it is necessary to offer clearly more advantages to start change rather than just offering close equivalents.

Blockchain also implies that humans should adopt automation, yet, we see daily evidence of a lack of trust toward automation. Consider this simple example: when calling a bank, an energy supplier, an airline, or a hospital, how many times do you actually think that talking to an automated voice system is sufficient? You probably try to ply your way through the entire digital menu to talk to the human operator, don't you? Under these circumstances, how can we expect that in supply chains humans will behave differently?

4.2 Reference Point Biases

Blockchain technology can enable visibility and transparency, exposing the user to much more data and information. Perfect visibility over a SC can backfire. In an experiment, Sterman and Dogan (2015) demonstrate that even with perfect visibility of the demand signal (which was known to be constant) as well as perfect knowledge of the orders at each instance of the supply chain, managers cannot resist the urge to hoard, ending up destabilizing the whole supply chain. On a larger scale, the behavioral effect is often cumulative and can throw an efficient supply chain out of balance (Croson and Donohue 2006; Mason-Jones and Towill 2000). Such behavior causes amplification of small variations (over-reactions and distorted information) e.g., the Forrester effect (Forrester 1958) also known as the bullwhip effect (Lee et al. 1997). This results in extra costs, e.g., inventory, markdowns, stock-outs or obsolescence (e.g., Niranjan et al. 2009). Additionally, observing demand and supply changes "on the fly" can create many dangers for planning processes and outcomes. For example, demand planners tend to make too many changes (they overreact to changes in demand) and, as a result, overstock (e.g., Kharlamov 2016). With the development of Blockchain, when information is delivered to planners not daily, but on a real-time basis, it is possible that supply chain systems will suffer from a more significant fault of planning.

In order to better understand the dangers of exposing managers to detailed and constantly up to date data, it is essential to consider the broader category of reference point biases. The main problem is that humans are more sensitive to changes rather than states (Kahneman 2012). Reference point category of biases listed in Table 2 gathers biases focused on the tendency to make judgements from a relative point of reference rather than absolute.

Reference point bias (Barkan et al. 2005; Boyle et al. 1998; Kahneman and Tversky 1979; McFadden 1999) is of major importance as it relates to the tendency to find an often arbitrary reference point, also known as anchor, and judge situations from that reference resulting in a highly subjective judgement influenced by the original starting point. This is also described as anchoring and adjustment or focalism bias (Epley and Gilovich 2001; Strack and Mussweiler 1997; Tversky and Kahneman 1974, 1992) which more specifically describes how people tend to

Table 2 Reference point biases

Decision bias	Description	References
Anchoring and Adjustment or focalism	Is the tendency to rely too heavily on one piece of information when making decisions, anchoring and adjusting OR Decision makers tend to take a reference point for evaluation and adjust accordingly to it	Epley and Gilovich (2001); Strack and Mussweiler (1997); Tversky and Kahneman (1974, 1992)
Bayesian likelihood	Tendency to conserve estimates of conditional probabilities	DuCharme (1970); Edwards (1968); Phillips and Edwards (1966); Phillips et al. (1966); Tversky and Kahneman (1974)
Certainty effect	Tendency to underestimate outcomes that are almost certain but not certain (usually from 95 to 99%)	Donkers et al. (2001); Kahneman (2012); McCord and de Neufville (1986); Weber and Milliman (1997)
Conservatism	Tendency not to revise or adjust estimates on the receipt of new significant information	Hogarth (1987); Nelson (1996)
Confidence	Tendency to make too extreme judgements depending on the level of confidence	Adams and Adams (1960); Fischer and Budescu (2005); Keren (1997); Lichtenstein and Fischhoff (1977); Tversky and Kahneman (1974)
Conservatism or regressive bias	Tendency to underestimate high values and high likelihoods, probabilities or frequencies and overestimate the low ones	Attneave (1953); Fiedler (1991); Fischhoff et al. (1977); Hilbert (2012); Kaufman et al. (1949); MacGregor et al. (1988); Tversky and Kahneman (1974)
Exaggerated expectation	Real-world evidence is less extreme than our expectations (inverse of conservatism bias)	Erev et al. (1994); Wagenaar and Keren (1985)
First impression	Tendency to fail to update the initial impression as additional information is available	McKinney et al. (1987); Nordstrom et al. (1998); Rabin and Schrag (1999)
Hard-easy	Tendency to conserve the confidence in judgements (inverse of confidence bias)	Juslin et al. (2000); Keren (1988); Lichtenstein and Fischhoff (1977); Merkle (2009); Suantak et al. (1996)
Non-linear extrapolation	Inability to extrapolate a non-linear growth process	Hogarth (1987); Mackinnon and Wearing (1991); Wagenaar and Timmers (1979)
Placement	Tendency to self-estimate better than estimates about others	Cooper et al. (1988); Kruger (1999); Kruger and Dunning (1999); Moore and Cain (2007); Moore and Healy (2008)

(continued)

Table 2 (continued)

Decision bias	Description	References
Possibility effect	Tendency to overweight the likelihood of barely possible events	Donkers et al. (2001); Kahneman (2012); McCord and de Neufville (1986); Weber and Milliman (1997)
Reference point	Tendency to find an anchor (reference point) randomly to judge situations	Barkan et al. (2005); Boyle et al. (1998); Kahneman and Tversky (1979); McFadden (1999)
Regression	Tendency to disregard that events regress to the mean in subsequent trials	Hogarth (1987); Joyce and Biddle (1981); Tversky and Kahneman (1973, 1974)
Subadditivity	Tendency to estimate the likelihood estimates less than the sum of its mutually exclusive components	Fox and Levav (2000); Neil Bearden and Wallsten (2004); Tversky and Koehler (1994)

find a reference point for evaluation and adjust accordingly to it, but relying too heavily on one piece of information when making decisions.

From debate regarding the man as intuitive statistician (Gigerenzer 1991; Peterson and Beach 1967), more specifically whether as a frequentist or probabilistic type of statistician, comes a large family of biases. Subadditivity bias (Fox and Levav 2000; Neil Bearden and Wallsten 2004; Tversky and Koehler 1994) refers to the tendency to estimate the likelihood of estimates less than the sum of its mutually exclusive components. Human inability to mentally deal with nonlinear progressions is reflected in the non-linear extrapolation bias (Hogarth 1987; Mackinnon and Wearing 1991; Wagenaar and Timmers 1979) that describes the difficulty to extrapolate a non-linear growth process. Regarding being able to update one's judgement when new relevant information arises, people generally exhibit a first impression bias (McKinney et al. 1987; Nordstrom et al. 1998; Rabin and Schrag 1999) and conservative bias (Hogarth 1987; Nelson 1996) when failing to revise and adjust estimates or the first impression with new information. This conservation of estimates is also called hard-easy bias (Juslin et al. 2000; Keren 1988; Lichtenstein and Fischhoff 1977; Merkle 2009; Suantak et al. 1996) which is the tendency to conserve the confidence in judgements. More specifically, the tendency to conserve estimates of conditional probabilities is called the Bayesian likelihood (DuCharme 1970; Edwards 1968; Phillips and Edwards 1966; Phillips et al. 1966; Tversky and Kahneman 1974).

Regarding judgements of expectations, exaggerated expectation bias (Erev et al. 1994; Wagenaar and Keren 1985) is the inverse of conservatism bias and describes the mismatch between extreme expectations and the real-world evidence. Despite a similar name, conservatism or regressive bias (Attneave 1953; Fiedler 1991; Fischhoff et al. 1977; Hilbert 2012; Kaufman et al. 1949; MacGregor et al. 1988; Tversky and Kahneman 1974) relates to the how people underestimate high values and likelihoods and overestimate the low ones, which is the inverse of the near possibility and near certainty effects. Near possibility and certainty effects (Donkers et al. 2001; Kahneman 2012; McCord and de Neufville 1986; Weber and Milliman 1997) refer to how people overweight small risks (fear of the possibility of a disaster), and underweight almost certain outcomes, e.g. disproportional investments to rise from 95% service level to 100% service level and going bankrupt because the gains are not proportional to the costs.

Confidence bias (Adams and Adams 1960; Fischer and Budescu 2005; Keren 1997; Lichtenstein and Fischhoff 1977; Tversky and Kahneman 1974) is how the high perceived confidence level leads to extreme judgement. Finally, placement bias (Cooper et al. 1988; Kruger 1999; Kruger and Dunning 1999; Moore and Cain 2007; Moore and Healy 2008) is a very particular type of point of reference bias as from an individual perspective, people self-estimate better than estimate others.

At this point, we identify some of the potential biases affecting human behavior. However, it is possible to eliminate or reduce the bias. Debiasing techniques can offer possible ways of improving decision making and the use of technology.

4.3 Debiasing

Debiasing (e.g. Keren 1990; Larrick et al. 2004; Idson and Chugh 2004; Kaufmann et al. 2009, 2010; Croskerry et al. 2013) is essentially the elimination or reduction of biases. Also, *debiasing* has been called as '*cognitive engineering*' by Fischhoff et al. (1982, p.427). The main assumption of debiasing is that the bias is real.

Regarding the processes for debiasing, Keren (1990, p. 523) suggests a simple framework for debiasing compromising three main steps as listed in Table 3. Firstly, identification of the potential bias (its nature and existence), its cognitive triggers and task structure and environment. Secondly, consideration of alternative means for reduction/elimination of the bias given the previously identified bias in form of cognitive triggers and task environment which divides in two types of ignorant user or active and aware of the task structure. And thirdly, monitoring and evaluation of the effectiveness of the adopted debiasing technique, where special attention should be given to the possibly induced side-effects of debiasing methods.

Table 3 Debiasing process proposed by Keren (1990, p.523)

	Step	Description
Step 1	Bias Identification	It is necessary to identify the existence of the potential bias and its nature. Both the cognitive triggers and the environment are important
Step 2	Alternatives for debiasing	Once the potential bias is identified, it is necessary to consider alternative means (strategies) for its reduction or elimination
Step 3	Monitoring and Evaluation	Once the debiasing technique is deployed, it is necessary to monitor and evaluate how effective it is. Particular attention should be given to possible negative side effects

The means for debiasing mentioned in Table 3 divides in two sets, one set is superficial concerning situations when the user is ignorant about the internal structure of the task and the other, much deeper is when the user is aware of the bias and the task can be manipulated. Concerning the effectiveness of different approaches to debiasing, whereas most research on debiasing has been of (superficial) procedural nature, the deep understanding of task environment and biases often (and quite logically) leads to better results (Croskerry et al. 2013).

Persistence biases are those linked with the natural tendency to make judgements and choose conserving past choices, relying on habits and aversion to change. The strategy of considering the alternative has been suggested to mitigate the both persistence bias (Anderson 1980, 1983; Lord et al. 1984) and status quo bias (Hammond et al. 1998).

Reference point category of biases gathers biases focused on the tendency to make judgements from a relative point of reference rather than absolute. This links into the idea that humans are more sensitive to changes rather than states (Kahneman 2012). Debiasing reference point biases has been mostly focused on anchoring and adjustment, and the suggested strategies for its debiasing are considering an alternative (Chapman and Johnson 1994, 1999; Hammond et al. 1998), considering the opposite (Mussweiler et al. 2000) and considering why the anchor is possibly wrong (Epley and Gilovich 2005) which is similar to the latter of considering the opposite. Providing instructions on how to avoid the bias (Friedlander and Phillips 1984) and warning about the possible bias (Epley and Gilovich 2005) are also suggested as effective debiasing methods.

Placement bias has been mitigated with the "*put yourself in the shoes of*" method (Faro and Rottenstreich 2006, p. 5s35), which is a particular form of consider the alternative perspective. Finally, regression bias can be reduced by warning the decision-maker about the possible bias (Faro and Rottenstreich 2006).

It is worth noting that debiasing methods do not necessary make the bias disappear. They do, however, help to correct for the bias. In some cases, debiasing can backfire. Cases when to correct for a bias, the debiasing method consists of inducing another "corrective" bias can lead to an even greater problem. Care must be taken when applying debiasing methods and ideally multiple methods should be used.

5 Concluding Remarks

The Blockchain technology is a great invention of the digital age with a multitude of possible applications in supply chains. If it works well, it will allow humanity to reach new horizons. Imagine a day from the future where your self-driving car will take you to work and automatically pay for parking and a charging station with cryptocurrency. Your company will use Blockchain to verify transactions and you do not need to worry about currency exchange as that will take fractions of a second. Your smart fridge will automatically replenish your home food supplies, linked to your diary of activities that ensures food and exercise maintains your health, whilst your smart home ensures waste and recycling is limited and scheduled. All this at a greater efficiency and reduced operational cost when compared to the traditional solutions.

Advanced supply chains with perfect visibility enabled by Blockchain technology is possible. It is, however, dependent on people. The list of biases is extensive with the respective debiasing methods that can potentially help to correct for the error. It is, however, not to discourage the progress. Its purpose is to make people aware of what humans are good at and what they are not. Therefore, any implementation of the Blockchain technology in the future should take into account the behavioral aspect in order to ease its implantation, acceptance and use.

Yet, much of the possible future supply chains depends on the readiness of human psychology to accept automated and decentralized systems. So, what does the future hold for Blockchains? We shall see...

References

Anderson CA (1980) Inoculation and counter-explanation: debiasing techniques in the perseverance of social theories. Soc Cogn 1(2):126–139

Anderson CA (1983) Abstract and concrete data in the perseverance of social theories: when weak data lead to unshakeable beliefs. J Exp Soc Psychol 19(2):93–108

Adams P, Adams J (1960) Confidence in the recognition and reproduction of words difficult to spell. Am J Psychol 73(4):544–552

Attneave F (1953) Psychological probability as a function of experienced frequency. J Exp Psychol 46(2):81–86

Barkan R, Danziger S, Ben-Bashat G, Busemeyer JR (2005) Framing reference points: the effect of integration and segregation on dynamic inconsistency. J Behav Decis Mak 18(3):213–226. https://doi.org/10.1002/bdm.496

Bayer D, Haber S, Stornetta WS (1993) Improving the efficiency and reliability of digital time-stamping. In: Capocelli R, De Santis A, Vaccaro U (eds) Sequences II: Methods in communication, security and computer science. Springer, New York, pp 329–334

Boldrin M, Christiano L, Fisher J (2001) Habit persistence, asset returns, and the business cycle. Am Econ Rev 91(1):149–166

Boyle B, Dahlstrom R, Kellaris J (1998) Points of reference and individual differences as sources of bias in ethical judgments. J Bus Ethics 17(5):517–525

Chapman GB, Johnson EJ (1994) The limits of anchoring. J Behav Decis Mak 7(4):223–242

Chapman G, Johnson E (1999) Anchoring, activation, and the construction of values. Organ Behav Hum Decis Process 79(2):115–153

Cooper WW (1964) New perspectives in organization research. Wiley, New York

Cooper AC, Woo CY, Dunkelberg WC (1988) Entrepreneurs' perceived chances for success. J Bus Venturing 3(2):97–108. https://doi.org/10.1016/0883-9026(88)90020-1

Croskerry P, Singhal G, Mamede S (2013) Cognitive debiasing 2: impediments to and strategies for change. BMJ Qual Saf 22(Suppl 2):ii65–ii72

Croson R, Donohue K (2006) Behavioral causes of the bullwhip effect and the observed value of inventory information. Manag Sci 52(3):323–336. https://doi.org/10.1287/mnsc.1050.0436

Donkers B, Melenberg B, Van Soest A (2001) Estimating risk attitudes using lotteries: a large sample approach. J Risk Uncertainty 22(2):165–195

DuCharme W (1970) Response bias explanation of conservative human inference. J Exp Psychol 85 (1):66–74
Edwards W (1968) Conservatism in human information processing. In: Kleinmuntz B (ed) Formal representation of human judgment. Wiley, New York
Epley N, Gilovich T (2001) Putting adjustment back in the anchoring and adjustment heuristic: differential processing of self-generated and experimenter-provided anchors. Psychol Sci 12 (5):391–396
Epley N, Gilovich T (2005) When effortful thinking influences judgmental anchoring: differential effects of forewarning and incentives on self-generated and externally provided anchors. J Behav Decis Mak 18(3):199–212
Erev I, Wallsten TS, Budescu DV (1994) Simultaneous over- and underconfidence: the role of error in judgment processes. Psychol Rev 101(3):519–527. https://doi.org/10.1037/0033-295X.101.3.519
Faro D, Rottenstreich Y (2006) Affect, empathy, and regressive mispredictions of others' preferences under risk. Manag Sci 52(4):529–541
Fernandez R, Rodrik D (1991) Resistance to reform: status quo bias in the presence of individual-specific uncertainty. Am Econ Rev 81(5):1146–1155
Fiedler K (1991) The tricky nature of skewed frequency tables: an information loss account of distinctiveness-based illusory correlations. J Pers Soc Psychol 60(1):24–36. https://doi.org/10.1037//0022-3514.60.1.24
Fischer I, Budescu DV (2005) When do those who know more also know more about how much they know? The development of confidence and performance in categorical decision tasks. Organ Behav Hum Decis Process 98(1):39–53. https://doi.org/10.1016/j.obhdp.2005.04.003
Fischhoff B, Slovic P, Lichtenstein S (1977) Knowing with certainty: the appropriateness of extreme confidence. J Exp Psychol Hum Percept Perform 3(4):552–564. https://doi.org/10.1037/0096-1523.3.4.552
Fischhoff B, Slovic P, Lichtenstein S (1982) Lay foibles and expert fables in judgments about risk. Am Stat 36(3b):240–255
Forrester JW (1958) Industrial dynamics: a major breakthrough for decision makers. Harv Bus Rev 36(4):37–66
Fox C, Levav J (2000) Familiarity bias and belief reversal in relative likelihood judgment. Organ Behav Hum Decis Process 82(2):268–292. https://doi.org/10.1006/obhd.2000.2898
Friedlander ML, Phillips SD (1984) Preventing anchoring errors in clinical judgment. J Consult Clin Psychol 52(3):366–371
Geerts GL, O'Leary DE (2014) A supply chain of things: the EAGLET ontology for highly visible supply chains. Decis Support Syst 63:3–22
Gigerenzer G (1991) How to make cognitive illusions disappear: beyond "heuristics and biases". Eur Rev Soc Psychol 2(1):83–115. https://doi.org/10.1080/14792779143000033
Haber S, Stornetta WS (1990) How to time-stamp a digital document. In: Conference on the theory and application of cryptography. Springer, pp 437–455
Hammond JJS, Keeney RRL, Raiffa H (1998) The hidden traps in decision making. Harv Bus Rev 76(5):47–58
Hilbert M (2012) Toward a synthesis of cognitive biases: how noisy information processing can bias human decision making. Psychol Bull 138(2):211–237. https://doi.org/10.1037/a0025940
Hogarth RM (1987) Judgement and choice: the psychology of decision, 2nd edn. Wiley, Chichester
Idson L, Chugh D (2004) Overcoming focusing failures in competitive environments. J Behav Decis Mak 17(3):159–172
Joyce E, Biddle G (1981) Are auditors' judgments sufficiently regressive? J Account Res 19 (2):323–349
Juslin P, Winman A, Olsson H (2000) Naive empiricism and dogmatism in confidence research: a critical examination of the hard-easy effect. Psychol Rev 107(2):384–396. https://doi.org/10.1037//0033-295X.107.2.384
Kahneman D (2012) Thinking, fast and slow. Studies in intelligence, vol 56. Penguin, New York

Kahneman D, Tversky A (1979) Prospect theory: an analysis of decision under risk. Econometrica 47(2):263–292

Kahneman D, Knetsch J, Thaler R (1991) Anomalies: the endowment effect, loss aversion, and status quo bias. J Econ Perspect 5(1):193–206

Kaplan S, Schoar A (2005) Private equity performance: returns, persistence, and capital flows. J Finance 60(4):1791–1823

Kaufman E, Lord M, Reese T, Volkmann J (1949) The discrimination of visual number. Am J Psychol 62(4):498–525

Kaufmann L, Michel A, Carter CR (2009) Debiasing strategies in supply management decision-making. J Bus Logist 30(1):85–106

Kaufmann L, Carter CR, Buhrmann C (2010) Debiasing the supplier selection decision: a taxonomy and conceptualization. Int J Phys Distrib Logist Manag 40(10):792–821

Keren G (1988) On the ability of monitoring non-veridical perceptions and uncertain knowledge: some calibration studies. Acta Psychol 67(2):95–119

Keren G (1990) Cognitive aids and debiasing methods: can cognitive pills cure cognitive ills? Adv Psychol 68:523–552

Keren G (1997) On the calibration of probability judgments: some critical comments and alternative perspectives. J Behav Decis Making 10(3):269–278. https://doi.org/10.1002/(SICI)1099-0771(199709)10:3<269::AID-BDM281>3.0.CO;2-L

Kharlamov AA (2016) Exploring the contribution of individual differences and planning policy parameters to demand planning performance. The University of Warwick

Kruger J (1999) Lake Wobegon be gone! The "below-average effect" and the egocentric nature of comparative ability judgments. J Pers Soc Psychol 77(2):221–232

Kruger J, Dunning D (1999) Unskilled and unaware of it: how difficulties in recognizing one's own incompetence lead to inflated self-assessments. J Pers Soc Psychol 77(6):1121–1134

Lamport L, Shostak R, Pease M (1982) The Byzantine generals problem. ACM Trans Program Lang Syst 4(3):382–401. https://doi.org/10.1145/357172.357176

Larrick RP, Koehler DJ, Harvey N (2004) Debiasing. In: Koehler DJ, Harvey N (eds) Blackwell handbook of judgment and decision making. Blackwell, Hoboken, pp 316–337

Lee HL, Padmanabhan V, Whang S (1997) The bullwhip effect in supply chains. Sloan Manag Rev 38(3):93

Lichtenstein S, Fischhoff B (1977) Do those who know more also know more about how much they know? Organ Behav Hum Perform 183(3052):159–183. https://doi.org/10.1016/0030-5073(77)90001-0

Lord C, Lepper M, Preston E (1984) Considering the opposite: a corrective strategy for social judgment. J Pers Soc Psychol 47(6):1231–1243

MacGregor D, Lichtenstein S, Slovic P (1988) Structuring knowledge retrieval: an analysis of decomposed quantitative judgments. Organ Behav Hum Decis Process 42(3):303–323

Mackinnon AJ, Wearing AJ (1991) Feedback and the forecasting of exponential change. Acta Psychol 76(2):177–191. https://doi.org/10.1016/0001-6918(91)90045-2

Mason-Jones R, Towill DR (2000) Coping with uncertainty: reducing "Bullwhip" behaviour in global supply chains

McCord M, de Neufville R (1986) "Lottery equivalents": reduction of the certainty effect problem in utility assessment. Manag Sci 32(1):56–60

McFadden D (1999) Rationality for economists? J Risk Uncertainty 19(1):73–106

McKinney K, Sprecher S, Orbuch T (1987) A person perception experiment examining the effects of contraceptive behavior on first impressions. Basic Appl Soc Psychol 8(3):235–248. https://doi.org/10.1207/s15324834basp0803

Merkle EC (2009) The disutility of the hard-easy effect in choice confidence. Psychonomic Bull Rev 16(1):204–213. https://doi.org/10.3758/PBR.16.1.204

Moore DA, Cain DM (2007) Overconfidence and underconfidence: when and why people underestimate (and overestimate) the competition. Organ Behav Hum Decis Process 103(2):197–213. https://doi.org/10.1016/j.obhdp.2006.09.002

Moore DA, Healy PJ (2008) The trouble with overconfidence. Psychol Rev 115(2):502–517. https://doi.org/10.1037/0033-295X.115.2.502

Mussweiler T, Strack F, Pfeiffer T (2000) Overcoming the inevitable anchoring effect: considering the opposite compensates for selective accessibility. Personal Soc Psychol Bull 26(9):1142–1150
Nakamoto S (2008) Bitcoin: a peer-to-peer electronic cash system
Neil Bearden J, Wallsten TS (2004) MINERVA-DM and subadditive frequency judgments. J Behav Decis Making 17(5):349–363. https://doi.org/10.1002/bdm.477
Nelson MW (1996) Context and the inverse base rate effect. J Behav Decis Making 9(1):23–40. https://doi.org/10.1002/(SICI)1099-0771(199603)9:1<23::AID-BDM210>3.0.CO;2-X
Niranjan TT, Metri BA, Aggarwal V (2009) Behavioral causes of bullwhip effect: breaking the mould. Int J Serv Oper Manag 5(3):350–374
Nordstrom CR, Hall RJ, Bartels LK (1998) First impressions versus good impressions: the effect of self-regulation on interview evaluations. J Psychol 132(5):477–491. https://doi.org/10.1080/00223989809599281
Parry G, Brax SA, Maull R, Ng I (2016) Visibility of consumer context: improving reverse supply with internet of things data. Supply Chain Manag Int J 21(2):228–244
Peterson CR, Beach LR (1967) Man as an intuitive statistician. Psychol Bull 68(1):29–46
Phillips LD, Edwards W (1966) Conservatism in a simple probability inference task. J Exp Psychol 72(3):346–354
Phillips LD, Hays WL, Edwards W (1966) Conservatism in complex probabilistic inference. IEEE Trans Hum Factors Electron HFE-7(1):7–18. https://doi.org/10.1109/THFE.1966.231978
Prodan M, Prodan A, Purcarea AA (2015) Three new dimensions to people, process, technology improvement model. In: Rocha A, Correia AM, Costanzo S, Reis LP (eds) New contributions in information systems and technologies, vol 1. Springer, Cham, pp 481–490. https://doi.org/10.1007/978-3-319-16486-1_47
Rabin M, Schrag J (1999) First impressions matter: a model of confirmatory bias. Q J Econ 114 (1):37–82
Sage AP (1981) Behaviour and organizational considerations in the design of information systems and processes for planning and decision support. IEEE Trans Syst Man Cybern SMC-11 (9):640–678
Samuelson W, Zeckhauser R (1988) Status quo bias in decision making. J Risk Uncertainty 1:7–59
Slovic P (1975) Choice between equally valued alternatives. J Exp Psychol Hum Percept Perform 1 (3):280–287. https://doi.org/10.1037/0096-1523.1.3.280
Sterman JD, Dogan G (2015) "I'm not hoarding, I'm just stocking up before the hoarders get here.": Behavioral causes of phantom ordering in supply chains. J Oper Manag 39–40:6–22. https://doi.org/10.1016/j.jom.2015.07.002
Strack F, Mussweiler T (1997) Explaining the enigmatic anchoring effect: mechanisms of selective accessibility. J Pers Soc Psychol 73(3):437–446
Suantak L, Bolger F, Ferrell W (1996) The hard–easy effect in subjective probability calibration. Organ Behav Hum Decis Process 67(2):201–221
Tversky A, Kahneman D (1973) Availability: a heuristic for judging frequency and probability. Cognit Psychol 5(2):207–232. https://doi.org/10.1016/0010-0285(73)90033-9
Tversky A, Kahneman D (1974) Judgment under uncertainty: heuristics and biases. Science 185 (4157):1124–1131. https://doi.org/10.1126/science.185.4157.1124
Tversky A, Kahneman D (1992) Advances in prospect theory: cumulative representation of uncertainty. J Risk Uncertainty 5(4):297–323. https://doi.org/10.1007/BF00122574
Tversky A, Koehler DJ (1994) Support theory: a nonextensional representation of subjective probability. Psychol Rev 101(4):547–567. https://doi.org/10.1037/0033-295X.101.4.547
Wagenaar W, Keren G (1985) Calibration of probability assessments by professional blackjack dealers, statistical experts, and lay people. Organ Behav Hum Decis Process 36(3):406–416
Wagenaar W, Timmers H (1979) The pond-and-duckweed problem; Three experiments on the misperception of exponential growth. Acta Psychol 43(1979):239–250
Weber EU, Milliman RA (1997) Perceived risk attitudes: relating risk perception to risky choice. Manag Sci 43(2):123–144

Alexander Kharlamov is a research fellow at the Faculty of Business and Law, UWE Bristol. Currently, he is working at the cutting edge of IoT and business model thinking, part of a multi-university research team (Cambridge, Warwick, Surrey and UWE) on the EPSRC Hub of All Things Living Lab project. Hub of All Things (HAT) is an enabler for personal data collection and use. The focus is on trust, identity, privacy, and security (TIPS) in the digital economy, centered on understanding and measuring user's perceived vulnerability to TIPS issues. Parallel areas of work include behavioral operations (decision making, errors and biases, individual differences in planning), supply chain segmentation, and integrating analytics in operations management.

Glenn Parry is Professor of Strategy and Operations Management at the University of the West of England, UK. He is primarily interested in understanding "good" business, where value in all its forms is the unit of measure. He is working on digital business models as CoI on the EPSRC "Hub of All Things [HAT]" research projects [EP/N028422/1 and EP/K039911/1], British Academy "Blockchain for Good," and AHRC "Bristol & Bath by Design" [AH/M005771/1]. His research is characterized by a strong industrial focus on process combined with the rigor of academic analysis. He has published in several international journals as well as the books *Build to Order: The Road to the 5-day Car*, *Complex Engineering Service Systems*, and *Service Design and Delivery* which was ranked in The IIJ top 20 upcoming books for innovators.